Molecular and Cell
BIOCHEMISTRY

Energy in BIOLOGICAL SYSTEMS

SMITH AND WOOD

CHAPMAN & HALL
University and Professional Division
London · New York · Tokyo · Melbourne · Madras

UK Chapman & Hall, 2–6 Boundary Row, London SE1 8HN

USA Chapman & Hall, 29 West 35th Street, New York NY100001

JAPAN Chapman & Hall Japan, Thomson Publishing Japan, Hirakawacho Nemoto Building, 7F, 1-7-11 Hirakawa-cho, Chiyoda-ku, Tokyo 102

AUSTRALIA Chapman & Hall Australia, Thomas Nelson Australia, 102 Dodds Street, South Melbourne, Victoria 3205

INDIA Chapman & Hall India, R. Seshadri, 32 Second Main Road, CIT East, Madras 600 035

First edition 1991
Reprinted 1993 (with corrections)

Typeset in 10/11½pt Palatino by EJS Chemical Composition, Midsomer Norton, Bath, Avon
Printed in Hong Kong

ISBN 0 412 40770 1

British Library Cataloguing in Publication Data

Energy in biological systems. – (Molecular and cell biochemistry)
I. Smith, C. II. Wood, E.J. III. Series
574.5

ISBN 0–412–40770–1

Library of Congress Cataloging-in-Publication Data
Available

Copy Editors: Sara Firman and Judith Ockenden
Sub-editor: Simon Armstrong
Production Controller: Marian Saville
Layout Designer: Geoffrey Wadsley (after an original design by Julia Denny)
Illustrators: Capricorn Graphics
Cover design: Amanda Barragry

Energy in BIOLOGICAL SYSTEMS

Contents

Editors' foreword	vii
Contributors	viii
Preface	ix
Abbreviations	x
Greek alphabet	xii
1 Energy and life	**1**
1.1 Introduction	1
1.2 Thermodynamics	2
1.3 Free energy, enthalpy and entropy	3
1.4 The biochemistry of adenosine triphosphate (ATP)	11
1.5 Overview	15
Answers to exercises	15
Questions	16
2 An overview of bioenergetics	**18**
2.1 Introduction	18
2.2 Organization of metabolism	20
2.3 What is catabolism for?	26
2.4 Acquiring energy from the environment	33
2.5 Strategies for generating ATP and NADPH in catabolism	40
2.6 Overview	44
Answers to exercises	44
Questions	45
3 Electron transport	**47**
3.1 Introduction	47
3.2 Electron transport	48
3.3 Oxidative phosphorylation	56
3.4 Synthesis of ATP	61
3.5 Photosynthesis	65
3.6 The mechanism of ATP synthesis	70
3.7 Overview	72
Answers to exercises	73
Questions	73
4 The tricarboxylic acid cycle	**75**
4.1 Introduction	75
4.2 How the cycle was elucidated	76
4.3 Some biochemical details	84
4.4 The TCA cycle in relation to other cellular processes	91
4.5 Overview	98
Answers to exercises	98
Questions	99

5 Glycolysis **101**
5.1 Introduction 101
5.2 Early history of the study of glucose metabolism 101
5.3 The Embden–Meyerhof pathway 104
5.4 Biological significance of the glycolytic pathway 107
5.5 Glycogen and other polysaccharides 110
5.6 Other monosaccharides 115
5.7 The pentose phosphate pathway 116
5.8 Overview 122
Answers to exercises 123
Questions 123

6 Lipids: breakdown of fatty acids; brown fat; ruminant metabolism **125**
6.1 Introduction 125
6.2 Triacylglycerols as energy reserves 125
6.3 Oxidation of fatty acids 126
6.4 Thermogenesis 129
6.5 Ruminant metabolism 133
6.6 Overview 137
Answers to exercises 137
Questions 138

7 Amino acid catabolism **139**
7.1 Introduction 139
7.2 Protein degradation 139
7.3 Removal of nitrogen from amino acids 141
7.4 Essential amino acids 144
7.5 Carbon chain catabolism 146
7.6 Interconversion of amino acids and transamination 150
7.7 Pathways for nitrogen excretion 155
7.8 Biosynthesis of urea 156
7.9 Overview 160
Answers to exercises 160
Questions 160

Answers to questions 162
Glossary 167
Index 169

Editors' Foreword

This book is one of a series of brief fundamental texts for junior undergraduates and diploma students in biological science. The series, Molecular and Cell Biochemistry, covers the whole of modern biochemistry, integrating animal, plant and microbial topics. The intention is to give the series special appeal to the many students who read biochemistry for only part of their course and who are looking for an all-encompassing and stimulating approach. Although all books in the series bear a distinct family likeness, each stands on its own as an independent text.

Many students, particularly those with less numerate backgrounds, find elements of their biochemistry courses daunting, and one of our principal concerns is to offer books which present the facts in a palatable style. Each chapter is prefaced by a list of learning objectives, with short overviews and revision aids at the ends of chapters. The text itself is informal, and the incorporation of marginal notes and information boxes to accompany the main text give a tutorial flavour, complementing and supporting the main narrative. The marginal notes and boxes relate facts in the text to applicable examples in everyday life, in industry, in other life sciences and in medicine, and provide a variety of other educational devices to assist, support, and reinforce learning. References are annotated to guide students towards effective and relevant additional reading.

Although students must start by learning the basic vocabulary of a subject, it is more important subsequently to promote understanding and the ability to solve problems than to present the facts alone. The provision of imaginative problems, examples, short answer questions and other exercises are designed to encourage such a problem-solving attitude.

A major challenge to both teacher and student is the pace at which biochemistry and molecular biology are advancing at the present time. For the teacher and textbook writer the challenge is to select, distill, highlight and exemplify, tasks which require a broad base of knowledge and indefatigable reading of the literature. For the student the challenge is not to be overwhelmed, to understand and ultimately to pass the examination! It is hoped that the present series will help by offering major aspects of biochemistry in digestible portions.

This vast corpus of accumulated knowledge is essentially valueless unless it can be used. Thus these texts carry frequent, simple exercises and problems. It is expected that students will be able to test their acquisition of knowledge but also be able to use this knowledge to solve problems. We believe that only in this way can students become familiar and comfortable with their knowledge. The fact that it is useful to them will mean that it is retained, beyond the last examination, into their future careers.

This series was written by lecturers in universities and polytechnics who have many years of experience in teaching, and who are also familiar with current developments through their research interests. They are, in addition, familiar with the difficulties and pressures faced by present-day students in the biological sciences area. The editors are grateful for the co-operation of all their authors in undergoing criticism and in meeting requests to re-write (and sometimes re-write again), shorten or extend what they originally wrote. They are also happy to record their grateful thanks to those many individuals who very willingly supplied illustrative material promptly and generously. These include many colleagues as well as total strangers whose response was positive and unstinting. Special thanks must go to the assessors who very carefully read the chapters and made valuable suggestions which gave rise to a more readable text. Grateful thanks are also due to the team at Chapman & Hall who saw the project through with good grace in spite, sometimes, of everything. These include Dominic Recaldin, Commissioning Editor, Jacqueline Curthoys, formerly Development Editor, Simon Armstrong, Sub-editor, and Marian Saville, Production Controller.

Finally, though, it is the editors themselves who must take the responsibility for errors and omissions, and for areas where the text is still not as clear as students deserve.

Contributors

DR P.S. AGUTTER *Department of Biological Sciences, Napier University, Edinburgh, UK. Chapters 4 and 5.*

DR J. ARONSON *Department of Molecular and Cellular Biology, University of Arizona, Tucson, Arizona, USA. Chapter 7.*

DR J.J. GAFFNEY *Department of Biological Sciences, The Manchester Metropolitan University, Manchester, UK. Chapters 3 and 6.*

PROFESSOR J.J.A. HEFFRON *Department of Biochemistry, University College, Cork, The Republic of Ireland. Chapter 1.*

DR S. SHAW *Department of Applied Sciences and Computing, University College Salford, Salford, UK. Chapter 2.*

DR C.A. SMITH *Department of Biological Sciences, The Manchester Metropolitan University, Manchester, UK. Chapters 5 and 6.*

Preface

The generation of biologically usable energy by oxidative processes, including electron transport, is a major area of importance to all life forms. It is a complex area and one that is still rather poorly understood despite many person-years of research and the accumulation of a plethora of data. For both student and teacher this presents problems. Many students, for example, find the idea of thermodynamics intrinsically difficult and are resistant to it as well as being poorly equipped mathematically to appreciate fully the concepts involved. Many teachers, too, are sceptical of the value of the rigorous application of classical thermodynamics to biological systems and may be less than enthusiastic in their teaching of it as a result. However, it has to be said that without a knowledge of energy relationships, the catabolic and anabolic pathways of metabolism are virtually meaningless.

The initial chapters of this book, therefore, offer a simple approach to, and an explanation of, relevant thermodynamic principles in familiar terms, and provide a basis for the rest of the material in the book. The next chapter gives the basic information required for understanding the ways in which biologically usable energy in the form of ATP is produced *via* oxidative pathways. It aims to give a clear and concise explanation of the structure and function of the coenzymes, and an up-to-date account, in simple terms, of how the transduction of oxidative energy is achieved in organelles such as the mitochondria. The pathways and mechanisms described form the metabolic hub into which reducing equivalents are fed from the breakdown of all food and storage materials. The four final chapters then describe these catabolic pathways—the TCA cycle, and carbohydrate, lipid and amino acid breakdown—in detail.

Athough this approach is the reverse of that taken in many textbooks, it is our aim to show students where they are going or where pathways lead. In our experience, students all too often 'start at the top' with little idea of why glucose, or a fatty acid, or amino acid, is undergoing catabolism in the first place. Therefore, an initial understanding of 'the final common pathway' of catabolism and energy production helps them orient their study of the vast numbers of catabolic routes.

In a number of cases we have attempted to give metabolism a new slant by discussing the pathways from a historical standpoint. We tend to forget rather too easily the years of painstaking pioneering work by dedicated biochemists, who produced brilliant advances without the help of the sophisticated equipment, automated devices and computers that are regarded as the *sine qua non* these days. The aim of this is to emphasise that it is the original, experimental scientific approach that is important in making advances rather than the number of pounds spent on sophisticated equipment.

We have also tried to stress the importance of linking chemical detail to biological function. From a student's point of view, nothing is worse than perceiving metabolic pathways as series of complex reactions to be learned by rote, without any comprehension of what they are. Hopefully, our approach will encourage students to take a step back and understand, rather than learn by heart because there might be a question in the examination!

The area of bioenergetics is poorly served by student texts: we hope that the present volume goes some way towards providing an accessible and enjoyable introduction to it.

Abbreviations

A	adenine (alanine)
ACP	acyl carrier protein
ACTH	adrenal corticotrophic hormone
ADP	adenosine diphosphate
Ala, A	alanine
AMP	adenosine monophosphate
cAMP	adenosine 3′,5′-cyclic monophosphate
Arg, R	arginine
Asn, N	asparagine
Asp, D	aspartic acid
ATP	adenosine triphosphate
ATPase	adenosine triphosphatase
C	cytosine (cysteine)
CDP	cytidine diphosphate
CMP	cytidine monophosphate
CTP	cytidine triphosphate
CoA, CoASH	coenzyme A
CoQ, Q	coenzyme Q, ubiquinone
Cys, C	cysteine
d-	2-deoxy-
D	aspartic acid
d-Rib	2-deoxyribose
DNA	deoxyribonucleic acid
cDNA	complementary DNA
e^-	electron
eV	electron-volt
E	glutamic acid
E	oxidation–reduction potential
F	phenylalanine
F	the Faraday (9.648×10^4 coulomb mol^{-1})
FAD	flavin adenine dinucleotide
Fd	ferredoxin
fMet	*N*-formyl methionine
FMN	flavin mononucleotide
Fru	fructose
g	gram
g	acceleration due to gravity
G	guanine (glycine)
G	free energy
Gal	galactose
Glc	glucose

Gln, Q	glutamine
Glu, E	glutamic acid
Gly, G	glycine
GDP	guanosine diphosphate
GMP	guanosine monophosphate
GTP	guanosine triphosphate
H	histidine
H	enthalpy
Hb	haemoglobin
His, H	histidine
Hyp	hydroxyproline (HOPro)
I	isoleucine
Ig G	immunoglobulin G
Ig M	immunoglobulin M
Ile, I	isoleucine
ITP	inosine triphosphate
J	Joule
K	degrees absolute (Kelvin)
K	lysine
L	leucine
Leu, L	leucine
ln x	natural logarithm of $x = 2.303 \log_{10} x$
Lys, K	lysine
M	methionine
M_r	relative molecular mass, molecular weight
Man	mannose
Mb	myoglobin
Met, M	methionine
N	asparagine
N	Avogadro's number (6.022×10^{23})
N	any nucleotide base (e.g. in NTP for nucleotide triphosphate)
NAD^+	nicotinamide adenine dinucleotide
$NADP^+$	nicotinamide adenine dinucleotide phosphate
P	proline
Pi	inorganic phosphate
PPi	inorganic pyrophosphate
Phe, F	phenylalanine
Pro, P	proline
Q	coenzyme Q, ubiquinone
Q	glutamine
R	arginine
R	the gas constant ($8.314\,J\,K^{-1}\,mol^{-1}$)
Rib	ribose
RNA	ribonucleic acid
mRNA	messenger RNA

rRNA	ribosomal RNA
tRNA	transfer RNA
s	second
s	sedimentation coefficient
S	svedberg unit (10^{-13} seconds)
S	serine
SDS	sodium dodecylsulphate
Ser, S	serine
T	thymine
Thr, T	threonine
TPP	thiamine pyrophosphate
Trp, W	tryptophan
TTP	thymidine triphosphate (dTTP)
Tyr, Y	tyrosine
U	uracil
UDP	uridine diphosphate
UDP-Glc	uridine diphosphoglucose
UMP	uridine monophosphate
UTP	uridine triphosphate
V	valine
V	volt
Val, V	valine
W	tryptophan
Y	tyrosine

Greek alphabet

Α	α	alpha
Β	β	beta
Γ	γ	gamma
Δ	δ	delta
Ε	ε	epsilon
Ζ	ζ	zeta
Η	η	eta
Θ	θ	theta
Ι	ι	iota
Κ	κ	kappa
Λ	λ	lambda
Μ	μ	mu
Ν	ν	nu
Ξ	ξ	xi
Ο	o	omicron
Π	π	pi
Ρ	ϱ	rho
Σ	σ	sigma
Τ	τ	tau
Υ	υ	upsilon
Φ	ϕ	phi
Χ	χ	chi
Ψ	w	psi
Ω	ω	omega

1

Energy and life

Objectives

After reading this chapter you should be able to:

□ examine which factors govern the direction of biochemical events in cells;

□ calculate the actual free energy changes of different types of biochemical reactions;

□ understand the structural basis of the high biochemical reactivity of ATP;

□ discuss how the free energy of ATP hydrolysis is transduced in biosynthetic reactions.

1.1 Introduction

One of the most characteristic features of organisms is their ability to capture, transduce, use and store energy for their continued existence. On our planet, life depends on the continuing flow of energy from the sun to the earth. The solar thermonuclear reactions produce energy which is emitted into space in the form of electromagnetic radiation. Upon reaching the earth this energy is captured by molecules of chlorophyll present in green plants and photosynthetic bacteria which employ it to reduce carbon dioxide to carbohydrates and other useful compounds (Fig. 1.1). It is estimated that the amount of energy reaching the earth each year is about 5×10^{24} joules. Only

□ The joule (J), is the derived unit of energy in the International System of Units (SI); it is related to the old unit, the calorie, thus,

$$4.184 \text{ J} = 1 \text{ calorie.}$$

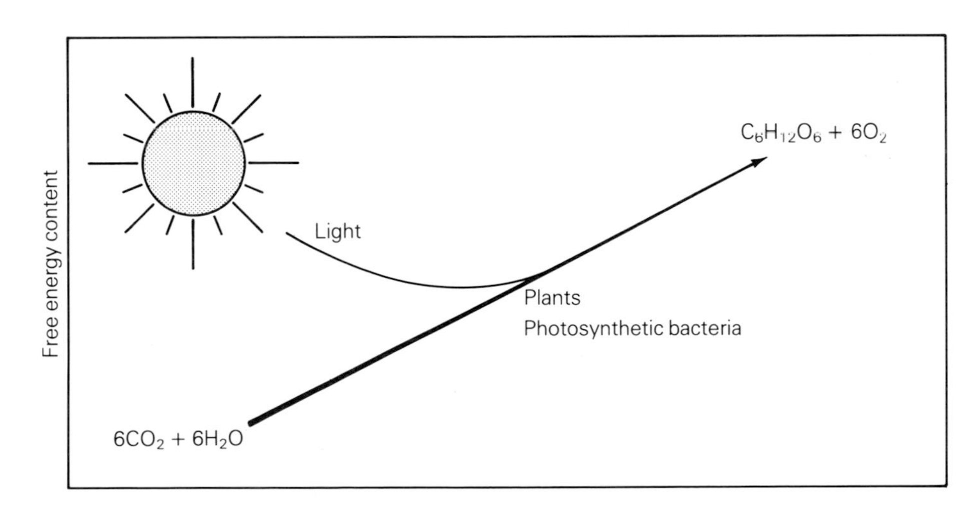

Fig. 1.1 A simplified diagram of photosynthesis. The sun generates radiant energy from nuclear fusion reactions. This energy is captured on earth by plants and photosynthetic bacteria and is used by them to convert carbon dioxide (CO_2) and water (H_2O) to glucose ($C_6H_{12}O_6$). Thus, the energy of the sun is used to drive the endergonic conversion of simple molecules to quite complex six-carbon molecules with a correspondingly greater free energy content.

a small fraction (about 0.025%) of this energy is converted into organic compounds by plants and other photosynthetic organisms. This captured energy drives the life processes of practically all other organisms, unicellular and multicellular, vertebrate and invertebrate. Thus, organisms can be viewed as biochemical machines that extract chemical energy from organic compounds by the process of ***catabolism*** and channel it into life-sustaining functions.

The flow of energy in biological systems is perhaps best understood by applying the principles of chemical thermodynamics, the branch of science which deals with energy transformations. It should be noted that the most fundamental difference between ordinary chemical systems and biochemical systems is that the latter are essentially ***isothermal*** in operation. They use not heat but chemical energy to drive and maintain their life processes.

Table 1.1 *Types of energy of importance in biological systems*

Energy form	SI Unit	'Old' unit
Thermal	joule (J)	calorie
Mechanical	joule	erg
Electrical	joule	erg
Light	joule	erg
Kinetic	joule	erg
Osmotic	joule	litre-atmosphere
Free energy	joule	calorie (kcal)
Entropy*	joule	calorie

* Entropy is regarded as a form of 'useless' energy because it is not available for the performance of work; under conditions of constant pressure and temperature, i.e. biological conditions. Thermal or heat energy cannot be converted into useful work.

1.2 *Thermodynamics*

The First Law

The various forms of energy found in biological systems are shown in Table 1.1. Although the different forms are interconvertible, energy can neither be created nor destroyed. This is a statement of the **First Law of Thermodynamics** or the principle of conservation of energy. When an amount of energy in one form is lost, an exactly equivalent amount of energy appears in another form. If heat is absorbed or work is done on any system, animate or inanimate, these forms of energy are conserved in the system as the internal energy, E. What the First Law does not indicate is whether natural processes in a system, for example the reactions in a metabolic pathway, can take place of their own accord in a given direction. To determine if a natural process or reaction can take place spontaneously a different principle is needed and this is embodied in the **Second Law of Thermodynamics**.

☐ Thermodynamics literally means 'movement of heat'.

The Second Law

In essence, this Law states that during any physical or chemical change a certain amount of energy is transformed into a random, disordered form

Box 1.1
Laws of thermodynamics

The First Law

Energy is neither created nor destroyed during any physical or chemical process in the universe but it may undergo change from one form to another.

The Second Law

All physical and chemical changes tend to proceed in that direction which leads to disorder of the system and its surroundings; that is, the disorder or entropy of the universe always tends to increase.

Several technical terms have specific meanings when applied to thermodynamic systems. These terms include:

System: a chemical reaction, natural process, a cell, the earth
Universe: the system and surroundings
Surroundings: the space that encloses the system
Open system: one that exchanges both energy and matter with its surroundings
Closed system: one that exchanges energy only with its surroundings

Catabolism: *that part of metabolism concerned with the chemical breakdown of organic molecules with the release of energy.*
Isothermal: *literally the same temperature (throughout the body).*

Reference Bray, W.S. (1978) *Physical Chemistry and its Biological Applications*, Academic Press, New York, USA. Chapters 3 and 4 give good accounts of the types of energy, their interconversions, units, entropy and free energy and their significance for living systems.

called **entropy** (S) and becomes unavailable to do work. Entropy is often spoken of as 'useless' energy to distinguish it from 'useful' or **free energy** which can be made to do useful work. Entropy is a measure of disorder on a molecular scale and can be visualized as the energy dissipated during a natural process by increasing the purely random motion of molecules and atoms in a system.

□ Entropy is an expression of the degree of disorder or randomness in a system. Like other thermodynamic quantities, such as temperature and pressure, entropy depends only on the state of the system and not on the path by which the state is reached. It is usually regarded as a quantity with an arbitrary value of zero; only changes in entropy are of significance generally. Entropy is sometimes spoken of as a 'low quality' energy because it represents a chaotic form of energy and is not available for doing useful work on its surroundings.

Some thermodynamic terminology

A more complete understanding of the applicability of the Second Law of Thermodynamics to biological systems requires the definition of a number of terms.

- A **system** consists of any piece of matter in the universe being observed; it might be a single reactant atom or molecule, a mitochondrion, a cell, an animal or even a planet as large as Jupiter.
- The **surroundings** refers to the space outside an imaginary surface or boundary around the system.
- The **universe** consists of the system and the surroundings (Fig. 1.2).

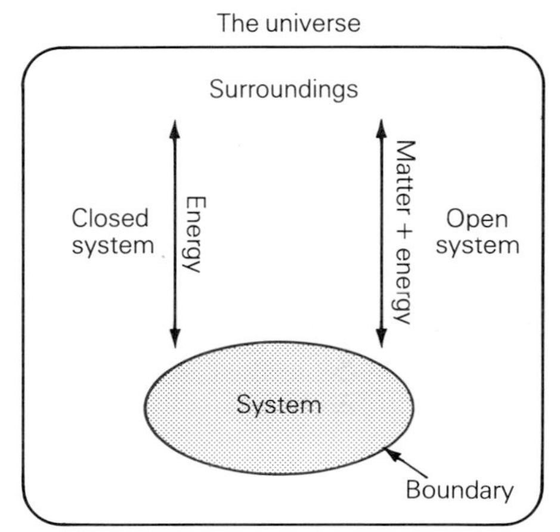

Fig. 1.2 Cells are open systems, both matter and energy may cross the boundary, the plasma membrane.

One might then ask the question, can matter and energy cross the boundary? This may be answered by defining three possible types of system:

- An **isolated system** is enclosed by a boundary which does not permit matter or energy to cross it.
- A **closed system** allows energy but not matter to cross the boundary.
- An **open system** allows passage of both matter and energy.

The earth may be regarded as a closed system because it does not exchange significant amounts of matter with its surroundings. The cell is a good example of an open system since it permits the movement of both matter and energy across its boundary, the plasma membrane. Isolated systems are hypothetical systems largely of theoretical interest. Now it is necessary to refer to useful energy again. There are two forms of useful energy:

- **Free energy**, usually denoted by the symbol G, can do work at *constant* temperature and pressure.
- **Heat energy** can do work by causing a *change* in temperature or pressure (Table 1.2).

Table 1.2 *Symbols of thermodynamic quantities*

$E^{0'}$	Standard redox potential at pH 7.0, in V or mV
E	Actual redox potential under non-standard conditions
F	The faraday or Faraday's constant: 96 485 coulombs mol^{-1} or 96.5 kJ V^{-1} mol^{-1}
R	Gas constant: 8.314 J mol^{-1} K^{-1}
G	(Gibbs) Free energy (kJ mol^{-1})
H	Enthalpy or heat content (kJ mol^{-1})
K'_{eq}	Apparent equilibrium constant at pH 7.0
$\Delta E^{0'}$	Difference in standard redox potential at pH 7.0
ΔE	Difference in actual redox potential
$\Delta G^{0'}$	Standard free energy change at pH 7.0
ΔG	Actual or physiological free energy change

1.3 *Free energy, enthalpy and entropy*

Organisms function at relatively uniform temperatures and pressures. Only minor variations in temperature and pressure occur between the various organs and tissues and any such differences are not used to drive biochemical reactions. For all practical purposes energy transformations in living systems take place under isothermal and constant pressure conditions. Under such conditions, a simple relationship exists between changes in free energy, enthalpy (also known as heat content) and entropy occurring during a biochemical reaction or process:

$$\Delta G = \Delta H - T\,\Delta S \qquad (1.1)$$

where ΔG is the change in free energy of the reaction (system), ΔH is the change in its enthalpy, T is the absolute temperature at which the reaction is taking place, and ΔS is the change in entropy of the universe. Note that the *absolute* values of the quantities such as free energy, enthalpy and entropy are not considered, but only their *changes*.

□ The Greek letter Δ (delta) is often used as an abbreviation to mean 'a change in'.

Reference Atkins, P.W. (1986) *Physical Chemistry*, 3rd edn, Oxford University Press, Oxford, UK, pp. 96–8. A very good qualitative account of the nature of entropy.

Reference Weast, R.C. (ed.) (1987) *Handbook of Physics and Chemistry*, 1st student edn, CRC Press, FL, USA, F-142–5. Excellent summary of the international systems of units (SI) and symbols for units.

Box 1.2
Functions of state

All the variables that are uniquely defined at an equilibrium state are called functions of state. Because changes in variables that are functions of state are not dependent on the intermediate states passed through in going from one arbitrary equilibrium state to another, it follows that the change is independent of how the process takes place. Such variables are the internal energy E, enthalpy H, entropy S and free energy G.

Example: The free energy yield of oxidation of glucose to CO_2 and H_2O is independent of whether it is oxidized by an ordinary chemical oxidation process or by the enzyme-catalysed reaction sequences of glycolysis and the tricarboxylic acid cycle Chapters 5 and 4.

- □ A system is at equilibrium when none of its properties change in time. Any system at equilibrium has exhausted its capacity for doing work on its surroundings. It follows that $\Delta G = 0$ when a system or reaction is at equilibrium.

- □ A spontaneous process or change is one that takes place as a consequence of the natural tendency of the universe towards greater chaos. Spontaneous processes are accompanied by an increase in entropy and a decrease in free energy. In biochemical terms, reactions are said to be spontaneous if they proceed with a large and negative ΔG value, e.g. < -20 kJ mol^{-1}. Although a given biochemical reaction may be said to be spontaneous in the thermodynamic sense, an enzyme is usually necessary if the reaction is to go at an appreciable rate.

As a reaction or system proceeds to chemical equilibrium, the entropy of the universe always increases. This is designated by a positive ΔS value for all spontaneous processes. This is yet another way of stating the Second Law of Thermodynamics. Because the universe consists of the *system* (e.g. chemical reactions) plus its *surroundings*, it follows from the Second Law that the change in entropy during a chemical or physical process can take place in the system or the surroundings. Normal biological systems do not undergo any change in their internal disorder when they metabolize their dietary foodstuffs. Organisms maintain, and indeed increase, their complexity or orderliness as they continue their metabolic processes. It follows that it is the entropy of the surroundings of living things that must increase during the continuance of the life processes (Fig. 1.3). Metabolism proceeds in order to maintain the structural orderliness of the organism by extracting free energy from nutrients and by giving off an amount of useless energy, principally as heat, which increases the randomness of the surroundings. In this way, biological systems satisfy the Second Law of Thermodynamics.

Organisms do not indefinitely postpone undergoing an increase in entropy, however. Net breakdown of cell structures, organs and tissues are well-known to occur during ageing and in disease processes, and would be associated with corresponding increases in entropy of the system; death of the organism brings about the ultimate change in entropy of the system.

Although the change in entropy may be used to predict if any biochemical process can proceed spontaneously (Fig. 1.4), the value of ΔS of such

Exercise 1

Consider a gas, a liquid and a crystal at the same temperature. Which system has the least entropy and why?

Exercise 2

Organisms continually create highly ordered structures in their cells and tissues from simple, less-ordered nutrient molecules. Thus, they appear to violate the Second Law of Thermodynamics. Explain why this is not so.

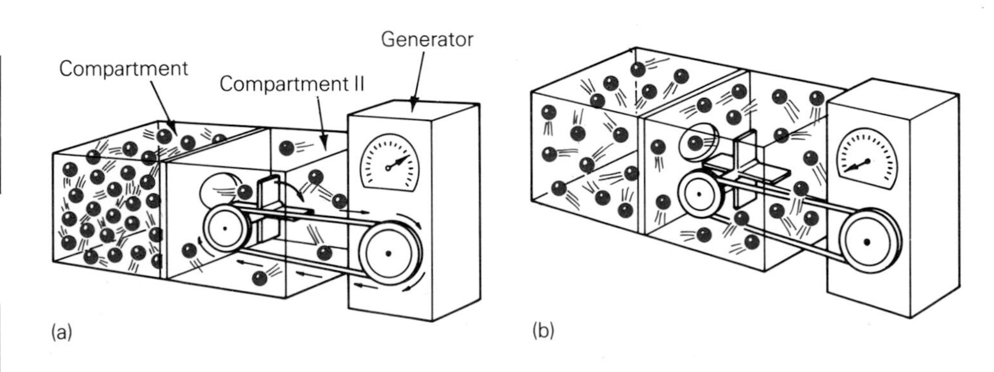

Fig. 1.3 Mechanical analogy of low- and high-entropy state. In (a) the situation shown represents an organized, high-energy state in which nearly all the molecules are in compartment I. As they are allowed to diffuse, the molecules enter compartment II, thereby increasing the entropy and lowering the free energy of the system until equilibrium is reached (b). The change from a low- to a high-entropy state releases free energy, which in this model is harnessed by the paddle wheel. The ability to do work approaches zero as the system comes into equilibrium. Redrawn from Eckert, R. and Randall, D. (1983) *Animal Physiology*, W.H. Freeman, San Francisco, USA, p. 49.

processes is not easily determined. Biochemists have found it more convenient to use the other function of state, G, to act as an indicator of reaction spontaneity and of equilibrium. The most important properties of G for biological systems are:

- G *decreases* during all spontaneous processes occurring at constant temperature and pressure.
- The *maximum amount of work* which can be obtained from a biochemical process occurring at constant temperature and pressure is given by the ΔG of the process; this is only true for reversible processes and the work done is always less in irreversible reactions.

Organisms function at essentially constant temperature and pressure, and the useful energy derivable from biochemical reactions under these conditions is the free energy, G. This is the form of energy that drives the biochemical machinery of all organisms. Heat *cannot* be used by organisms to do work because heat can only do work as it passes from one system to another system at a *lower* temperature. It is for this reason that heat is considered to be *useless* energy when examining the driving forces of biochemical machinery. Although this is true, organisms use the heat generated as a by-product of metabolic reactions to maintain body temperature at a level which ensures that enzyme-catalysed reactions proceed at appropriate rates.

□ Reaction types based on ΔG changes are: endergonic, $\Delta G > 0$; exergonic, $\Delta G < 0$. Exergonic reactions proceed spontaneously (in the presence of a catalyst) while endergonic ones require the input of free energy if they are to proceed. In biological systems this means coupling them to another process such that the *overall* process still has a negative ΔG.

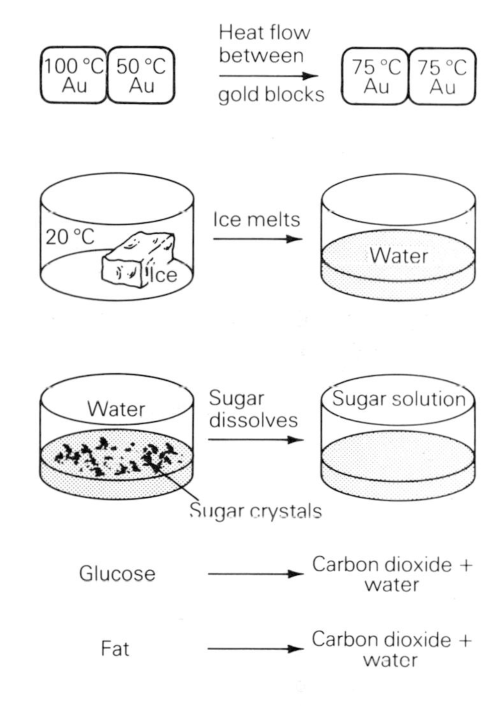

Fig. 1.4 Examples of some spontaneous or natural processes.

Free energy changes in biochemical reactions

Free energy changes in biochemical reactions may be calculated, given that the free energy change of a reaction is related to the changes in heat content and entropy (Eqn 1.1). Every reaction has a characteristic ΔG under defined specific conditions: ΔG under the appropriate specific conditions, **standard conditions**, and is denoted as ΔG^0. 'Standard conditions' are 25°C or 298 K, a pressure of 101 kPa, while maintaining reactant and product concentrations at 1 mol dm^{-3}, and a pH of 0.0. These are the standard conditions as used by chemists and physicists. They need to be altered so that meaningful calculations of free energy changes under biochemical conditions may be carried out. Standard biochemical conditions are identical to standard chemical conditions with the single exception that the pH is 7.0 rather than 0.0 (most metabolism proceeds at about pH 7.0). The standard biochemical free energy change is then denoted by $\Delta G^{0'}$

The most straightforward definition of the $\Delta G^{0'}$ of a biochemical reaction is as the difference in free energy content between the reactants and products under standard conditions. Using a reaction type as an example:

$$A + B = C + D$$

$\Delta G^{0'}$ is less than zero when the sum of the free energy contents of C and D is less than that of A and B under standard conditions. A common difficulty in understanding the definition of $\Delta G^{0'}$ arises from the fact that both reactant and product concentrations are all 1 mol dm^{-3}. However, for a reaction with $\Delta G^{0'} < 0$ the equilibrium point lies to the right, that is, the concentrations of products will be greater than those of the reactants. Some of the confusion in this regard may arise from the relationship between $\Delta G^{0'}$ and the equilibrium constant (K'_{eq}) outlined below. The equilibrium constant for the above reaction type is:

$$K'_{eq} = [C][D]/[A][B]$$

and the value of $\Delta G^{0'}$ may be found from

$$\Delta G = \Delta G^{0'} + R\,T\,\ln[C][D]/[A][B]$$

□ The pH of the cytoplasm of most cells lies in the range 6.9 to 7.0. A sustained fall in intracellular pH is an indicator of the onset of disease.

□ Strictly speaking, the concentrations of reactants and products in equations should not be used, rather, *activity* is the appropriate quantity. The activity of compound X, denoted a, is equal to the product of the activity coefficient, f, and the concentration of X, i.e. $a = f[X]$. At low concentrations of X, its activity will closely approximate its concentration.

Reference Harold, F.M. (1986) *The Vital Force: A study of Bioenergetics*, W.H. Freeman, New York, USA. An excellent account of energy, work and order is found in Chapter 1.

where R is the gas constant, T is the absolute temperature and ΔG is the *actual* free energy change of the reaction. Since $\Delta G = 0$ at equilibrium, it is possible to equate the terms on the right-hand side of the equation to zero and write:

$$\Delta G^{0'} + RT \ln[C][D]/[A][B] = 0$$

then

$$\Delta G^{0'} = -RT \ln[C][D]/[A][B]$$

or

$$\Delta G^{0'} = -RT \ln K'_{eq} \quad (1.2)$$

Exercise 3

Phosphoglucomutase catalyses the reaction:

glucose 1-phosphate ⇌ glucose 6-phosphate

Experimentally, it was found that the reaction mixture at equilibrium contained 5.0% glucose 1-phosphate at 25°C. Calculate the equilibrium constant and the $\Delta G^{0'}$ for this enzyme-catalysed reaction.

Thus, $\Delta G^{0'}$ is directly proportional to the natural logarithm of the equilibrium constant but $\Delta G^{0'}$ is *not* the free energy change of a reaction at equilibrium. The most important point to note is that under biochemical (physiological) conditions in organisms, it is not the $\Delta G^{0'}$ of a reaction in a metabolic pathway which is relevant but rather, ΔG, the actual free energy yield under defined but non-standard physiological conditions. The appropriate equation for calculating the free energy yield of a biochemical reaction is:

$$\Delta G = \Delta G^{0'} + RT \ln[C][D]/[A][B] \quad (1.3)$$

Thus, to calculate the physiological free energy yield of a cellular reaction it is only necessary to know the equilibrium constant (usually available in biochemical data books) and the actual concentrations of reactants and products. These concentrations in living cells are usually in the range 10^{-5}–10^{-2} mol dm^{-3}, far removed from the 1 mol dm^{-3} concentrations of the standard state. Equation 1.3 shows that in biological systems reactions can be driven by the relative concentrations of reactants and products. The equation may also indicate the *directionality* of the reaction.

Box 1.3 Calculation of the free energy of hydrolysis of ATP under physiological conditions

The $\Delta G^{0'}$ of hydrolysis of ATP is −31 kJ mol^{-1} but this value refers to 1 mol dm^{-3} reactant and products, while in living cells the relevant concentrations are in the mmol dm^{-3} concentration range. For example, it is found that intracellular concentrations of ATP, ADP and inorganic phosphate (P_i) in frog skeletal muscle are 8.5, 0.25 and 2.6 mmol dm^{-3}, respectively, assuming that the muscle temperature and pH are 25°C and 7.0, respectively. The equation for ATP hydrolysis can be written:

$$ATP + H_2O \rightarrow ADP + P_i$$

Any change in the concentration of water is minimal and can be ignored. Thus, the following relationship for the physiological free energy yield of ATP hydrolysis can be given:

$$\begin{aligned}
\Delta G &= \Delta G^{0'} + RT \ln [ADP][P_i]/[ATP] \\
&= -31 \text{ kJ mol}^{-1} + 2.303 \times 8.314 \text{ J mol}^{-1} \text{ K}^{-1} \times 298 \text{ K} \times \log (2.5 \times 10^{-4}) \\
&\quad (2.6 \times 10^{-3})/(8.5 \times 10^{-3}) \\
&= -31 \text{ kJ mol}^{-1} + 5706 \text{ J mol}^{-1} \log(7.65 \times 10^{-5}) \\
&= -31 \text{ kJ mol}^{-1} + 5706 \text{ J mol}^{-1} \times (-4.12) \\
&= -31 \text{ kJ mol}^{-1} - 24 \text{ K mol}^{-1} \\
&= -55 \text{ kJ mol}^{-1}
\end{aligned}$$

Therefore, the *actual* free energy of hydrolysis of ATP in skeletal muscle is about 76% greater than the standard free energy of hydrolysis. Since the concentrations of ATP, ADP and P_i vary considerably in different tissues, it is clear that ΔG will also vary with the tissue examined.

Reference Banks, B.E.C. and Vernon, C.A. (1978) Biochemical abuse of standard equilibrium constants. *Trends in Biochemical Sciences*, July, **3**, N156–8. An interesting critique of the biochemical approach to thermodynamics.

REDOX POTENTIALS. The free energy of, for example, the nutrients in human diets is released in a controlled manner through **oxidation–reduction** or **redox reactions** occurring mainly in the mitochondria (Fig. 1.5). These are reactions in which electrons or hydrogen atoms are transferred from one molecule to another. The donating molecule is called the **reductant** and the accepting one the **oxidant**; they operate in pairs known as ***redox pairs*** or ***couples***. In cells, redox reactions may involve electrons *only*, as in the case of the cytochromes, hydride ions, or a combination of both, as for the $NAD^+/NADH$ redox pair.

The ability of redox pairs to donate and accept electrons or hydrogen varies widely but may be expressed quantitatively by a constant, the **standard redox potential,** $E^{0'}$. A redox potential is measured as the ***electromotive force***, in volts, of a ***half-cell*** consisting of both members of the redox couple when compared with the standard reference half-cell by convention referred to as

See Chapter 3

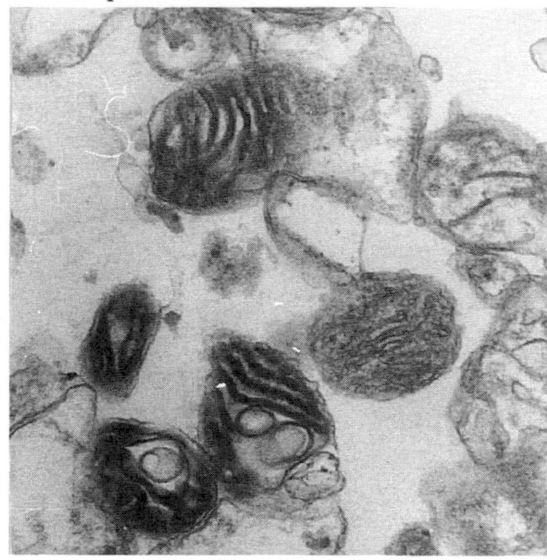

Fig. 1.5 Mitochondria isolated from disrupted cells and viewed by electron microscopy.

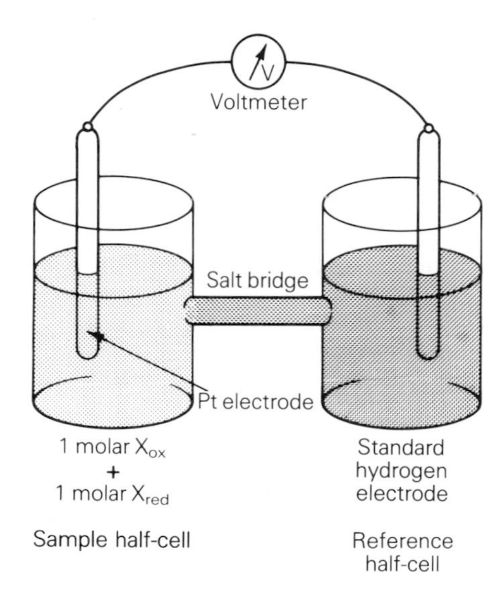

Fig. 1.6 Apparatus for measuring oxidation–reduction pontentials. The beaker on the left contains a solution of 1 mol dm^{-3} of the oxidized (ox) and reduced (red) species of compound X. A salt bridge containing saturated KCl solution provides electrical continuity between the reference half-cell and the sample cell and the two inert platinum electrodes are connected to a voltmeter. The standard hydrogen electrode half-cell contains a solution of 1 mol dm^{-3} H^+ in equilibrium with hydrogen as at a pressure of 101 kPa (old unit, 1 atm). The potential of the hydrogen electrode is conventionally set at 0.0 V at 25°C. The difference in potential between the two half-cells is measured by the voltmeter. In practice, it is customary to measure potentials against a more convenient electrode than the hydrogen electrode, such as the calomel electrode. Since the latter has a constant potential difference compared with the hydrogen electrode, it is a simple matter to calculate the absolute value of the standard redox potential.

Box 1.4 The nucleotides NAD⁺ and NADP

These two coenzymes, nicotinamide adenine dinucleotide (NAD^+) and nicotinamide adenine dinucleotide phosphate ($NADP^+$) like the flavin coenzymes, are involved in oxidation–reduction reactions. NAD^+ and $NADP^+$ contain a tetravalent nitrogen atom in the nicotinamide ring, which is positively charged. It is conventional therefore to show the oxidized forms as having a positive charge. NAD^+ contains adenosine linked through a pyrophosphate group to *N*-ribosyl nicotinamide. $NADP^+$ has an additional phosphate esterified to the 2′-hydroxyl of adenosine. Both NAD^+ and $NADP^+$ contain a nucleotide as part of their structure. They act as coenzymes for a variety of dehydrogenases.

The types of reaction in which the nicotinamide nucleotides function are as follows:

$$AH_2 + NAD^+ \rightarrow A + NADH + H^+$$

and

$$AH_2 + NADP^+ \rightarrow A + NADPH + H^+$$

The substrate AH_2 is oxidized by transfer of a hydride ion (H^-, that is, $H^+ + 2e^-$) to position 4 of the nicotinamide ring with the simultaneous release of H^+ to the medium. The coenzymes are loosely bound to the enzyme and can easily dissociate, carrying hydrogen in the process. The coenzymes may be thought of as dissociable electron carriers or as reusable substrates.

Redox pair/couple: *an electron-donating molecule and its oxidized form.*

Half-cell: *one electrode of an electrolytic cell and the solution or electrolyte with which it is in contact.*

Electromotive force: *the free energy, available to do work, arising from an unequal distribution of charged particles.*

Table 1.3 Standard redox potentials of some biochemically important substrates

Electrode equation	n	$E^{0'}$ (V)
acetate + $2H^+ + 2e^-$ = acetaldehyde	2	−0.58
$2H^+ + 2e^- = H_2$	2	−0.42
$NAD^+ + 2H^+ + 2e^-$ = $NADH + H^+$	2	−0.32
pyruvate + $2H^+ + 2e^-$ = lactate	2	−0.19
cytochrome *c* (Fe^{3+}) + e^- = cytochrome *c* (Fe^{2+})	1	+0.22
$\frac{1}{2}O_2 + 2H^+ + 2e^- = H_2O$	2	+0.82

The standard redox potentials are expressed under biochemical standard conditions, i.e. at pH 7.0 and 25°C. n = number of electrons transferred. The standard redox potentials of systems such as the cytochromes vary, depending on whether they are in the isolated state or present in the mitochondrial inner membrane to which the values in the table refer. The guiding principle in interpreting $E^{0'}$ values of redox systems is the *more negative* or the *less positive* the value the stronger the reducing power of the system. Under non-standard conditions, such as those which prevail in all living cells, it is essential to know the actual concentrations of oxidants and reductants before one can make a decision on whether one redox system will actually reduce another under physiological conditions. (See also Table 2.1 and 3.4)

the **hydrogen electrode** (Fig. 1.6). The potential of the latter is set by convention at 0.0 V at pH 0.0 and in the presence of 101 kPa of H_2 gas at 25°C. At pH 7.0, the standard hydrogen electrode potential becomes −0.42 V and is the reference potential used in biochemistry for characterization of the tendency of substrates to undergo oxidation. The $E^{0'}$ of a redox pair is then measured in a half-cell containing 1 mol dm^{-3} oxidant and reductant at pH 7.0 and 25°C against the standard hydrogen electrode half-cell. It is worth noting that the standard conditions for determination of $E^{0'}$ are the same as those used in measuring standard free energy changes discussed previously. A list of redox potentials of metabolically important compounds is shown in Table 1.3.

The important principle in the listing of $E^{0'}$ is that the redox pairs are set out in order of increasing standard potentials or increasing oxidizing ability. For example, the NAD^+/NADH redox couple with a large negative $E^{0'}$ will be readily oxidized by the coenzyme Q_{ox}/coenzyme Q_{red} system. Likewise, the latter will be readily oxidized by the $\frac{1}{2}O_2/H_2O$ couple with the large positive potential. It should now be evident that the $E^{0'}$ values of the various redox pairs enable one to predict the direction of electron flow from one redox pair to another, when both are under standard conditions in the presence of an appropriate catalyst: the direction will always be from negative to more positive potential values.

The standard free energy change of a redox reaction can be calculated readily from the difference in redox potentials of the two pairs undergoing reaction using the equation:

$$\Delta G^{0'} = -n\ F\ \Delta E^{0'} \qquad (1.4)$$

where F is the **Faraday**, n is the number of electrons transferred and $\Delta E^{0'}$ is the difference in standard redox potential of the electron-donating and electron-accepting pairs. For example, in the reduction of pyruvate by NADH to lactate, a reaction catalysed by the enzyme lactate dehydrogenase,

$$\text{pyruvate} + \text{NADH} + H^+ \rightleftharpoons \text{lactate} + NAD^+$$

The NAD^+/NADH pair will be oxidized by the pyruvate/lactate pair since their $E^{0'}$ values are −0.32 V and −0.19 V, respectively. $\Delta E^{0'}$ is obtained by subtracting $E^{0'}_{NAD^+/NADH}$ from $E^{0'}_{pyruvate/lactate}$

$$\Delta E^{0'} = -0.19\ V - (-0.32\ V) = 0.13\ V$$

The free energy yield corresponding to this $\Delta E^{0'}$ is shown in Box 1.5. Positive $\Delta E^{0'}$ values indicate an exergonic reaction in contrast with the notation for $\Delta G^{0'}$ values referred to above.

So far only the terms standard redox potential and potential differences of biological redox reactions have been used. It is probably evident, having considered the situation with free energy calculations, that redox reactions in cells do not take place under standard conditions. Indeed, the concentrations of the individual redox systems in cells are in the range 10^{-6}–10^{-2} mol dm^{-3}. Thus if redox potentials are to be of real value in biochemistry, the potential must be calculated for physiological conditions.

Box 1.5 Calculation of $\Delta G^{0'}$ from $\Delta E^{0'}$

The $\Delta E^{0'}$ for the oxidation of the NADH by pyruvate (see text) is 0.13 V; two electrons are transferred in the process. Since $\Delta G^{0'} = -n\ F\ \Delta E^{0'}$ in this case:

$$\begin{aligned}\Delta G^{0'} &= -2 \times 96.5\ kJ\ V^{-1}\ mol^{-1} \times 0.13\ V\\ &= -193\ kJ\ mol^{-1} \times 0.13\\ &= -25.1\ kJ\ mol^{-1}\end{aligned}$$

This may be done by using the **Nernst equation**:

$$E = E^{0'} + (RT/nF)\ \ln[A_{ox}]/[A_{red}]$$

where E is the *physiological* redox potential of the redox pair A_{ox} and A_{red}, R is the gas constant, T is the absolute temperature, n is the number of electrons transferred and F is the Faraday. Based on this equation, the physiological redox potential difference of the redox reaction between the redox systems A and B:

$$A_{ox} + B_{red} \rightleftharpoons A_{red} + B_{ox}$$

can be shown to be:

$$\Delta E = \Delta E^{0'} - (RT/nF)\ \ln[A_{red}][B_{ox}]/[A_{ox}][B_{red}] \quad (1.5)$$

It is noteworthy that the equation is similar in form to the expression for the change in free energy of a chemical reaction under non-standard conditions.

Exercise 4

The redox pair X_{ox}/X_{red} has a more negative standard redox potential than that of Y_{ox}/Y_{red}. Does it follow that system X will reduce system Y under intracellular conditions? Give reasons for your answer.

TRANSPORT OF MOLECULES ACROSS CELL MEMBRANES. All cells are separated from their immediate environment by a surface or plasma membrane. In eukaryotic animal and plant cells, there are also numerous intracellular membrane-bound systems or ***organelles*** such as the mitochondria, endoplasmic reticulum, lysosomes, peroxisomes and nucleus. A property of biological membranes is their selective permeability to the various molecules and ions dissolved in the intracellular and extracellular fluids. Generally, the greater the lipid solubility of a molecule, the greater its permeability through a membrane. Thus hydrophobic compounds cross cell membranes by simple diffusion from the compartment where their concentration is greatest.

See *Cell Biology*, Chapter 1

The simple sugars glucose, fructose and galactose can cross membranes by binding to protein carrier molecules in the membrane. This is commonly known as facilitated diffusion. In both simple and facilitated diffusion processes, the driving force for transport of the solute is its concentration gradient across the membrane. The free energy of the hydrolysis of ATP is not required for simple or facilitated diffusion. Cell membranes are usually less permeable to charged molecules and metal ions, and are impermeable to molecules of high M_r such as proteins, polysaccharides and nucleic acids.

Most cells are also capable of transporting an ion or molecule across a membrane to another compartment where its concentration is higher. Cells doing this are said to be carrying out **active transport**. Active transport processes usually require the free energy of hydrolysis of ATP to drive them. The best-known example of an active transport process is the transport of Na^+ and K^+ across the cell membrane of red blood, nerve and muscle cells. In muscle cells, the K^+ concentration in the cytoplasm is 124 mmol dm^{-3}, whereas its concentration in the serum is about 3 mmol dm^{-3}. The Na^+ concentration in a muscle cell is only 4 mmol dm^{-3} compared with about 140 mmol dm^{-3} in the serum. The gradients of these ions across the muscle cell membrane are maintained by an enzyme in the cell membrane, the **Na^+/K^+-transporting adenosine triphosphatase (Na^+/K^+-ATPase)**. This enzyme acts as a molecular pump. The free energy of hydrolysis of ATP drives Na^+ out of the muscle cell against a relative concentration gradient of 35 and K^+ into the cell against a relative concentration gradient of 41.

☐ Active transport is the process by which a molecule is transported against its concentration gradient across a semi-permeable membrane such as the plasma membrane. Free energy is always required to drive active transport of both uncharged molecules and ions and is derived from ATP hydrolysis or directly from redox reactions such as those taking place in the electron transport chain of the mitochondrion.

A similar ATP-driven molecular pump for translocating H^+ exists in the cell membranes of the acid-secreting parietal cells of the lining of the stomach (Fig. 1.7). This pump generates the acid conditions in the stomach that are necessary for the activation of pepsinogen and action of pepsin in the first phase of digestion of proteins. Mitochondria also possess H^+ pumps on their inner membranes which actively transport H^+ across this membrane using

Organelle: *a membrane-bound structure in cells which performs specific biochemical functions. Examples: the nucleus, endoplasmic reticulum, mitochondrion.*

Reference Keynes, R.D. and Aidley, D.J. (1981) Cambridge University Press, Cambridge, UK. A good short account of ion transport and basic mechanisms underlying muscular contraction and nerve impulse conduction.

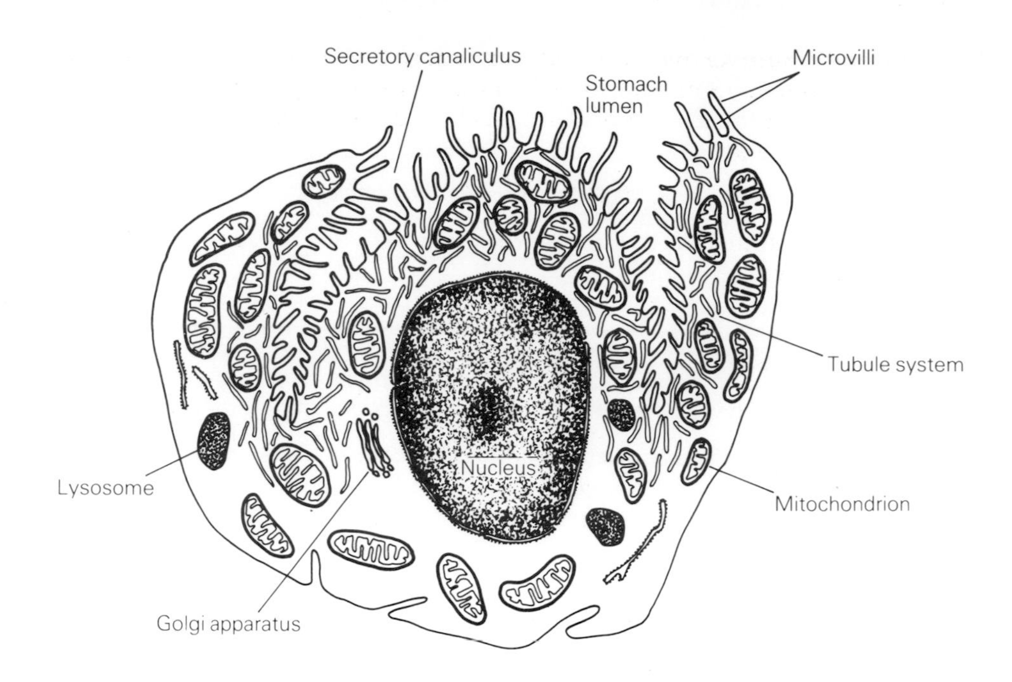

Fig. 1.7 Diagram of a parietal (or oxyntic) cell from the mammalian stomach. This type of cell secretes HCl into the stomach lumen. Secretion is facilitated by the increased surface area of the cell due to the numerous infoldings (canaliculi) and microvilli of the surface membrane. The heavy metabolic energy load associated with secretion is satisfied by the large number of mitochondria. The extensive tubules within the cytoplasm are thought to be involved in the release of ions from the cell, although their exact role is uncertain. Redrawn from Holtzman, E. and Novikoff, A.B. Saunders, (1984) *Cells and Organelles*, 3rd edn, Saunders, p. 407.

the redox energy liberated by the oxidative reactions of the electron transport chain.

The amount of free energy required for the transfer of one mole of an uncharged solute molecule across the cell membrane from a concentration $[S]_i$ inside the cell to a concentration $[S]_o$ outside the cell can be calculated using the equation:

$$\Delta G = -R\,T \ln [S]_o/[S]_i \qquad (1.6)$$

If the molecule is charged or is a metal ion, an additional electrical term has to be added to this equation. This arises because a small potential difference exists across cell membranes, varying from about −10 to −90 mV, depending on the cell type. This potential difference is commonly called the **membrane potential** and is often denoted by $\Delta\psi$. When an ion L^{m+} is transported down an electrical potential difference of $\Delta\psi$ mV in the absence of a concentration gradient the free energy change is given by the equation:

$$\Delta G = -m\,F\,\Delta\psi$$

□ Membrane potential is the potential difference that exists across the inner and outer surfaces of plasma and most intracellular membranes. It is a localized potential difference and is not a property of the cytoplasm as a whole, i.e. it is not a bulk phase property of cells or organelles. In excitable cells such as muscle fibres and neurons, it can change very rapidly during cell activation. In the mitochondrion, the membrane potential is a major driving force in the synthesis of ATP. In most cells, the value of the membrane potential lies in the range −10 to −90 mV.

Box 1.6
Calculation of the free energy required to transfer non-ionic materials between biological compartments

The concentration of creatine in urine is about 40-fold greater than that in serum. It is possible to calculate the free energy per mole required for its transfer from blood to urine at 37°C, since $\Delta G^{0'} = RT \ln[S]_o/[S]_i$, in this case:

$$\begin{aligned} G^{0'} &= 8.314 \text{ J mol}^{-1}\text{ K}^{-1} \times 310 \text{ K} \times \ln 40/1 \\ &= 9507.5 \text{ J mol}^{-1} \\ &= 9.5 \text{ kJ mol}^{-1} \end{aligned}$$

Reference Nicholls, D.G. (1982) *Bioenergetics: An Introduction to Chemiosmotic Theory*, Academic Press, New York, USA. The standard monograph on mitochondrial proton pumping and ATP synthesis.

where m is the number of unit charges on the ion. In the more general case, for example the transport of Na^+ and K^+, movement of the ions will be affected by both a concentration gradient and an electrical gradient. The free energy change is then given by:

$$\Delta G = - m F \Delta\psi + R T \ln[L^{m+}]_o/[L^{m+}]_i \qquad (1.7)$$

the subscripts o and i refer to the concentrations of L^{m+} outside and inside the cell, respectively.

1.4 The biochemistry of adenosine triphosphate (ATP)

It will be shown in Chapter 3 that the free energy liberated from dietary nutrients by the redox reactions occurring in the electron transport chain of

Box 1.7 ATP and the 'high-energy' phosphate bond

It is still frequently asserted that the two terminal *phosphoanhydride* bonds of ATP are 'high-energy' phosphate bonds; indeed that of phosphoenolpyruvate has been called a super high-energy phosphate bond. Such bonds have been denoted for many years by the symbol ~P and this is a useful way of illustrating the reactive part of the ATP molecule. Nevertheless, the terminology is misleading because it conveys the idea that there is energy in the bond that is released as free energy when it is cleaved. It appears to imply that energy is packed into that particular bond even though it is known that free energy is an *extensive property of the molecule as a whole*. This is clearly incorrect since energy is *required* to break chemical bonds.

There are three principal reasons why the hydrolysis of the ATP molecule has a more negative $\Delta G^{0'}$ than most other organic phosphates in the cell:

1. At the physiological pH of about 7.0, the ATP molecule will have four closely-spaced negative charges on the O atoms as shown below:

$$\text{adenosine}-O-\overset{\overset{O^-}{|}\ \alpha}{\underset{\underset{O}{\|}}{P}}-O-\overset{\overset{O^-}{|}\ \beta}{\underset{\underset{O}{\|}}{P}}-O-\overset{\overset{O^-}{|}\ \gamma}{\underset{\underset{O}{\|}}{P}}-O^-$$

   These charges will strongly repel each other and cause steric strain within the molecule which can only be partially relieved by hydrolysis to ADP and P_i.
2. When the reaction for ATP hydrolysis is considered:

$$ATP + H_2O \rightarrow ADP + P_i + H^+$$

   it is clear that under biochemical conditions the H^+ ion concentration will be very low. Thus, by the law of mass action, the equilibrium for ATP hydrolysis will lie far to the right compared with ordinary chemical standard conditions.
3. The third reason resides in the fact that the products of ATP hydrolysis, ADP and P_i, are more ***resonance stabilized*** than ATP. That is, there occurs greater electron delocalization in the products than in ATP itself. Mechanistically, it is possible to write more canonical forms for ADP plus P_i than for ATP, indicating that the greater reactivity of ATP is related to the lower degree of resonance stabilization.

Overall, ATP can be considered a highly reactive nucleoside triphosphate with two functional parts. One part, the nucleoside, enables it to recognize and bind to specific enzymes. The other part, the triphosphate arm, confers upon the molecule its high chemical reactivity. The role of ATP should then be seen as increasing the chemical reactivity of the various substrates that it phosphorylates even though very transiently. One may continue to use the flowery 'high-energy' phosphate language when describing the general metabolic role of ATP but with the proviso that there is more to it than meets the eye.

Pauling, L. (1970) Structure of high-energy molecules. *Chemistry in Britain*, **6**, 468–72. A definitive descricption of the structure of ATP by the twice Nobel prizewinner.

Srere, P.A. (1988) On the origins of the squiggle, in *The Roots of Modern Biochemistry* (eds H. Kleinkauf, H. von Dahren and L. Jaenicke), de Gruyter, Berlin. Impressive series of articles on the history of biochemistry, in addition to the specific paper indicated.

'Squiggle' or ~: *the symbol introduced by Lipmann in 1941 to denote a 'high-energy' bond in biochemical compounds.*
Resonance stabilization: *the stabilization of organic molecules mediated by the delocalization or 'sinking' of certain electrons to a lower energy level, thus forming a hybrid structure between two or more canonical forms.*
Phosphoanhydride bond: *the P–O–P linkage formed when two phosphate molecules or groups condense with the elimination of a water molecule.*

the mitochondria is not dissipated as heat but rather is conserved in the compound **adenosine triphosphate (ATP)**. ATP is centrally involved in energy exchanges and transformations in virtually all biological systems. It has been likened to a chemical battery by one author and this is an apt description of its role in metabolism.

STRUCTURE OF ATP. The partial structure of ATP is shown in Box 1.7. It is a nucleotide consisting of the purine base adenine, a five-carbon sugar D-ribose and a triphosphate unit. It also provides one of the basic building blocks of the genetic material, DNA. Generally, the metabolically active form of ATP is its magnesium salt, $MgATP^{2+}$. The divalent Mg^{2+} is complexed with the negatively-charged oxygen atoms of the triphosphate arm. Although the structure looks complex it is only necessary initially to focus on the triphosphate arm since it is this part of the molecule that undergoes structural change during the myriad of reactions in which ATP participates. Nevertheless, the relatively high potential energy which is made available on hydrolysis of the two terminal ***phosphoanhydride bonds*** is a property of the reaction *as a whole* and it is not merely confined to the phosphate bond cleaved on hydrolysis (Box 1.7).

Table 1.4 *Standard free energies of hydrolysis of some phosphorylated compounds*

Compound	$\Delta G^{0'}$ (kJ mol)
glucose 1-phosphate	−20.9
glucose 6-phosphate	−13.8
fructose 6-phosphate	−15.9
$ATP \rightarrow ADP + P_i$	−30.5
$ATP \rightarrow AMP + P_i$	−30.5
$AMP \rightarrow$ adenosine $+ P_i$	−14.2
phosphocreatine	−43.1
phosphoenolpyruvate	−61.9

The reactions in which the γ- and β-phosphoanhydride bonds of ATP are hydrolysed have high K_{eq} values and release large amounts of free energy as indicated below (see also Table 1.4):

1. $ATP + H_2O \rightarrow ADP + P_i + H^+ \quad \Delta G^{0'} = -31 \text{ kJ mol}^{-1}$
2. $ADP + H_2O \rightarrow AMP + PP_i + H^+ \quad \Delta G^{0'} = -31 \text{ kJ mol}^{-1}$

This free energy is harnessed or **transduced** to drive a variety of endergonic reactions. For example, those of the biosynthetic pathways and energy-requiring processes of muscular contraction, ion transport and nerve impulse.

Exercise 5

Is the reaction $ATP + H_2O \rightarrow ADP + P_i$ at equilibrium in the cell?

When ATP is cleaved to ADP and P_i, as happens during muscular contraction, ion transport and nerve impulse conduction, the stored chemical energy pool in the ATP battery tends to run down or become discharged. In normal cells the ATP concentration is kept at a fairly constant concentration through the synchronous rephosphorylation of ADP at the expense of redox reactions taking place in the mitochondrial electron transport chain. Should the two terminal phosphoanhydride bonds of ATP be hydrolysed without doing any useful work, such as energizing biosynthetic reactions, the discharge of the ATP battery would result in heat production. Indeed, this happens in brown adipose tissue as well as in certain diseases where the coupling mechanism between ATP hydrolysis and the energy-requiring systems is defective.

See Chapter 3

COUPLED REACTIONS. The free energy contained in organic molecules is extracted by various redox processes during catabolism. This energy is harnessed in the ATP molecule and is available for driving biosynthetic and other endergonic processes. All these processes represent forms of chemical work that are accomplished through the agency of **coupled reactions**, a general example of which is:

$$A + B \rightarrow C + \mathbf{D}$$

$$\mathbf{D} + E \rightarrow F + G$$

Such reactions are catalysed by specific enzymes. Note that the two reactions are linked by the **common intermediate, D** whose free energy content must be sufficiently high to drive the second reaction. ATP is most often the common intermediate as illustrated in the following coupled reactions leading to the

Reference Staunton, J. (1978) *Primary Metabolism: A Mechanistic Approach*, Oxford University Press, Oxford, UK. Contains a highly original approach to the role of ATP in metabolism and its chemical reactivity.

Canonical forms: *two or more contributing structures which approximate to the real hybrid molecule and are written according to certain rules.*

Box 1.8
ATP and the energy status of the cell

The free energy status of the cell can be expressed in several ways. Perhaps the most useful is that known as the adenylate *energy charge* which was originally proposed by Atkinson. It is an index of the extent to which the total adenine nucleotide concentration of a cell is 'filled' with 'high-energy' phosphate bonds and an indicator of the status of the major energy-yielding pathways. It is defined as:

$$\text{adenylate energy charge (AEC)} = \frac{[\text{ATP}] + 0.5[\text{ADP}]}{[\text{ATP}] + [\text{ADP}] + [\text{AMP}]}$$

The energy charge is 1.0 when all the adenine nucleotide exists as ATP and is zero when all the β- and γ-phosphoanhydride bonds of ADP and ATP have been hydrolysed. In healthy cells the value of the energy charge ranges from 0.90 to 0.95.

Example: The inherited skeletal muscle disease, malignant hyperthermia, which occurs in humans and pigs, is characterized by a very rapid increase in body temperature, skeletal muscular spasm and accelerated muscle metabolism. It is fatal if not promptly treated by the specific muscle relaxant dantrolene.

Sodium salt of dantrolene

The value of the energy charge of skeletal muscle has been measured before the onset and just before death of a pig suffering from the disease. This involved taking small samples of muscle from the pig under anaesthesia with a special biopsy device, followed by estimation of the levels of ATP, ADP and AMP in the samples. The results were:

	ATP	ADP (μmol/g tissue)	AMP
Before onset	4.2	0.37	0.029
Before death	2.1	0.66	0.19

The energy charge before the onset of the disease is given by:

$$\text{AEC} = \frac{4.2 + (0.5 \times 0.37)}{4.2 + 0.37 + 0.029} = 0.95$$

Before death AEC = 0.82

These values indicate that a comparatively small decrease in the energy charge from 0.95 to 0.82 signals the onset of death. It is generally believed that energy charge values less than 0.90 are indicative of the onset of disease in any given tissue.

Atkinson, D.E. (1977) *Cellular Energy Metabolism and its Regulation*, Academic Press, New York, USA. The author shows that energy change is a linear measure of useful energy stored in the adenine nucleotides and presents evidence that changes in the ratios of the adenylates are of greater significance in metabolic control than changes in their individual concentrations.

Ellis, F.R. and Heffron, J.J.A. (1985) Clinical and biochemical aspects of malignant hyperthermia, in *Analgesia* (eds R.S. Atkinson and A.P. Adams), Churchill Livingstone, Edinburgh, UK, pp. 173–207. Useful background reading.

synthesis of glucose 6-phosphate:

1. phosphoenolpyruvate + ADP → pyruvate + **ATP**
2. **ATP** + glucose → ADP + glucose 6-phosphate

A somewhat more complex set of coupled reactions is seen in a process essential for protein synthesis:

3. phosphoenolpyruvate + ADP → pyruvate + **ATP**
4. **ATP** + amino acid → aminoacyl-AMP + PP_i
5. **aminoacyl-AMP** + tRNA → **aminoacyl-tRNA** + AMP

Reference Hochachka, P.W. and Somero, G.N. (1984) *Biochemical Adaptation*, Princeton University Press, New Jersey, USA. An excellent short account of the thermogenic role of brown adipose tissue and its mechanism of thermogenesis is given in pp. 364–8.

Reference McPartland, A.A. and Segel, I.H. (1986) Equilibrium constants, free energy changes and coupled reactions: concepts and misconcepts, *Biochemical Education*, **14**, 137–41. A very good treatment of the energetics and efficiency of multistep processes.

In this example, there are two common intermediates, ATP and aminoacyl-AMP.

These reactions illustrate the principal ways in which free energy can be transferred from one reactant to another in metabolic pathways, under the isothermal and isobaric conditions that prevail in all organisms (Fig. 1.8). ATP fulfills the role of a free energy-carrying intermediate in that it links the wide variety of energy-producing reactions with those reactions that require energy. Other nucleoside triphosphates (Fig. 1.9) can also act as common intermediates between some catabolic and anabolic processes but are of much less significance than ATP in this fundamental strategy.

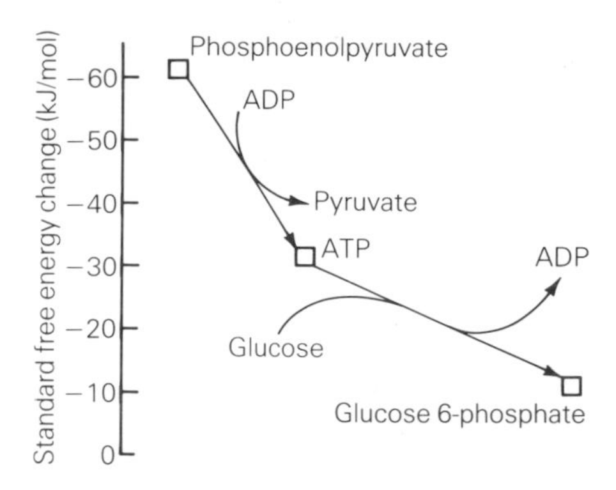

Fig. 1.8 Free energy diagram showing the intermediate position of ATP in phosphate group transfer reactions in metabolism. Phosphoenolpyruvate, produced in glycolysis, has the greatest free energy content and transfers it to ADP by phosphate group transfer to form ATP which has a free energy content intermediate to such compounds as glucose 6-phosphate. The latter is often referred to as a 'low energy' compound but, as may be judged from the diagram, this is an arbitrary distinction.

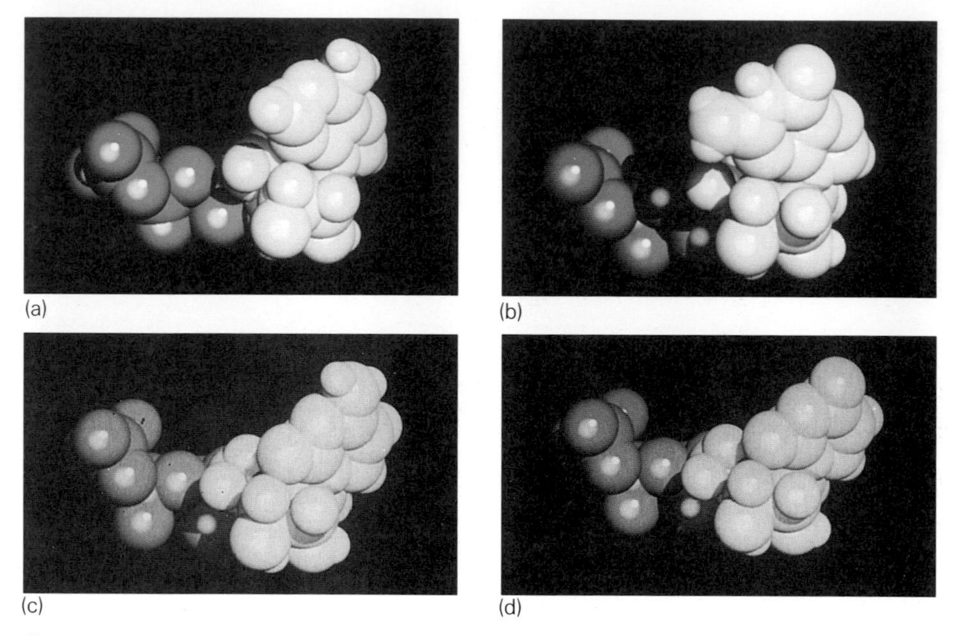

Fig. 1.9 Molecular models of (a) ATP, (b) GTP, (c) CTP and (d) UTP. Courtesy of Dr C. Freeman, Polygen, University of York.

□ The involvement of ATP as a common intermediate in a metabolic reaction is often expressed in the abbreviated form:

$$S \xrightarrow{ATP \quad ADP + P_i} P$$

Thus, the conversion of the substrate S to the product P is 'driven' by the hydrolysis of ATP to ADP.

It is evident therefore that the most important feature of such coupled reactions is the existence of a **common intermediate** without which the driving force, ΔG, cannot be communicated to the reactant(s). By analogy with electronics, the common intermediate is the energy ***transducer*** of biological endergonic processes. Of course, ΔG for the combined reactions must also obey the principle derived from the Second Law of Thermodynamics that ΔG must be less than zero if the coupled reaction process is to proceed spontaneously. For example, in the case of the following reactions:

$$6.\ A + B \rightarrow C + D \quad \Delta G_{(1)} > 0$$

$$7.\ D + E \rightarrow F + G \quad \Delta G_{(2)} < 0$$

The spontaneous conversion of A, B and E to C, F and G can only occur if the sum of $\Delta G_{(6)}$ and $\Delta G_{(7)}$ is less than zero. Usually, $\Delta G_{(6)}$ is also negative in coupled reactions. In reactions 1 and 2 above, the respective ΔG values are -31 and -17 kJ mol^{-1}.

This statement can be extended to the generalization that the net free energy change of a metabolic pathway or reaction sequence may be obtained by taking the sum of the free energy changes of the individual reactions. More simply, the ΔG values of a sequence of reactions are additive.

Transducer: *a device for transmitting the energy from one system to another. This often involves changing the form of the energy.*

Other nucleoside triphosphates may also be used to activate endergonic metabolic reactions (Fig. 1.9). For instance, guanosine triphosphate (GTP), uridine triphosphate (UTP) and cytidine triphosphate (CTP) are essential free energy sources in the biosyntheses of proteins, glycogen and phospholipids respectively. Nucleoside triphosphates are also required in the biosynthesis of DNA and RNA where they act both as substrates and as sources of free energy in the formation of the nucleic acids.

Exercise 6

Given the following two reactions and their $\Delta G^{0'}$ values, determine the standard free energy of ATP hydrolysis.

1. glucose + ATP → glucose 6-P + ADP
   $\Delta G^{0'} = -16.7 \text{ kJ mol}^{-1}$
2. glucose 6-P + H_2O → glucose + P_i
   $\Delta G^{0'} = -13.8 \text{ kJ mol}^{-1}$

1.5 Overview

Energy is required to enable all biochemical reactions to proceed in organisms. Of the many interconvertible forms of energy, free energy is the most useful for predicting whether a given reaction can take place in the cell. A reaction will take place spontaneously only if the change in free energy is negative. Although changes in entropy occurring during reactions may also be used to predict the direction of reaction, they are not readily calculated. However, changes in free energy are easily calculated from the equilibrium constant and reactant and product concentrations and from redox potentials. Even more importantly, free energy changes may be calculated from other thermodynamic quantities in cases where the equilibrium constants are very large or very small and are difficult to determine experimentally. Free energy is defined as the energy that can be used to do useful work at constant temperature and pressure. Biological systems use it to energize such important processes as muscle contraction, the transport of ions across membranes and the biosyntheses of the major constituents of living matter. All cells have metabolic processes which operate according to the fundamental laws of thermodynamics under non-standard reaction conditions.

ATP, the so-called universal biochemical energy currency, drives endergonic processes by forming common or activated intermediates. The specific high chemical reactivity of ATP resides in the triphosphate part of the molecule. It is incorrect to imagine that the free energy of ATP is stored in the terminal phosphoanhydride bond, rather the energy change is a feature of the whole reaction. A good structural, chemical basis for the high reactivity of ATP now exists.

Answers to exercises

1. The crystal, because it is the most ordered structure.
2. Organisms use the free energy of materials drawn from their surroundings (i.e. food) to make complex, highly ordered protein and nucleic acid molecules. In so doing, they also degrade some of the food energy into entropy of the surroundings. Thus, the entropy of the universe increases while that of the system, in this case, the organisms, remains constant. As far as organisms are concerned, the production of entropy in their surroundings is by no means a totally useless activity. Indeed, without doing so, organisms would lose the splendid order of their cells, tissues and organs; disease processes would soon set in and would be followed later by death. Organisms do not violate the Second Law.
3.

$$\text{glucose 1-phosphate } (5\%) \rightleftharpoons \text{glucose 6-phosphate } (95\%)$$

$$K'_{eq} = \frac{[\text{glucose 6-phosphate}]}{[\text{glucose 1-phosphate}]} = \frac{0.95}{0.05} = 19$$

Since the relation between K'_{eq} and $\Delta G^{0'}$ is given by the equation:

$$\Delta G^{0'} = -RT \ln K'_{eq}$$

the $\Delta G^{0'}$ of the above reaction is equal to $-8.3 \text{ J mol}^{-1} \text{ K}^{-1} \times 298 \text{ K} \times 19$
$= -7282 \text{ J mol}^{-1}$
$= -7.3 \text{ kJ mol}^{-1}$

4. No. It will depend on the concentration ratios $[X_{ox}]/[X_{red}]$ and $[Y_{ox}]/[Y_{red}]$ in the cell.
5. No. If it were, the ΔG would be zero and it would be impossible to derive useful work from ATP. The unique cellular role of ATP depends on maintaining this reaction very far from equilibrium.
6. Reactions 1 and 2 describe the formation of glucose 6-phosphate and its hydrolysis, respectively. The above reactions and $\Delta G^{0'}$ values may be added in the conventional algebraic manner to give:

3. ATP + H_2O → ADP + P_i
   $\Delta G^{0'} = -30.5 \text{ kJ mol}^{-1}$

This is an example of the principle that the standard free energy for a reaction may be determined by adding (or subtracting) the standard free energies of two other reactions which will combine to give the desired reaction.

Reference Edsall, J.T. and Gutfreund, H. (1983) *Biothermodynamics: The Study of Biochemical Processes at Equilibrium*, Wiley, New York, USA. An advanced treatment of biochemical thermodynamics.

QUESTIONS

FILL IN THE BLANKS

1. In a redox reaction the molecule that loses electrons becomes ______ and the molecule that gains electrons becomes ______ .

At pH 7.0, ATP occurs as a multiple ______ ______ because its ______ groups are almost completely ______ at this pH.

One of the principal reasons why the hydrolysis of ATP has a high ______ is that at ______ ______ ATP molecules have ______ closely spaced ______ charges.

ATP functions as an energy-carrying ______ ______ in ______ ______ .

ATP provides ______ ______ for the major forms of ______ work, namely, ______ of cell components, ______ contraction and ______ transport of numerous molecules and ions.

Choose from: active, biochemical, biosynthesis, charged anion, common intermediate, four, free energy, ΔG^0, ionized, living cells, muscular, negative, oxidized, pH 7.0, phosphate, reduced.

MULTIPLE-CHOICE QUESTIONS

2. Which one of the following equations is used to evaluate free energy changes in cells under physiological conditions?

A. $\Delta G = RT \ln K'_{eq}$

B. $\Delta G = \Delta G^{0'} + RT \ln \dfrac{[\text{products}]}{[\text{reactants}]}$

C. $\Delta G = RT \ln \dfrac{[\text{products}]}{[\text{reactants}]}$

D. $\Delta G = \Delta H - T\Delta S$

E. $\Delta G = \Delta G^{0'} + RT \dfrac{[\text{products}]}{[\text{reactants}]}$

3. Which one of the following equations relates the free energy change to redox potential difference under standard conditions?

A. $\Delta G^{0'} = nF\Delta E^{0'}$
B. $\Delta G^{0'} = RT \ln K'_{eq}$
C. $\Delta G = \Delta H + T\Delta S$
D. $\Delta G^{0'} = \Delta E^{0'} + \dfrac{RT}{nF} \ln \dfrac{[\text{oxidant}]}{[\text{reductant}]}$

E. None of the above

4. Which of the following redox pairs is the strongest reducing agent (the standard redox potential of each couple is given in parenthesis)?

A. ubiquinone/hydroquinone (0.10 V)
B. pyruvate/lactate (−0.19 V)
C. oxidized glutathione/reduced glutathione (−0.23 V)
D. NADP/NADPH (−0.32 V)
E. oxidized ferredoxin/reduced ferredoxin (−0.43 V)

5. Which of the following represents an *endergonic* reaction?

A. $ADP + P_i \rightarrow ATP + H_2O$
B. $ATP + H_2O \rightarrow ADP + P_i$
C. $6CO_2 + 6H_2O \rightarrow$ glucose $+ 6O_2$
D. glucose $+ 6O_2 \rightarrow 6CO_2 + 6H_2O$
E. fructose 1,6-bisphosphate $+ H_2O \rightarrow$ fructose 6-P $+ P_i$

6. State whether the following are true or false.

A. An isolated system is one that does not exchange matter or energy with its surroundings.
B. A closed system is one that exchanges matter but does not exchange energy with its surroundings.
C. An open system is one that exchanges both matter and energy with its surroundings.
D. A cell is a closed system.
E. The Second Law of Thermodynamics states that not all forms of energy are equivalent.
F. At equilibrium there is no free energy change and $\Delta G = O$.
G. All of the free energy released in an exergonic reaction in the cell is used to perform mechanical, electrical or osmotic work.
H. Spontaneous reactions have large, positive ΔG^0 values.
I. When $\Delta G^{0'}$ is positive, K'_{eq} is less than one.
J. Spontaneous reactions have large, positive ΔS values.

SHORT-ANSWER QUESTIONS

7. Consider the chemical reaction below whose K'_{eq} is 1×10^{-3}.

$$A \longrightarrow B$$

(a) Calculate the $\Delta G^{0'}$ of the reaction.
(b) Is this a spontaneous or non-spontaneous reaction?

8. Mg^{2+} ions will interact with ATP under intracellular conditions. Predict the effect of the interaction on the ΔG of hydrolysis of ATP.

9. What is the difference between $\Delta G^{0'}$ and ΔG?

10. Although animals continually generate heat from their metabolic processes, they cannot generally use this form of energy to perform any kind of work. Why not?

11. When one mole of the fatty acid, palmitate is oxidized to CO_2 and H_2O by the β-oxidation pathway in the liver at pH 7.0, 129 moles of ATP are formed from ADP and P_i. Calculate the minimum amount of free energy stored as ATP under physiological conditions when one mole of palmitate is oxidized.

12. The following reaction in the glycolytic pathway has a $\Delta G^{0'} = 24\ \text{kJ mol}^{-1}$. It is catalysed by the enzyme aldolase.
D-fructose 1,6-bisphosphate $\rightleftharpoons$ dihydroxyacetone phosphate +
D-glyceraldehyde 3-phosphate
Will this reaction go in the forward direction in a cell?

13. If the cytoplasmic concentration of K^+ of a typical mammalian cell is $140\ \text{mmol dm}^{-3}$, while the concentration in the extracellular fluid is only $5\ \text{mmol dm}^{-3}$, calculate the energy required to transfer one mole of K^+ into the cell if the membrane potential is -110 mV at 37°C.

2

An overview of bioenergetics

Objectives

After reading this chapter you should be able to:

- ☐ describe the ways in which organisms extract energy and raw materials from the environment;
- ☐ explain the biochemical basis of photoautotrophic, chemoautotrophic and heterotrophic modes of nutrition;
- ☐ evaluate the various roles of catabolic processes in cells of both autotrophs and heterotrophs;
- ☐ relate the broad strategies of catabolism adopted by cells to generate heat and make available free energy for useful work and produce precursor molecules for biosynthesis.

2.1 Introduction

Life is varied and diverse. Millions of different species of organisms are in existence and many types are now extinct. However, despite this enormous diversity of form and function (Fig. 2.1), all organisms contain compounds that have common molecular structures and operate their processes by the same sorts of chemical reactions.

Organisms require energy to do useful work and also need to obtain specific elements and molecules from the environment for their survival, growth and reproduction. The sun is the ultimate source of energy for almost all life on earth. This energy is used by photosynthetic organisms to convert carbon dioxide and other simple molecules into complex organic compounds, which make up the organism. Non-photosynthetic organisms, such as animals and many microorganisms, obtain their energy and chemical requirements when they consume and degrade the ready-made organic compounds of other organisms.

Fig. 2.1 (a) Photomicrograph of the filamentous bacterium *Bacillus cereus* (× 975). (b) Photomicrograph of cyanobacterium, *Nostoc* sp. filaments (× 1090). (c) Photomicrograph of *Amoeba proteus* (× 230). (d) Photomicrograph of the freshwater green alga, *Nitella* sp. (× 37). (Figs (a) to (d) courtesy of M.J. Hoult, Department of Biological Sciences, The Manchester Metropolitan University, UK.) (e) A community of lichens on a maritime rock surface (courtesy of Dr. A.H. Fielding, Department of Biological Sciences, The Manchester Metropolitan University, UK). (f) Common passion flower, *Passiflora caerulea*. (g) Purple moor grass, *Molinia caerulea*. (h) A group of Scots pine (*Pinus silvestris*) in the Forest of Marr, Scotland, UK. (i) Intertidal sessile acorn barnacle, *Chthamalus stellatus*. (j) Horseshoe crab, *Limulus* sp. (courtesy of Dr E.J. Wood, University of Leeds, UK). (k) Scanning electron micrograph of the grain weevil *Sitophilus infestans* (× 16) (courtesy of P. Carter, Department of Biological Sciences, The Manchester Metropolitan University, UK). (l) Young alligator in Shark Valley, Florida, USA (courtesy of Dr C.A. Smith, Department of Biological Sciences, The Manchester Metropolitan University, UK). (m) Barbary doves (*Streptopelia risoria*) (courtesy of Dr B. Stevens-Wood, Department of Biological Sciences, The Manchester Metropolitan University, UK). (n) Red deer (*Cervus elaphus*) (courtesy of Dr C.R. Goldspink, Department of Biological Sciences, The Manchester Metropolitan University, UK).

Reference Attenborough, D. (1979) *Life on Earth*, Collins, London, UK, 319 pp. Also Attenborough, D. (1984) *The Living Planet*, Collins, London, UK, 330 pp. These books are based on the BBC television series of the same names. Both provide pictorial accounts of the diversity of life on earth. There is a profusion of photographs of organisms from bacteria to human beings set in a range of natural and human-made habitats.

(a)
(b)
(c)
(d)
(e)
(f)
(g)
(h)
(i)
(j)
(k)
(l)
(m)
(n)

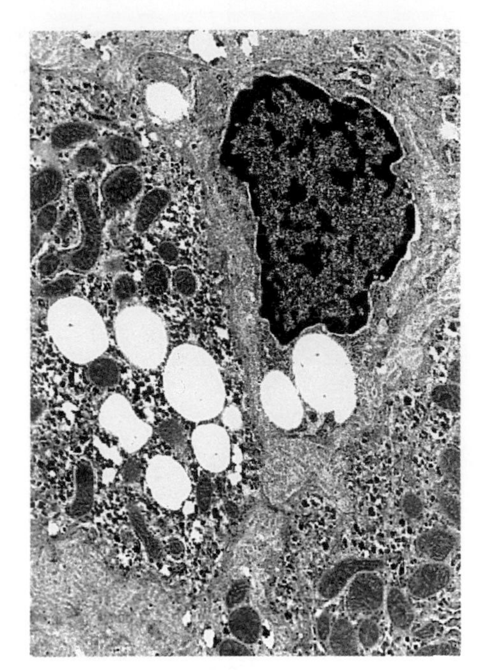

Fig. 2.2 Transmission electron micrograph of an animal cell showing membrane-bound compartments within the cell (× 4440).

The ability to turn energy into useful work and to degrade or build up cellular materials depends upon a relatively small number of co-ordinated biochemical reactions, which are common to all organisms. These reactions are controlled by enzymes, the catalysts of biological systems. The strategies of design, operation and control of biochemical processes are fundamentally the same throughout the ***biosphere***.

2.2 Organization of metabolism

Metabolism is the sum of all the chemical reactions that take place within organisms. Biochemical transformations in cells involve hundreds of thousands of chemical reactions. Reactions are organized into **metabolic pathways** in which the product of one reaction forms the reactant of another. Pathways are usually linear but some are cyclical. Pathways may also be branched so that intermediates in one pathway form the starting points of others. Metabolism can be subdivided into ***catabolism*** and ***anabolism***.

ANABOLISM is the biosynthesis of more complex molecules from simpler ones. Photosynthesis is an example of an anabolic process, which produces complex organic molecules from carbon dioxide and water. Other examples of anabolism include the biosynthesis of proteins, lipids and nucleic acids. During anabolism the substance of cells is built up from simple molecules or ***nutrients***, which are derived ultimately from the environment.

CATABOLISM is the network of chemical reactions in which complex molecules are broken down into smaller molecules. This process releases the chemical energy necessary for anabolism and the performance of work such as transport of materials across membranes and muscular movement. These degradations also form molecules that may be precursors for biosynthesis. During catabolism, waste products are produced, which must be excreted. Examples of catabolic processes include reactions that degrade carbohydrates, fatty acids and proteins and amino acids.

Catabolism and anabolism are part of a continuous recycling and renewal of cell constituents. Molecules within cells are continually being broken down and replaced by newly synthesized molecules.

Reduced nicotinamide adenine dinucleotide — Oxidized nicotinamide adenine dinucleotide

NADH → NAD^+

Ethanal CH_3CHO → Ethanol CH_3CH_2OH

(a)

Cytochrome *c* (Fe^{2+}) → Cytochrome *c* (Fe^{3+})

Cytochrome *a* (Fe^{3+}) → Cytochrome *a* (Fe^{2+})

(b)

Fig. 2.3 Examples of oxidation–reduction reactions in which: (a) hydrogen atoms are transferred between the redox couples NAD^+/NADH and acetaldehyde/ethanol; and (b) electrons are transferred between cytochrome *c* and cytochrome *a*. The reduced form of each redox couple is shown in red.

Cell structure and metabolism

Cell structure imposes organization on cellular chemistry. This is particularly true of eukaryotic cells (Fig. 2.2). Catabolic and anabolic pathways are physically as well as chemically separate. In the cells of plants and animals, for example, the catabolism of fatty acids takes place in mitochondria, whereas fatty acid biosynthesis involves a different set of reactions catalysed by different enzymes and occurs in the cytosol and endoplasmic reticulum.

OXIDATION–REDUCTION reactions, those involving the transfer of electrons or hydrogen atoms, are commonly associated with the consumption and production of metabolic energy, because the major part of the energy release occurs by oxidative processes.

Oxidations and reductions occur when one molecule receives electrons or hydrogen atoms from another molecule. The direction of transfer depends on the electron affinity of the oxidant of the couple. A couple with a low electron affinity (for example, NAD^+/NADH, Fig. 2.3) transfers electrons or hydrogen atoms to the couple with higher affinity (CH_3CHO/CH_3CH_2OH). The ability of a redox couple to donate electrons may be expressed quantitatively as the standard redox potential, $E^{0'}$. The difference between the $E^{0'}$ values of two

See Chapter 3

Biosphere: *that part of the earth which is living.*
Metabolism: *from the Greek* metabole, *change. Cells are in a state of continual chemical change.*

Catabolism: *from the Greek* katabole, *to throw down.*
Anabolism: *from the Greek* anabole, *to heap up.*
Nutrients: *environmental chemicals from which organisms are built. From the Latin* nutrire, *to nourish.*

Table 2.1 *Standard redox potentials of some redox couples; couples with more negative values readily donate electrons to couples lower in the table with less negative values* (see also Tables 1.3 and 3.4)

Couple (oxidant/reductant)	Redox potential $E^{0'}$ (V)	Biological importance
SO_4^{2-}/HSO_3^-	−0.52	Environmental source of electrons
Ferredoxin oxidized/reduced	−0.39	An electron acceptor in oxygenic photosynthesis
$NAD^+/NADH$	−0.32	Coenzymes
$NADP^+/NADPH$		Coenzymes
S/H_2S	−0.28	Environmental source of electrons
SO_4^{2-}/H_2S	−0.22	Environmental source of electrons
$FAD/FADH_2$	−0.22	Coenzyme involved in cellular oxidation and reduction
Pyruvate/lactate	−0.19	Metabolites in fermentation
Cytochrome *c* oxidized/reduced	+0.22	An electron transport protein
NO_3^-/NO_2^-	+0.43	Environmental source of electrons
Fe^{3+}/Fe^{2+}	+0.77	Environmental source of electrons
$\frac{1}{2}O_2/H_2O$	+0.82	Environmental source of electrons
Chlorophyll (P680) oxidized/reduced	+0.90	Chlorophyll reaction centre of photosystem II

$E^{0'}$ is the standard redox potential (pH 7, 25°C); it refers to the partial reaction oxidant + $e^- \rightarrow$ reductant.

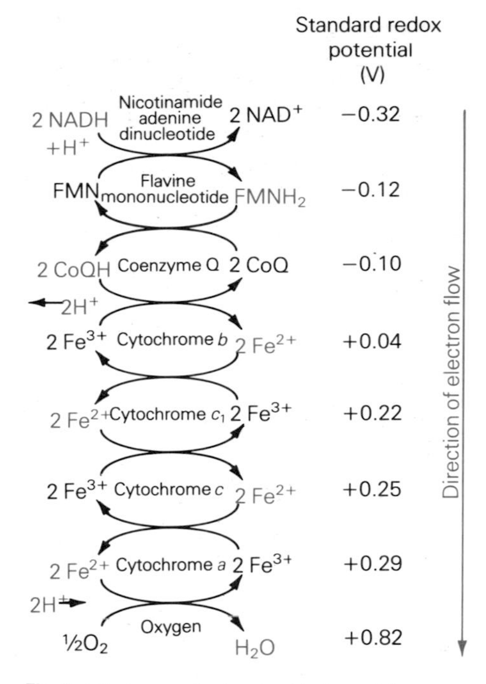

Fig. 2.4 Redox couples involved in the cellular oxidation of NADH. The reduced form of each couple is shown in red. The direction of electron flow is shown in red.

Table 2.2 *Examples of group-transfer molecules involved in metabolism*

	Group transferred	
ATP	Phosphate (HPO_4^{2-})	ADP
Creatine phosphate	Phosphate	Creatine
NADPH	Hydrogen (2H)	$NADP^+$
NADH	Hydrogen (2H)	NAD^+
$FADH_2$	Hydrogen (2H)	FAD
$FMNH_2$	Hydrogen (2H)	FMN
Acetyl coenzyme A	Acetyl (CH_3CO–)	Coenzyme A
Pyridoxamine	Amino group (–NH_3^+)	Pyridoxal

couples is a measure of the amount of energy involved in transferring electrons between them.

Table 2.1 shows the standard redox potentials of some redox couples. The reduced form of molecules in couples with negative reducing potentials are **strong** reductants, that is, they have a low electron affinity. Water is a weak reductant while NADH is a strong reductant. Figure 2.4 shows how a series of oxidation–reduction reactions can span the large energy difference between $NAD^+/NADH$ and $\frac{1}{2}O_2/H_2O$ with electrons being transferred from couples with relatively low electron affinity to those with a higher affinity.

Not every step of a catabolic pathway involves oxidation, nor does every step of an anabolic pathway involve reduction. However, where oxidation and reduction reactions do occur, **group-transfer molecules** or **coenzymes**, for example, NADH and NADPH (Table 2.2) are common participants. The reactions of catabolism generate NADH and NADPH from NAD^+ and $NADP^+$, respectively, during the oxidation of carbohydrates and fatty acids and the degradation of amino acids. The NADPH is used in anabolism regenerating $NADP^+$. Photosynthetic organisms are also able to produce NADPH using the energy of sunlight.

PROTON PUMPING harnesses the energy of electrons during the transfer of electrons from NADH to oxygen. This electron transfer takes place in the inner mitochondrial (Fig. 2.5) membrane. During transfer, hydrogen ions (H^+) are moved across the inner membrane. This pumping establishes a concentration gradient with more H^+ in the space outside the inner mitochondrial membrane than inside, and an electrical gradient with more

Exercise 1

The following reactions involve oxidation and reduction. Draw a table to show: (a) the redox couples involved; (b) which molecules are in the reduced and which are in the oxidized state; and (c) the group-transfer molecules involved.

(a) glucose 6-phosphate + $NADP^+$ ⇌ 6-phosphogluconolactone + NADPH + H^+

(b) succinate + FAD ⇌ fumarate + $FADH_2$

(c) pyruvate + CoASH + NAD^+ ⇌ acetyl CoA + CO_2 + NADH

Reference Stryer, L. (1988) W.H. Freeman, New York, 1089 pp. One of the modern 'bibles' of biochemistry, which gives a readable and fairly detailed treatment of metabolism.

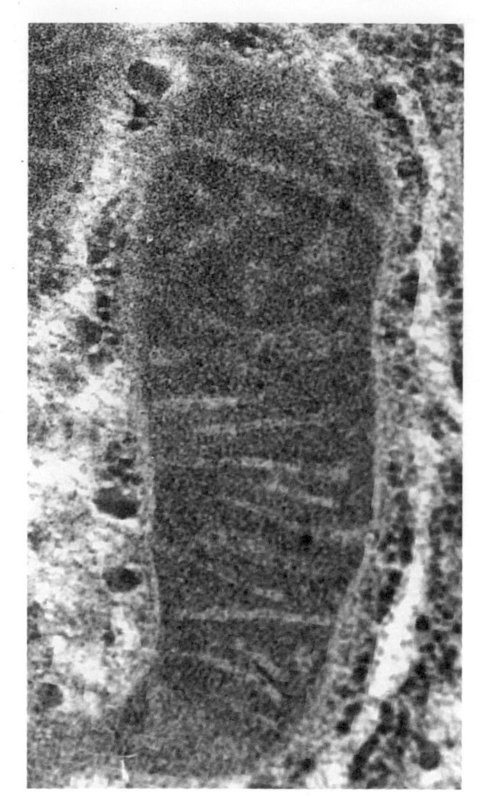

Fig. 2.5 Electron micrograph of a mitochondrion (× 47 500) shown in section so that the cristae are visible.

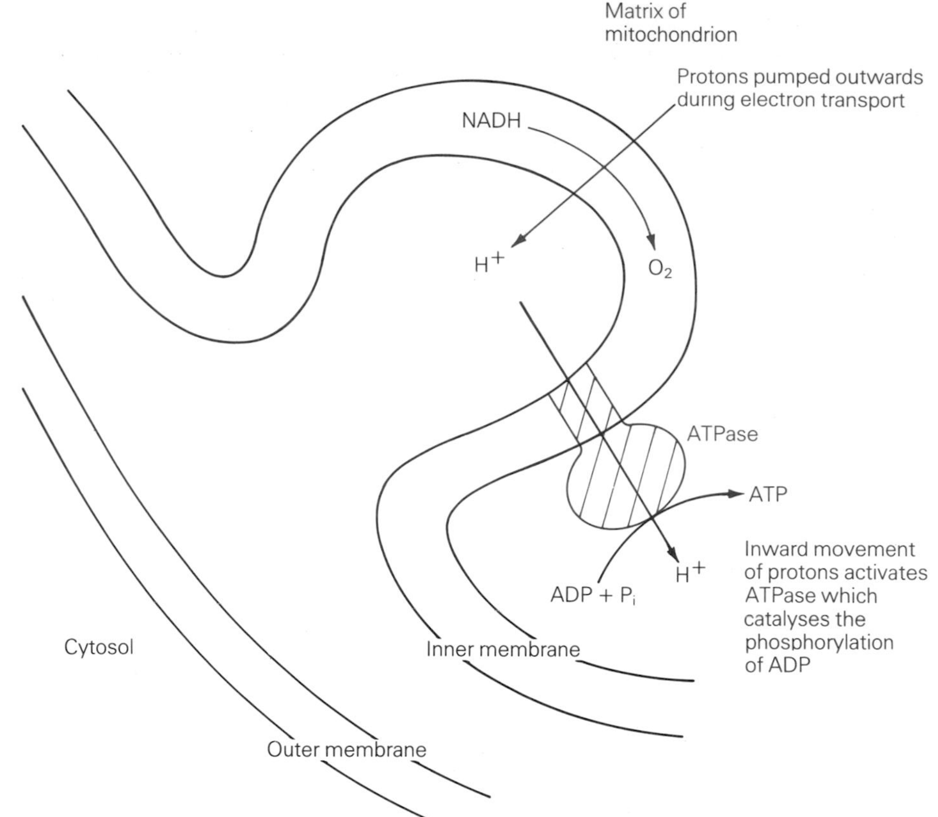

Fig. 2.6 Overview indicating the movement of H^+ in the mitochondrion during electron transfer and ATP generation.

See Chapter 3

positive charges outside than inside. These gradients of concentration and electrical potential are a form of stored energy and there is a strong tendency for H^+ to diffuse inwards across the membrane. The flow of H^+ back into the inner mitochondrial space or matrix activates an enzyme complex called ATPase, which catalyses the formation of ATP from ADP and P_i (Fig. 2.6). This generation of ATP during electron transport is called ***oxidative phosphorylation***. Very similar mechanisms occur during photosynthesis in a process called **photosynthetic phosphorylation**. *ATP* (Fig. 2.7) plays a central role in the transfer of P_i between molecules in the cell.

THE METABOLIC WEB of chemical reactions in the cell is organized into energy-consuming, reductive syntheses (anabolism) and energy-realizing, oxidative degradations (catabolism). The contrasting features of catabolism and anabolism are summarized in Table 2.3 and Figure 2.8.

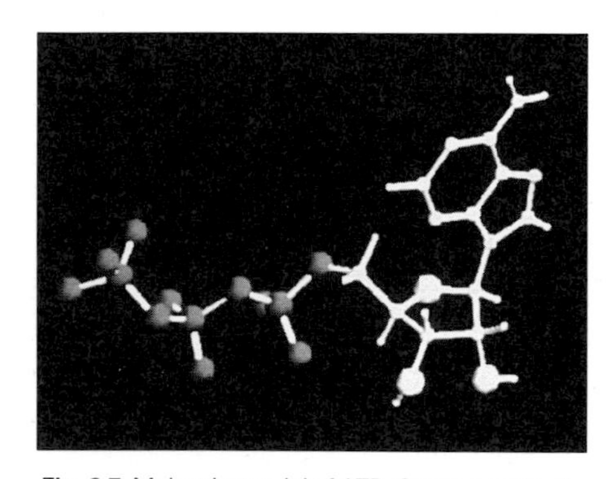

Fig. 2.7 Molecular model of ATP. Courtesy of Dr C. Freeman, Polygen, University of York, UK.

Heat and work in catabolism

When one mole of glucose is burned completely in oxygen, 2816 kJ of heat are released. The reaction may be expressed:

$$C_6H_{12}O_6 + 6O_2 \rightarrow 6CO_2 + 6H_2O \quad \Delta H = -2816 \text{ kJ mol}^{-1}$$

The heat of combustion is a measure of the change in the thermodynamic quantity called **enthalpy** (ΔH). The negative value indicates that heat is produced in this reaction.

In biological systems, not all the energy from the complete oxidation of one

Oxidative phosphorylation: *the production of ATP coupled to an electron transport chain transferring electrons from a reductant to an oxidant (O_2).*

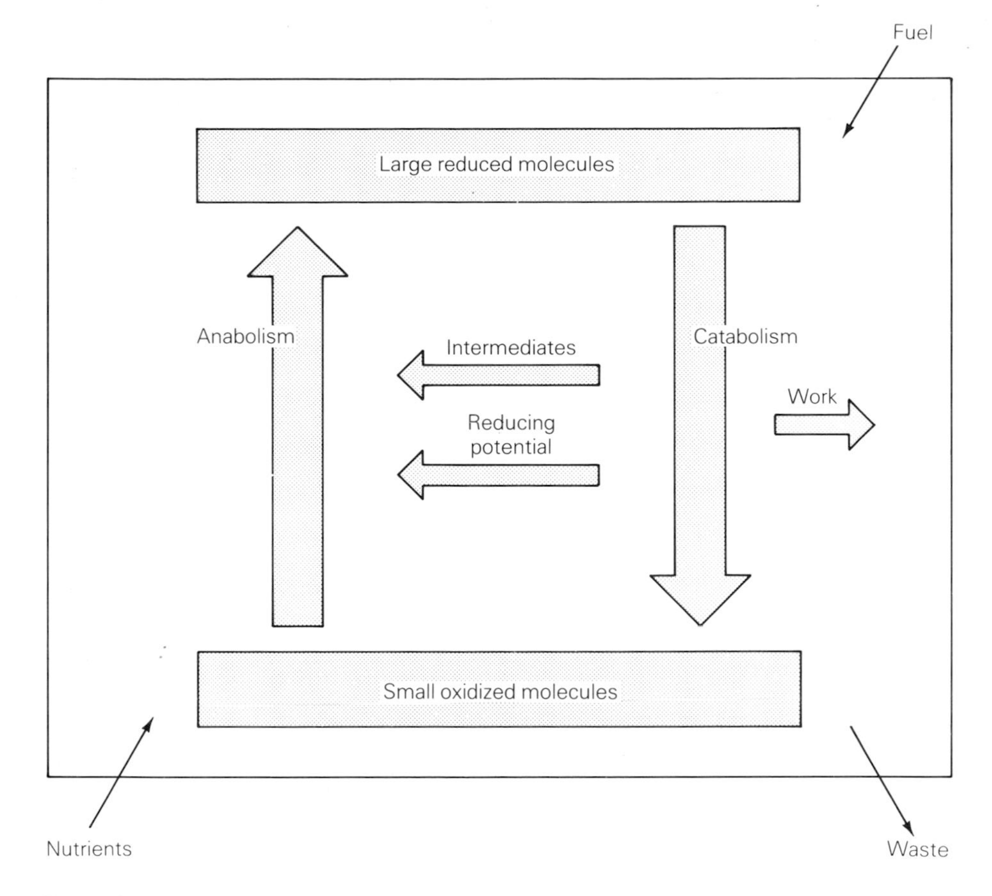

Fig. 2.8 Overview of the main processes of cell chemistry.

Table 2.3 *Comparison of the main features of catabolism and anabolism*

Catabolism	**Anabolism**
Degradative	Synthetic
Usually oxidative	Usually reductive
Energy yielding, produces ATP	Energy utilizing, consumes ATP
End-products and intermediates act as starting materials for anabolism	End-products act as starting materials for catabolism
Generates waste excreted to the environment	Utilizes nutrients from the environment

molecule of glucose is liberated directly as heat. Some of it is recovered or conserved in a form that can do work and drive biosynthesis. An example is the generation of an H^+ gradient during electron transfer. The production of a gradient of concentration and charge requires the expenditure of energy. This usable energy can **only** be generated in a system which also moves spontaneously towards equilibrium as in the diffusion of H^+ resulting in the dissipation of the concentration and electrical gradients.

See Chapter 1

The maximum amount of useful work which can be obtained from a biological system is that specified by the change in the **free energy** of the system, ΔG. ΔG accounts for both the change in enthalpy (ΔH) and the change in entropy, ΔS. An increase in S (that is, ΔS is positive) defines the direction in which a reaction moves spontaneously to equilibrium. The value

□ The free energy (G) is sometimes called Gibbs free energy after Josiah Willard Gibbs. As Professor of Mathematical Physics at Yale University during the nineteenth century, Gibbs devised much of the thermodynamic theory used in biochemistry today (Chapter 1).

Box 2.1
The bomb calorimeter

The bomb calorimeter provides a means of assessing the amount of energy made available during the catabolic degradation of food. The instrument measures the heat of combustion (calorific value) of materials. A sample of food is ignited and burned in an atmosphere of oxygen. Combustion takes place under pressure inside a thick-walled stainless steel container (the 'bomb'). The energy content or calorific value is determined by comparing the rise in temperature of the bomb resulting from combustion of the sample with that of a standard material of known energy content. A commonly used standard is benzoic acid, which has a known calorific value of 264.47 kJ g^{-1}. A chocolate bar might have an energy content of 20.05 kJ g^{-1}, a packet of salted peanuts, 26.64 kJ g^{-1}.

Exercise 2

Calculate the energy value of one gram of food, which contains 0.27 g protein, 0.28 g lipid and 0.38 g carbohydrate given that:

ΔH for carbohydrate $\rightarrow$ $CO_2 + H_2O = 17.6$ kJ g^{-1}
ΔH for lipid $\rightarrow$ $CO_2 + H_2O = 39.7$ kJ g^{-1}
ΔH for protein $\rightarrow$ urea + $CO_2 + H_2O = 18.0$ kJ g^{-1}

□ The Law of Mass Action states that the rate of a reaction depends upon the concentration of the reactants. Thus, in the reaction $A + B \rightleftharpoons C + D$ the reaction rate from left to right is directly proportional to [A] × [B]. The reaction rate from right to left is proportional to [C] × [D].

Exercise 3

The equilibrium constant (K_{eq}) for the reaction $X \rightleftharpoons Y + Z$ taking place in a cell is given by [Y] [Z]/[X] = 10^{-5} mol dm^{-3}.
(a) In a normal cell the steady-state concentrations of X, Y and Z are:
[X] = 10^{-4} mol dm^{-3};
[Y] = 10^{-5} mol dm^{-3};
[Z] = 10^{-6} mol dm^{-3}:
Is the reaction at equilibrium? If not, in which direction will the reaction tend to move to restore equilibrium?
(b) What would happen if a reaction utilizing Z as a precursor could not take place as might be the case in an abnormal cell?

See *Biological Molecules*, Chapter 6

See Chapter 5

of ΔG indicates the maximum amount of useful work which can be done *by* a reaction, and a reaction must have a negative ΔG in order to proceed spontaneously. A positive value of ΔG is sometimes said to indicate that work must be done *on* a reaction to displace it from equilibrium. In practice, reactions with a positive ΔG do not proceed and can only be made to proceed by combining them into a reaction sequence whose overall ΔG has a negative value. The ΔG for the complete oxidation of glucose:

$$C_6H_{12}O_6 + 6O_2 \rightarrow 6CO_2 + 6H_2O$$

is $\Delta G = -2879$ kJ mol^{-1}. Thus for every mole of glucose converted to carbon dioxide and water, 2879 kJ are potentially available for useful work. The difference between the useful work possible ($\Delta G = -2879$ kJ mol^{-1}) and the heat generated by burning glucose ($\Delta H = -2816$ kJ mol^{-1}) represents the amount of useful work possible from the increase in entropy. The maximum amount of work done by living systems in practice is unlikely to be greater than $-\Delta H$.

Control of metabolic reactions

The rates of reactions in metabolism depend upon the relative concentration of reactants and products. Thus, in the reaction:

$$A \rightleftharpoons B$$

a high concentration of A relative to B favours the formation of B, whereas high concentrations of B relative to A favour the formation of A. At equilibrium, the rate of formation of A is the same as that of B. The ratio of concentrations of A and B at equilibrium ([B]/[A]) is called the equilibrium constant K_{eq}. A value of K_{eq} much less than 1.0 indicates that [A] is large relative to [B] at equilibrium. Any displacement of [A] or [B] from equilibrium results in a tendency for the reaction to restore its equilibrium position. Thus, even a reaction which has an equilibrium favouring a high concentration of A can be made to produce B continuously if B is constantly removed, maintaining the ratio of [B]/[A] at a smaller value than that of K_{eq}. Reactions arranged in pathways can, therefore, proceed readily within the cell with some reactants maintained at steady-state concentrations far removed from the equilibrium ones.

ENZYMES CONTROL the rate of metabolism. Some enzymes, termed regulatory or allosteric enzymes, are sensitive to inhibition or activation by compounds other than their substrate or products, frequently metabolites that bear no particular structural relationship to the substrates of the enzymes. Such enzymes are typically situated at key points, frequently at an early reaction in a metabolic pathway. They are usually activated by the precursors of the pathway or inhibited by end-products or both. These enzymes often catalyse the slowest or rate-limiting step in a reaction sequence so that they exert fine control over the whole pathway.

GLUCOSE CATABOLISM provides an example of metabolic control. What can be achieved in one step by igniting glucose in the presence of oxygen and producing heat in a burst, takes 21 individual reactions within the cell. The first nine reactions of glucose catabolism (***glycolysis***) result in the cleavage of the C_6 glucose molecule and the formation of two C_3 molecules of pyruvate.

Table 2.4 shows the values of ΔG for each reaction of glycolysis calculated from *physiological* concentrations of metabolites. Cells maintain the concentrations of metabolites so that ΔG is usually close to 0 and negative. Most reactions therefore operate close to equilibrium ($\Delta G = 0$). Equilibria are

Glycolysis: *literally the breakdown of sugar. It refers specifically to the process where one molecule of glucoses is cleaved to produce two molecules of pyruvate.*

Reference Jones, M.N. (ed.) (1979) *Studies in Modern Thermodynamics.1.Biochemical Thermodynamics*, Elsevier Science Publishers, Oxford, UK, 389 pp. A collection of review papers. Chapter 11 is concerned with the energetic strategies of catabolism and anabolism.

Box 2.2 Metabolic disorders

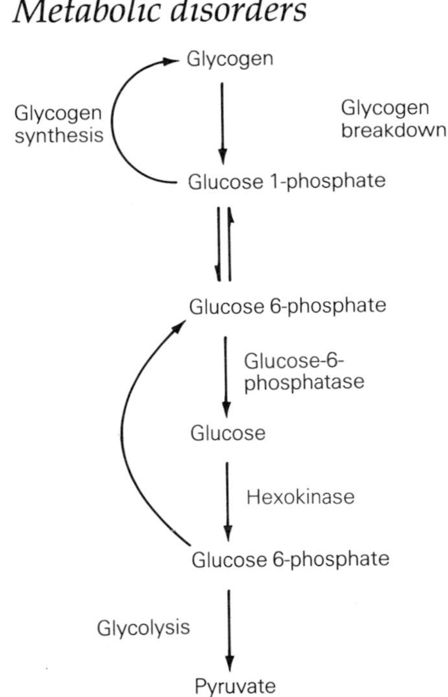

In human beings, and other organisms, a number of metabolic disorders are due primarily to the presence of a defective enzyme so that a normal metabolic reaction takes place at a much-reduced rate. Clinical problems arise when compounds that are normally metabolized accumulate in the cell. For example, the final step in the release of glucose from the liver is the conversion of glucose 6-phosphate to glucose by glucose 6-phosphatase (see figure). A defect in this enzyme results in the inability of the liver to release glucose into the blood. However, glycogen synthesis can still take place because the enzyme hexokinase catalyses the formation of glucose 6-phosphate from glucose. Thus, glycogen can be synthesized and accumulates in the liver because of the defect in a reaction essential for its breakdown. The clinical condition is characterized by an enlarged liver and low concentration of blood glucose.

Overview of the reactions of glycogen storage and breakdown.

Table 2.4 *Reactions of glycolysis*

Step	Reaction	Enzyme	*ΔG (kJ mol^{-1})
1	$C_6 \rightarrow C_6$6-P	Hexokinase	−29.2
2	$C_6$6-P $\rightleftharpoons$ $C_6$6-P	Phosphoglucose isomerase	−1.6
3	$C_6$6-P $\rightarrow$ $C_6$1,6-P_2	Phosphofructokinase	−22.5
4	$C_6$1,6-P_2 $\rightleftharpoons$ 2($C_3$3-P)	Aldolase	−9.2
5	$C_3$3-P $\rightleftharpoons$ $C_3$1,3-P_2	Glyceraldehyde-3-phosphate dehydrogenase	−6.9
6	$C_3$1,3-P_2 $\rightleftharpoons$ $C_3$3-P	Phosphoglycerate kinase	
7	$C_3$3-P $\rightleftharpoons$ $C_3$2-P	Phosphoglyceromutase	−1.8
8	$C_2$2-P $\rightleftharpoons$ $C_3$2-P	Enolase	−0.4
9	$C_3$2-P $\rightarrow$ C_3	Pyruvate kinase	−20.5

Net free energy change is −92.1 kJ mol^{-1} (note that for every one glucose molecule two C_3 molecules participate in steps 5 to 9).
* ΔG is calculated from cellular concentrations of metabolites (see p. 10).

maintained by mass action. Reactions close to equilibrium are accompanied by small changes in free energy. Thus, waste heat generated at each step is, in most cases, small.

Only three reactions of glycolysis (Table 2.4) have large negative values of ΔG, namely those catalysed by hexokinase, phosphofructokinase, and pyruvate kinase. The negative values of ΔG indicate that these reactions move spontaneously in the direction of glucose catabolism. The reactions are **not** easily reversible under the conditions existing in the cell and they provide sites for the enzymic control of the whole metabolic pathway. For example, hexokinase is a regulatory enzyme. It is inhibited by glucose 6-phosphate, which is the product of the reaction. All the kinetically reversible reactions in the pathway therefore depend on the activity of hexokinase.

□ As long ago as 1780, Lavoisier and Laplace measured the heat evolved by a guinea pig placed in a calorimeter and compared it with the amount of oxygen consumed by the animal. They concluded that respiration is a form of slow combustion.

2.3 *What is catabolism for?*

Catabolism generates:

- precursor molecules for biosynthesis;
- reducing potential, necessary for biosynthesis, usually in the form of the coenzyme NADPH;
- chemical potential energy, in the form of ATP, necessary for the performance of work and biosynthesis;
- heat.

Catabolism provides precursors for biosynthesis

There are numerous catabolic reactions and the stepwise degradation of fuel molecules leads to the formation of a seemingly vast array of intermediate

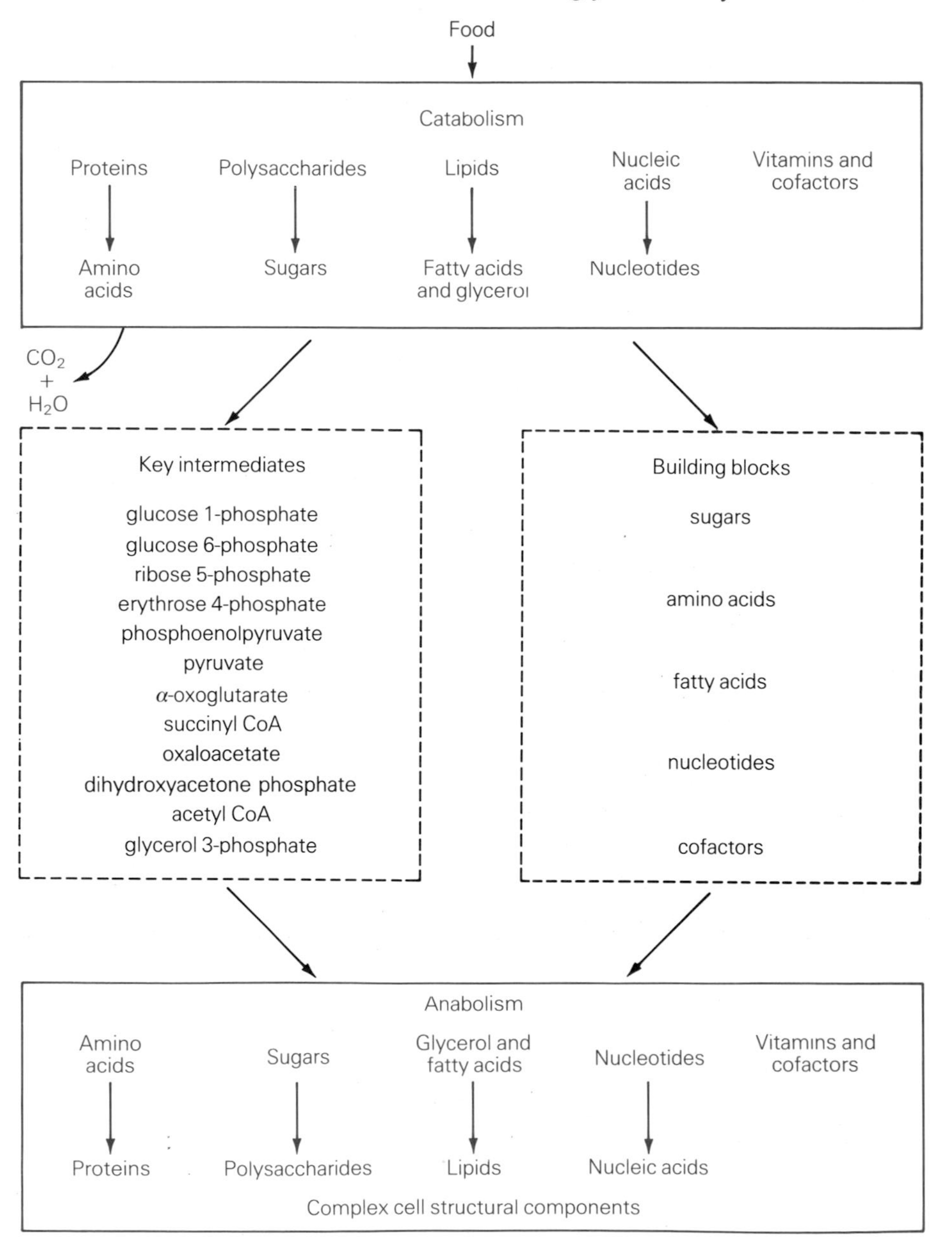

Fig. 2.9 Some of the key intermediate compounds generated by catabolism and used in anabolism.

compounds. However, an examination of the structures of these reveals that many of them are related. They include monosaccharides from polysaccharides, amino acids from proteins and fatty acids and acetyl CoA from lipids for example. These intermediates, along with the basic monomeric building blocks such as glucose, amino acids and nucleotides, act as precursors in biosynthesis (Fig. 2.9).

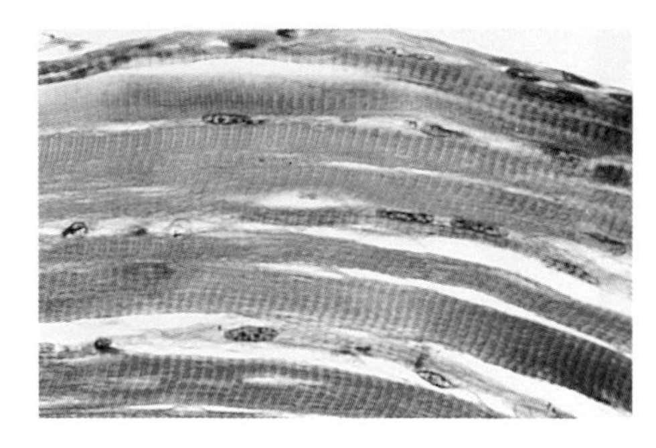

Fig. 2.10 Photomicrograph of vertebrate skeletal muscle (× 1100).

Catabolism realizes energy for useful work and biosynthesis

Various kinds of work are performed in biological systems. Mechanical work leads to movement. Perhaps the most obvious example of mechanical work is that done during muscular contraction (Fig. 2.10). Other examples include protoplasmic streaming, chromosome migration during cell division and the action of cilia and flagella. Work is done during the transport and accumulation (against a concentration gradient) of molecules and ions.

Box 2.3 Metabolism of ingested toxins

Many thousands of chemicals, such as pesticides, food additives, industrial chemicals and drugs, are toxic to cells. Mammals, including human beings, catabolize many ingested toxins and convert them into less poisonous compounds. Many of the detoxification mechanisms involve the addition of a hydroxyl group (hydroxylation) into the toxin. This reaction requires oxygen and electrons from NADPH. The reactions are generally mediated by cytochrome P450 (see figure) present in a complex bound to the endoplasmic reticulum of liver cells (*Cell Biology*, Chapter 1). The reaction, involving electron transfer in which oxygen and NADPH participate, is as follows:

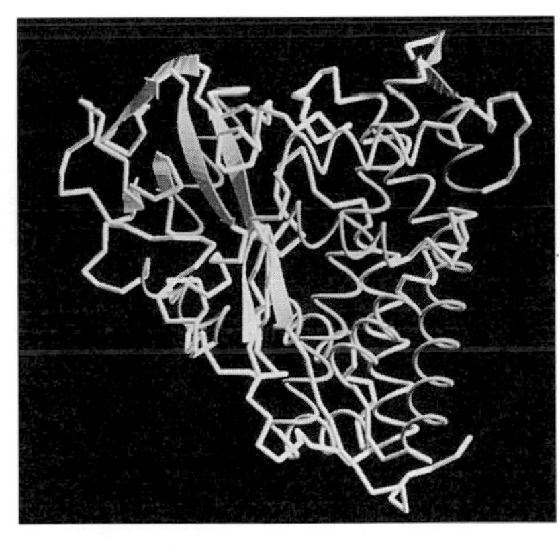

Computer-generated tertiary structure of cytochrome P450. Courtesy of Dr J.M. Burridge, IBM, UK.

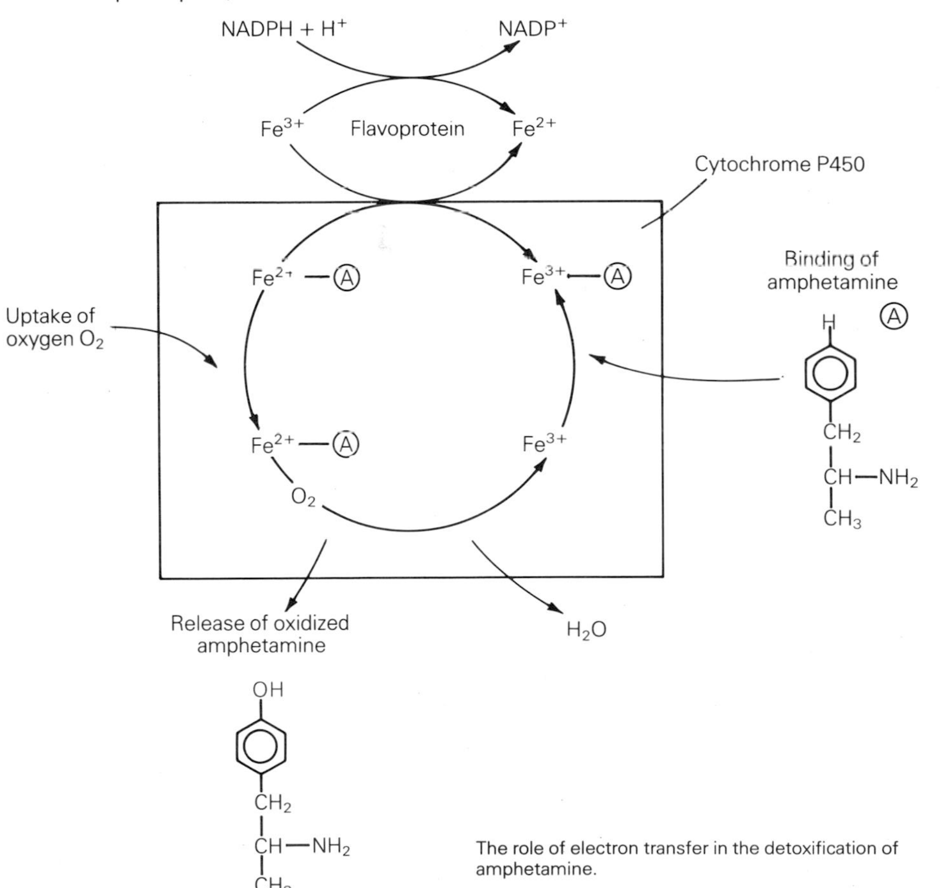

The role of electron transfer in the detoxification of amphetamine.

See Box 3.6

Transport systems are the key to several important biological processes, which include: the regulation of cell volume and the maintenance of intracellular pH and ionic composition; the acquisition of raw materials from the environment and excretion of toxic subtances; and the generation of electrical impulses (Fig. 2.11). All living systems do work in synthesizing macromolecules and other biomolecules from simple precursor molecules. Work is also done in the generation of light by bioluminescent organisms.

☐ Emitting light is probably useful to animals in recognizing mates and luring prey.

The performance of work depends on the catabolic generation of interconvertible, but alternative, forms of usable energy: reducing potential in the form of NADPH and NADH; chemical energy in the form of ATP; and concentration gradients of ions. Membrane-bound ATPases make possible the interconversion of these forms of cellular energy. In photosynthetic organisms light, not catabolism, is the ultimate energy source that similarly leads to the formation of NADPH and ATP.

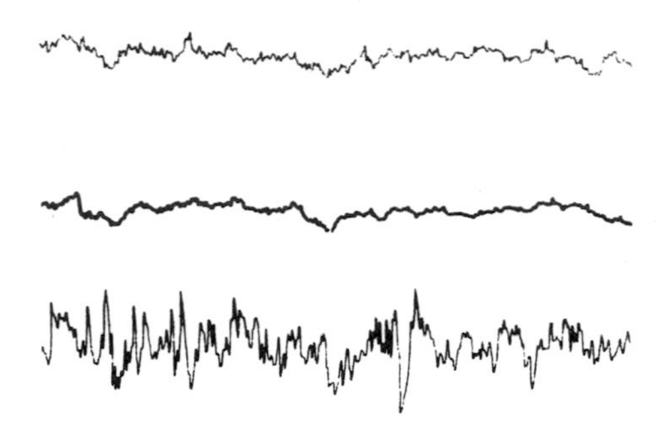

Fig. 2.11 Electrical work is performed, for example, in the brain. An electroencephalogram (EEG) indicating brain activity in a sleeping human being. Courtesy of Dr K. Hume, Department of Biological Sciences, The Manchester Metropolitan University, UK.

ATP production during electron transport is coupled, via an ATPase, to the diffusion of protons, as indicated in Fig. 2.12. Alternatively, ATP hydrolysis may be coupled via different ATPases to processes such as movement, transport of ions and molecules against diffusion gradients, generation of electrical impulses, and bioluminescence. ATP has, therefore, a variety of roles and biochemists attach much importance to measuring the concentration of ATP in cells.

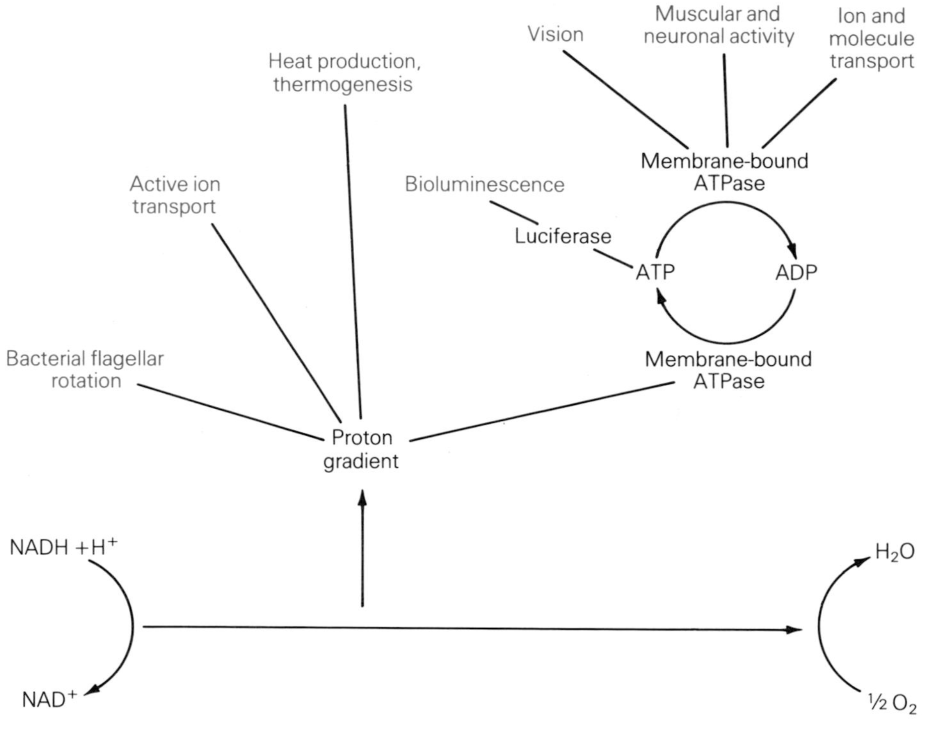

Fig. 2.12 Ion gradients are a universal link betweeen chemical processes in the cell and physical processes of work.

Catabolism maintains the concentration of ATP in cells

One of the major roles of catabolism is the maintenance of a high ratio of ATP to ADP and AMP in cells. These three forms of phosphorylated adenosine differ from each other in the number of bound phosphate groups present. The molecules are readily interconvertible through the action of **adenylate kinase**, which catalyses the reaction:

See Chapters 4 and 5

$$\text{ATP} + \text{AMP} \rightleftharpoons 2\text{ADP}$$

Reference Harold, F.M. (1986) *The Vital Force: A Study of Bioenergetics*. W.H. Freeman, New York, 577pp. An overview of the energetics of metabolism, particularly the generation and roles of ion diffusion gradients.

Reference Campbell, A.K. (1989), Living light: biochemistry, function and biomedical applications, *Essays in Biochemistry*, **24**, 41–81. An illustrated review of aspects of bioluminescence.

The summed concentration of ATP, ADP and AMP is known as the **adenylate pool** of the cell. The relative proportions of the three components of the adenylate pool are a measure of the capacity of the cell to do work. Thus, if *all* the adenylate pool is in the form of ATP, the cell will have a *maximum* free-energy content. Therefore, the total free energy in the form of adenylate is proportional to the *mean number* of bound phosphate groups per adenine nucleotide. The mean number of such groups would be a minimum of zero if all the cellular adenylate was in the form of AMP and a maximum of two if all the bound phosphate groups were in the form of ATP.

See *Cell Biology*, Chapter 2

THE ADENYLATE ENERGY CHARGE expresses quantitatively the mean number of phosphate groups per adenine nucleotide. The mean number of phosphate groups can be calculated from the following relationship:

$$\text{Mean number} = ([\text{ADP}] + 2[\text{ATP}])/([\text{AMP}] + [\text{ADP}] + [\text{ATP}])$$

where [AMP], [ADP] and [ATP] represent the cellular concentrations of AMP, ADP and ATP respectively; [AMP] + [ADP] + [ATP] is the total adenylate concentration; [ADP] is the number of phosphates carried by ADP and [ATP] × 2 is the total number carried by ATP. The values calculated for the mean number of bound phosphates per adenine nucleotide range from 0 to 2. The adenylate energy charge may have any value between 0 and 1 and is calculated from the relationship:

$$\text{Adenylate energy charge} = 0.5 \times (\text{mean number of phosphates per adenine nucleotide})$$

It has been argued that the adenylate energy charge is the key factor in the regulation of catabolic and anabolic pathways, which are coupled through the ATP/ADP/AMP system. The relationship between the adenylate energy charge and the rates of ATP-generating and ATP-utilizing reactions is shown in Fig. 2.13. The rate of ATP production is the same as that of ATP utilization where the two curves intersect. If the adenylate charge rises or falls from a value characteristic of this metabolic steady state, ATP, ADP and AMP act as allosteric activators or inhibitors of regulatory enzymes. Thus, ATP-utilizing pathways are stimulated by a high energy charge, while ATP-generating systems are activated by low energy charge. Active cells typically have an adenylate charge of about 0.9.

The adenylate energy charge is useful in that it can be calculated from data of relative rather than absolute values of the cellular concentrations of ATP, ADP and AMP.

Exercise 4

Examine the data presented in the figure, which shows the separation of AMP, ADP and ATP using high-performance liquid chromatography (HPLC). The heights of the peaks are proportional to the concentrations.
(a) Use the data to determine the adenylate energy charge. (b) How does the result compare with the value of 0.9 quoted in the text for active cells? Account for any differences.

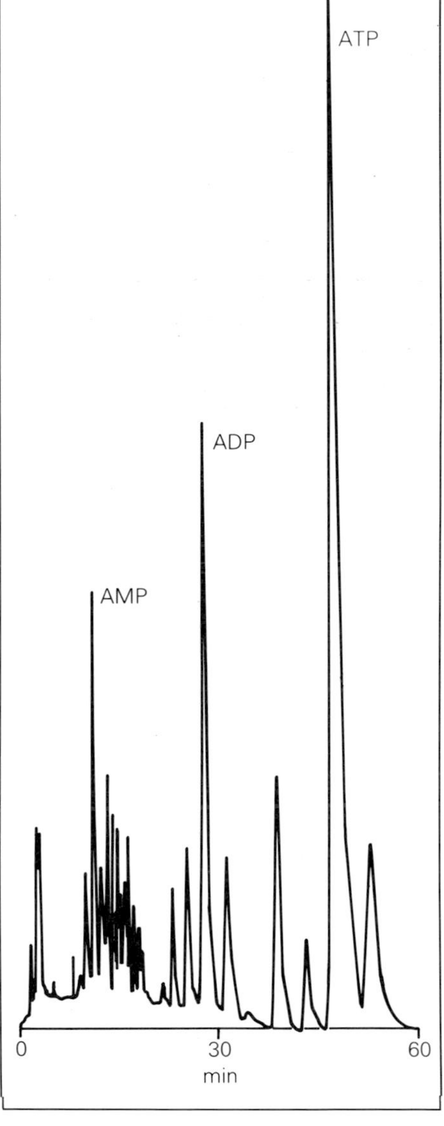

High-performance liquid chromatography (HPLC) separation of a perchloric acid-soluble extract of hybridoma cells. The retention times for AMP, ADP and ATP are indicated. The amount of each adenine nucleotide is proportional to its peak height. Courtesy of M. Costello, Department of Biological Sciences, The Manchester Metropolitan University, UK.

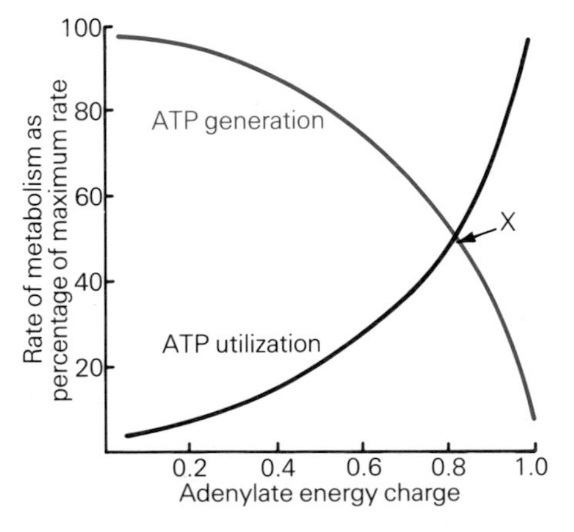

Fig. 2.13 The rate of ATP utilization and ATP generation in the cell plotted as a function of adenylate energy charge. At the point of intersection (X), the rate of production of ATP is the same as that of its utilization.

Box 2.4 Bioluminescence and the bioassay of ATP

A number of assays based on bioluminescence have been developed for determining the concentration of ATP. One method uses the reactions catalysed by the enzyme **luciferase**, extracted from the firefly (*Photinus* sp.). Extracts from *Photinus* contain both luciferase and compounds called luciferins, which can be oxidized in an ATP-dependent manner releasing light. The reaction takes place in two steps.

Step 1 is the ATP-driven, reversible formation of luciferase–luciferyl adenylate complex and pyrophosphate (PP_i).

$$\text{luciferin} + \text{ATP} \xrightleftharpoons{\text{luciferase}} \text{luciferin} + PP_i + \text{AMP}$$

Step 2 is the irreversible reaction of the complex with molecular oxygen to produce light with a peak wavelength of 560 nm.

$$\text{AMP} + \text{luciferin} + O_2 \xrightarrow{\text{luciferase}} \text{oxyluciferin} + \text{ATP} + \text{light}$$

The amount of light emitted is proportional to the concentration of ATP and can be measured accurately with a sensitive instrument called a luminometer. Concentrations as low as 10^{-15} mol dm^{-3} of ATP may be detected.

The concentration of ADP and AMP in samples can be determined by first converting them to ATP using appropriate enzymes and phosphate donor substrates:

$$\text{AMP} + \text{ATP} \xrightarrow[\text{kinase}]{\text{adenylate}} 2\text{ADP}$$

$$\text{ADP} + \text{PEP} \xrightarrow[\text{kinase}]{\text{pyruvate}} \text{ATP} + \text{pyruvate}$$

and then measuring the 'new' concentrations of ATP by luminometry. Knowledge of the initial concentrations of ATP allows the concentrations of ADP and AMP to be determined. Hence the adenylate energy charge of the cells or tissues can be calculated. Values of 0.8–0.9 are found in normal cells; lower values are associated with non-viable cells. The technique can be also be used to determine the continuous production of ATP in suspensions of isolated mitochondria.

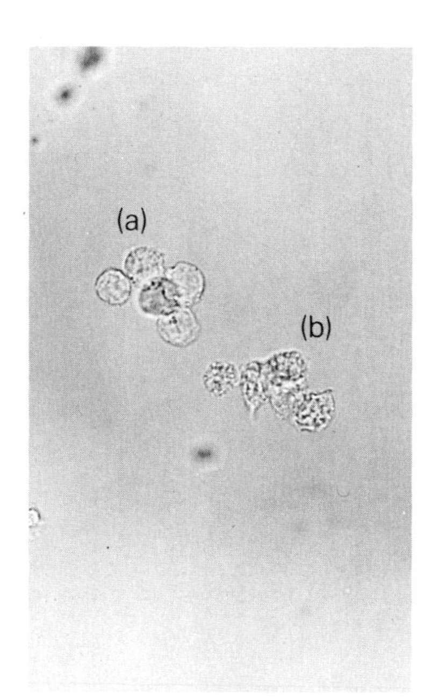

Hybridoma cells with (a) a high adenylate energy charge (~ 0.8) and (b) a low adenylate energy charge (~ 0.4). Note the differing appearance between the healthy and the non-viable cells (× 370). Courtesy of Dr H. Jenkins, Department of Biological Sciences, The Manchester Metropolitan University, UK.

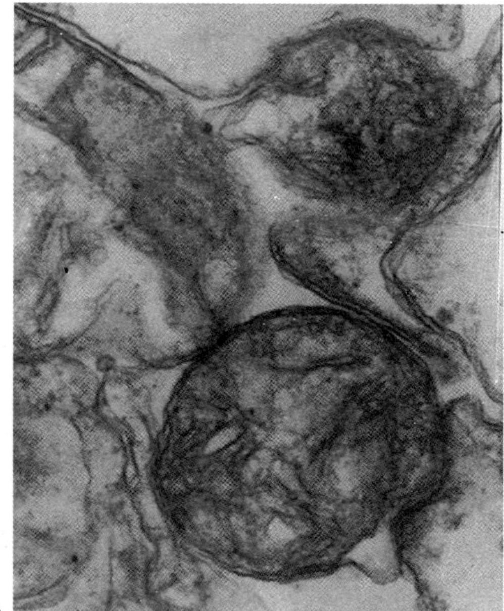

Isolated mitochondria (× 18 000).

Deluca, M.A. (ed.) (1978). *Bioluminescence and Chemoluminescence. Methods in Enzymology*, Vol. LVII, Academic Press, New York, 653 pp. The first of two advanced manuals in the 'Methods' series devoted to bioluminescence. The early part of the text dealing with underlying principles is most useful.

Box 2.5 Nuclear magnetic resonance spectroscopy (NMR)

NMR has been used to measure the concentrations of bound and free phosphate directly in tissues without the technical difficulties, inaccuracies and errors associated with conventional extraction, purification and characterization procedures.

In NMR, a powerful external homogeneous magnetic field is applied to the sample under study. Some atomic nuclei such as that of the phosphorus isotope ^{31}P possess a permanent magnetic moment due to nuclear spin. On application of the external field, there is an alignment of the nuclear magnets producing net orientation of spins of the nuclei in the direction of the magnetic field (lower energy level). Using radio-frequency electromagnetic radiation it is possible to excite the nuclei so that they align against the external field at a higher energy level. The radio frequency required to excite the nuclei (that is, the resonance frequency) is constant for a given nucleus in a given magnetic field strength. The resonance frequency depends on the precise chemical environment of the nucleus. Thus, ^{31}P in the γ-bonded phosphate (terminal phosphate) of ATP has a resonance frequency different from that of the α- and β-bonded phosphate.

The NMR spectrum is expressed as intensity, which is related to concentration of the nuclei, against the resonance frequency.

The figure shows ^{31}P-NMR spectra of P_i, ATP and ADP in maize root tips under aerobic and anoxic conditions. NMR can distinguish between the P_i present in the vacuole and that in the cytosol, enabling the cytosolic phosphorylation potential to be determined. The resonance frequency of the P_i in the vacuole and in the cytosol is different because of differences in pH in the two compartments.

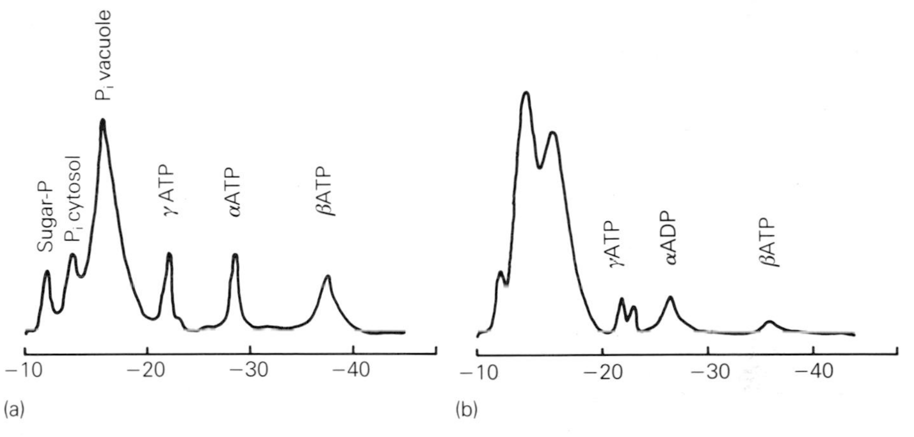

^{31}P-NMR spectra of maize root tips perfused with either (a) oxygen or (b) nitrogen. The peaks corresponding to P_i in the cytosol, P_i in the vacuole, and the phosphate in ATP and ADP are indicated. Redrawn from J.K.M. Roberts (1984) *Annual Review of Plant Physiology*, **35**, 375–86.

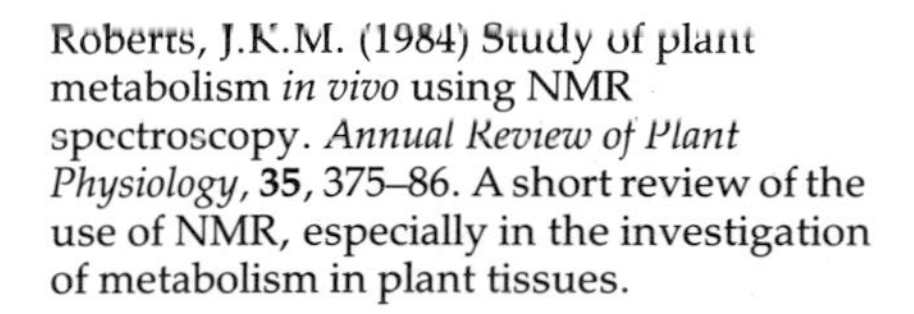

Roberts, J.K.M. (1984) Study of plant metabolism *in vivo* using NMR spectroscopy. *Annual Review of Plant Physiology*, **35**, 375–86. A short review of the use of NMR, especially in the investigation of metabolism in plant tissues.

THE PHOSPHORYLATION POTENTIAL is an index of metabolic energy which expresses the relative concentration of ATP in the cell according to the following relationship.

$$\text{Phosphorylation potential} = [\text{ATP}]/([\text{ADP}] + [\text{P}_i])$$

THE RATIO OF ATP TO ADP CONCENTRATION has also been used to indicate the free-energy status of cells. For example, Fig. 2.14 shows the relationship between the motility of bull spermatozoa as a function of the ratio of [ATP] to [ADP]. The spermatozoa were treated with a range of concentrations of KCN. KCN inhibits aerobic catabolism of glucose and so less ATP is produced. In this way, the [ATP]/[ADP] ratio was controlled. It is evident that motility of spermatozoa requires high concentrations of ATP relative to ADP.

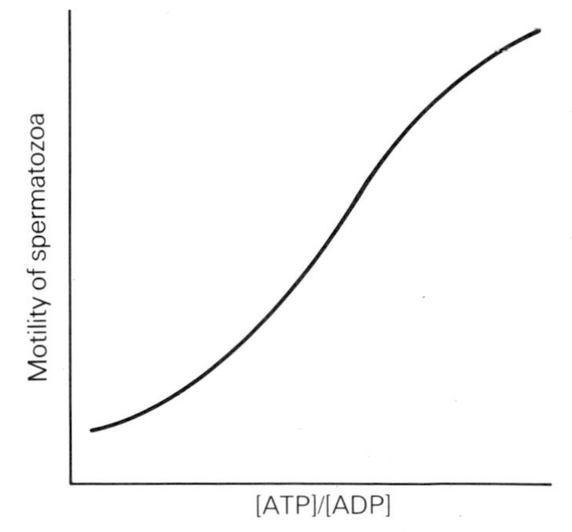

Fig. 2.14 Motility of bull spermatozoa (turbidity) plotted as a function of [ATP]/[ADP] ratio (controlled by the addition of KCN).

Exercise 5

In this chapter examples are given of organisms that generate heat. What are the biochemical advantages of regulating body temperature? Is there any evidence to suggest that cells might be heat engines?

☐ Animals classified as **ectotherms** use environmental sources of heat in the regulation of their body temperatures. For example, lizards bask in sunshine to raise their body temperatures.

☐ Under average conditions, a human male of mass 70 kg catabolizes fuel compounds at a rate of about 9000 kJ day^{-1}. All of this energy is eventually dissipated as heat and so contributes to the maintenance of body temperature. In a marathon race, a runner can generate heat at a rate of 1.4 kW, which is equivalent to approximately 12 000 kJ during a race lasting just a few hours. If the weather is hot and the sun is high, dissipation of this body heat can pose severe problems of dehydration and hyperthermia, possibly leading to death. Olympic marathons are never run at midday!

A marathon race. Courtesy of *Stockport Express, Advertiser and Times*, Stockport, UK.

Fig. 2.15 Respiratory heat production in the skunk cabbage *Symplocarpus foetidus*. Enough heat has been generated to melt the snow in the immediate vicinity of the plant. Courtesy of Professor R. Knutson, Luther College, Iowa, USA

Catabolism generates heat

All the energy in fuel compounds is eventually lost as heat which is dissipated in the environment. Although heat is often considered to be a waste by-product, many animals and some plants make use of the heat generated during the catabolism of fuel compounds in the regulation of their temperature.

In many ***endothermic*** animals, muscular movement, which consumes ATP and stimulates respiration, is a major way of producing the heat necessary for maintaining body temperatures above that of the environment. Such muscular action includes shivering in some mammals and wing vibrations in bumble bees. Some mammals supplement the process of shivering by oxidizing stored fat through a mechanism that uncouples the production of ATP from the diffusion of H^+ down a gradient established during electron transport. ATP is not produced but heat is generated. This mechanism of **thermogenesis** occurs in the mitochondria of brown adipose tissue of many cold-adapted animals as well as newly born mammals, including humans.

Some plants are able to produce heat directly by oxidizing NADH via an alternative electron transport pathway. Thermogenesis in Araceae plants (Fig. 2.15) is a classic example. In *Arum maculatum* (Lords and Ladies) shown diagrammatically in Fig. 2.16, heat is generated rapidly in the upper spadix (appendix) at certain times of the day (Fig. 2.17). Volatile compounds are evaporated, attracting pollinating insects to the flower. The reduction of oxygen by NADH in tissue extracts of the spadix of *Arum* is not inhibited by cyanide or azide, which block electron transport and oxidative phosphorylation. NADH must, therefore, be reducing oxygen by a pathway that directly generates heat.

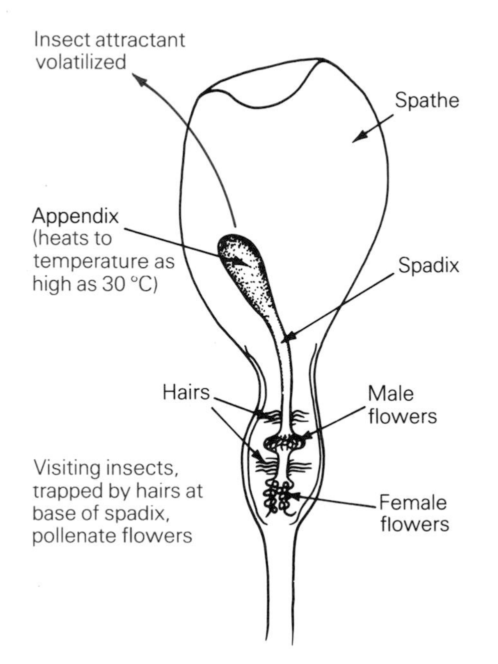

Fig. 2.16 The flower of *Arum maculatum* showing details of the flower structure associated with trapping pollinating insects.

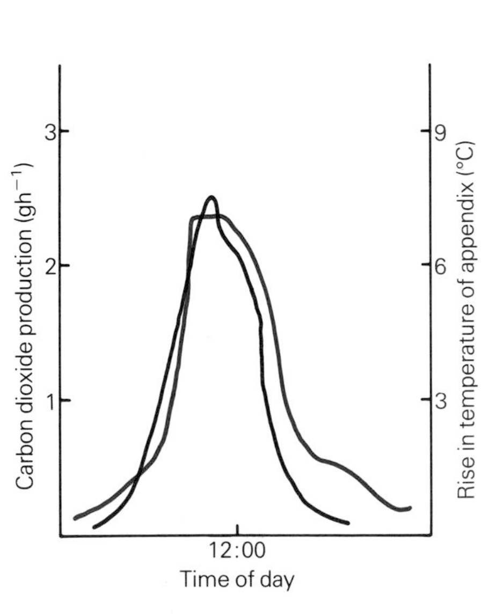

Fig. 2.17 Relationship between catabolic rate (carbon dioxide production) and rise in temperature of the spadix (appendix) of *Arum maculatum*. Redrawn from B.J.D. Meeuse (1966), *Scientific American*, **25(1)**, 80.

Endotherm: *an organism that uses an internal (metabolic) source of heat to maintain body temperature.*

Reference Aked, J. (1989) Arum – still a hot topic. *Plants Today*, **2**, 181–3. A short account of *Arum* species, with particular reference to the heat-generating capacity of plant species of this type.

2.4 *Acquiring energy from the environment*

Organisms can be classified broadly into two major nutritional types, depending on how they feed (Table 2.5). Those able to utilize environmental energy directly in the synthesis of their fuel compounds are called ***autotrophs***. All green plants belong to the largest group of autotrophs, the photosynthesizers. Photosynthetic autotrophs (***photoautotrophs***) use light energy in the conversion of carbon dioxide and inorganic compounds into cellular components, such as carbohydrates, proteins and lipids. A relatively small but significant group of organisms, the ***lithotrophs*** or ***chemo-autotrophs***, are able to perform similar biosynthetic processes but use energy derived from the oxidation of reduced inorganic compounds. In contrast to autotrophs, which synthesize their organic molecules from inorganic sources, **heterotrophs** obtain ready-made fuel compounds from the environment. Heterotrophs thus depend ultimately on autotrophs for their food supply (Table 2.6). Figure 2.18 shows the nutritional relationships between heterotrophs and autotrophs. The chemical potential energy originally derived from the environment by the autotrophs is eventually dissipated as heat.

Exercise 6

The hydrolysis of one mole of sucrose to glucose and fructose involves an enthalpy change of 29.3 kJ. To what extent should this reaction be taken into account when designing a calorie-controlled slimming diet?

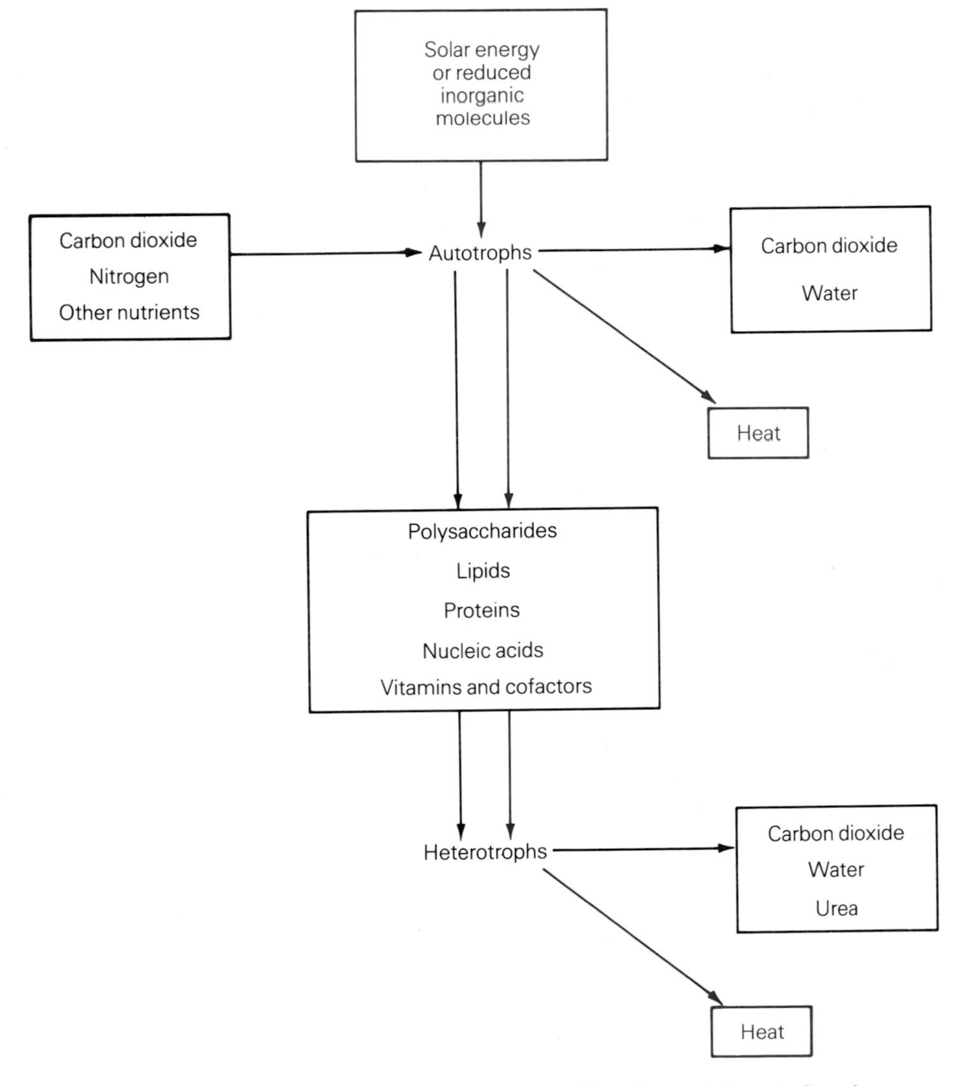

Fig. 2.18 The flow of carbon, energy and nutrients in the biosphere. The red arrows indicate the flow of energy, largely in the form of reducing potential. The dark arrows indicate the flow of nutrients and wastes.

Autotroph: *from the Greek* troph, *to feed, and* auto, *self. 'Self-feeders' need an environmental source of energy. Heterotroph means feeding from sources other than oneself.*

Photoautotroph: *autotroph that uses light as an environmental source of energy. From the Greek* photo, *light.*

Lithotroph *(chemoautotrophs): autotroph that uses inorganic sources of energy from the environment. From the Greek* litho, *rock.*

Table 2.5 Classification of organisms by cell type and principal modes of nutrition

	Autotrophs	Heterotrophs	
Prokaryotes	Aerobic and anaerobic photosynthesizers	Aerobic and anaerobic decomposers	
	Aerobic and anaerobic chemolithotrophs	Lithotrophic heterotrophs	
Eukaryotes	Plants: aerobic photosynthesizers	Fungi: aerobic decomposers	Animals: herbivores, carnivores, detritivores

Table 2.6 Classification of organisms according to their mode of acquiring energy and carbon compounds from the environment

Energy source	Carbon source	
	Inorganic (CO_2)	Organic
Light	Photoautotrophs	Photoheterotrophs
Inorganic chemical	Chemolithotrophs or chemoautotrophs	Lithotrophic heterotrophs
Organic chemical	–	Heterotrophs

Autotrophic nutrition

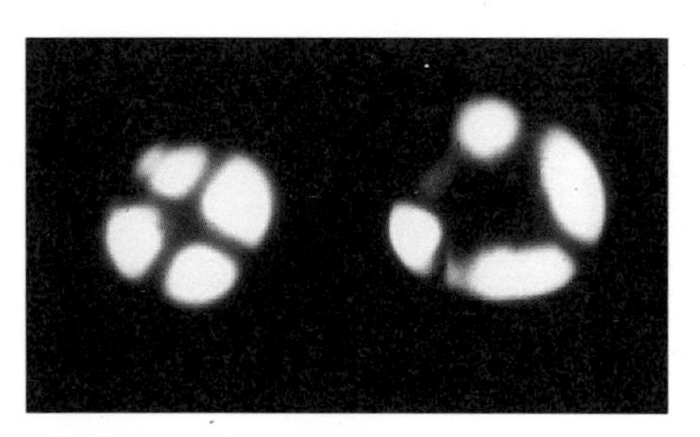

Fig. 2.19 Starch grains, as observed by polarized light microscopy, showing typical 'Maltese cross' Courtesy of M.J. Hoult, Department of Biological Sciences, The Manchester Metropolitan University, UK.

The primary products of light-driven and chemically-driven reductions in all autotrophs are ATP and NADPH. These compounds are used in the formation of carbohydrate in the Calvin cycle of photosynthesis. The enzymes of this cycle are common to virtually all autotrophs. Autotrophs typically polymerize the products of the Calvin cycle to produce fuel storage compounds such as starch (Fig. 2.19) in eukaryotic plants and cyanobacteria, or glycogen or poly-β-hydroxybutyrate in other autotrophs.

A feature common to virtually all autotrophs is the generation of ATP coupled to electron transfer. This type of ATP production is called **electron transfer phosphorylation.** Electron transfer phosphorylation in photosynthesis is called **photophosphorylation**.

CYANOBACTERIA AND UNICELLULAR AND MULTICELLULAR PLANTS (Figs 2.1 and 2.20) extract electrons from water and reduce carbon dioxide to carbohydrate. The oxygen of water is liberated as free oxygen.

The first part of the reaction involves the capture of light by pigments such as chlorophylls producing a redox couple with a higher affinity for electrons than water. For example, in the photosystem II (Fig. 2.21) part of the photosynthetic apparatus (see below), the oxidized/reduced chlorophyll couple has a more positive standard redox potential than that of the oxygen/water couple (Table 2.1). Electrons are thus lost from water, which dissociates to liberate molecular oxygen.

$$2H_2O \rightarrow 4H^+ + 4e^- + O_2$$

The electrons and protons (H^+) released eventually reduce $NADP^+$ to NADPH.

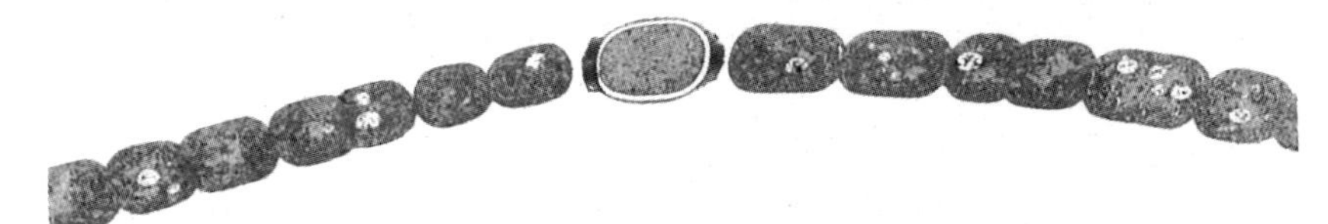

Fig. 2.20 Transmission electron micrograph of a filament of *Anabaena cylindrica* showing a heterocyst with its thick wall and vegetative cells (× 3600). Courtesy of Professor G.A. Codd and Ms G.A. Alexandre, Department of Biological Sciences, University of Dundee, UK.

Reference Blaxter, K. (1989) *Energy Metabolism in Animals and Man*, Cambridge University Press, Cambridge, UK, 336 pp. A readable text providing a refreshing treatment of energetics and integrating aspects of biochemistry, biophysics, physiology and environmental biology. The book sets biochemical energetics into the context of the environmental physiology of human beings and other animals.

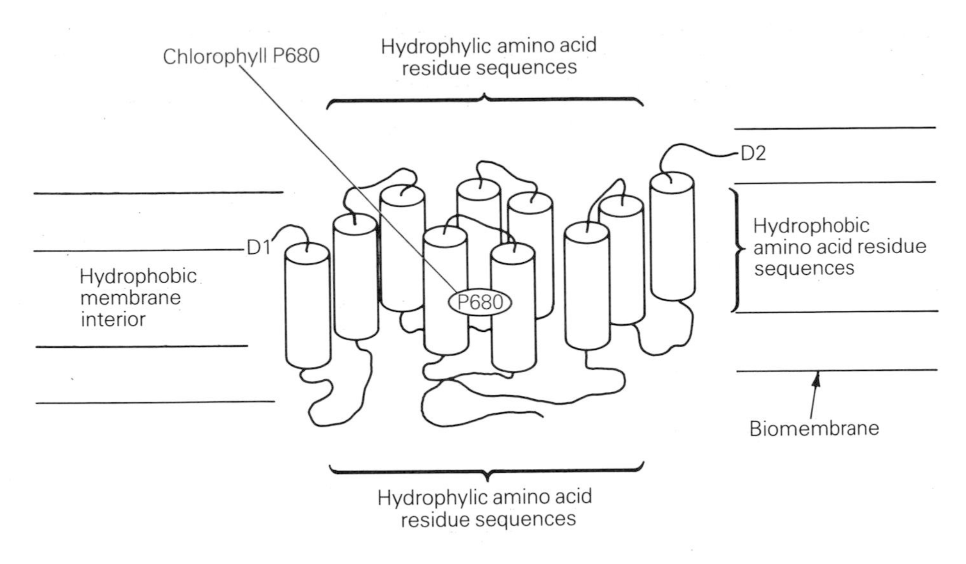

Fig. 2.21 A model of the organization of the reaction centre of photosystem II. Two main polypeptides, D1 and D2, constitute the reaction centre. Each polypeptide has five hydrophobic sequences of amino acid residues (represented by cylinders) embedded in the core of the chloroplast membrane. Linking hydrophilic portions are located on each side of the membrane. Chlorophyll P680 is attached to the middle of two of the hydrophobic sequences. Redrawn from D.S. Bendall, (1989) *Plants Today*, **2**, 188–92.

The second part of the reaction involves the reduction of CO_2 in reactions of the Calvin cycle. These reactions do not depend on light.

$$4H^+ + 4e^- + CO_2 \rightarrow CH_2O + H_2O$$

CH_2O represents the synthesized carbohydrate. Photosynthesis, which uses water as a source of electrons, is described as ***aerobic* or *oxygenic***. Aerobic photosynthesis accounts for the major part of the annual production of organic matter, estimated at 10^{17} g year^{-1} (Fig. 2.22).

Fig. 2.22 Photograph of an area of the Big Thicket National Preserve of southeast Texas, USA (Landsat satellite). The shaded squares represent different vegetation types, such as hardwood or pine forest, water, swamp, and natural and managed grassland. NASA's National Space Technology Laboratories is using Landsat data to assess the effects of natural and human-made changes on the vegetation cover of the earth. Courtesy of National Aeronautics and Space Administration (NASA), Washington D.C., USA.

***Aerobic** (oxygenic): metabolic processes that take place in the presence of oxygen.*
***Anaerobic** (anoxic): metabolic processes that take place in the absence of oxygen.*

PHOTOSYNTHETIC BACTERIA other than the cyanobacteria extract electrons from sources other than water. These sources include molecular hydrogen, hydrogen sulphide, and a very wide variety of organic compounds. The process may be summarized by the following general reaction:

$$CO_2 + 2H_2A \rightarrow CH_2O + H_2O + 2A$$

where H_2A represents the source of electrons and A the product remaining after its oxidation. Thus in, for example, the green and purple sulphur bacteria (e.g. *Chlorobium* sp. and *Thiocystis* sp. respectively), H_2S is the primary source of electrons. Consequently sulphur and not oxygen is liberated. These photosynthetic bacteria are obligate anaerobes and the process is called **anaerobic** or **anoxic** photosynthesis. The pigments involved in anoxic photosynthesis include the bacteriochlorophylls, which differ from chlorophylls a and b of green plants.

One major difference between aerobic and anaerobic photosynthesis is that the latter requires less energy input to convert CO_2 into organic compounds because hydrogen and hydrogen sulphide, etc. are stronger reductants than water and transfer their electrons more readily (Table 2.1). The maximum amount of work required to convert CO_2 into carbohydrate using H_2S as an electron source can be expressed as the change in the standard free energy, $\Delta G^{0'}$ in the following overall reaction:

$$6CO_2 + 12H_2S \rightarrow C_6H_{12}O_6 + 6H_2O + 12S \quad \Delta G^{0'} = +406 \text{ kJ mol}^{-1}$$

□ In evolutionary terms it has been argued that the energetic disadvantage of using water as an electron source is outweighed by the widespread availability of water.

The free-energy change for the fixation of CO_2 in aerobic photosynthesis may be similarly represented as follows:

$$6CO_2 + 12H_2O \rightarrow C_6H_{12}O_6 + 6H_2O + 6O_2 \quad \Delta G^{0'} = +469 \text{ kJ mol}^{-1}$$

The positive values for $\Delta G^{0'}$ mean that work must be done to carry out photosynthesis. Clearly, anoxic photosynthesis has a smaller energy demand than aerobic photosynthesis.

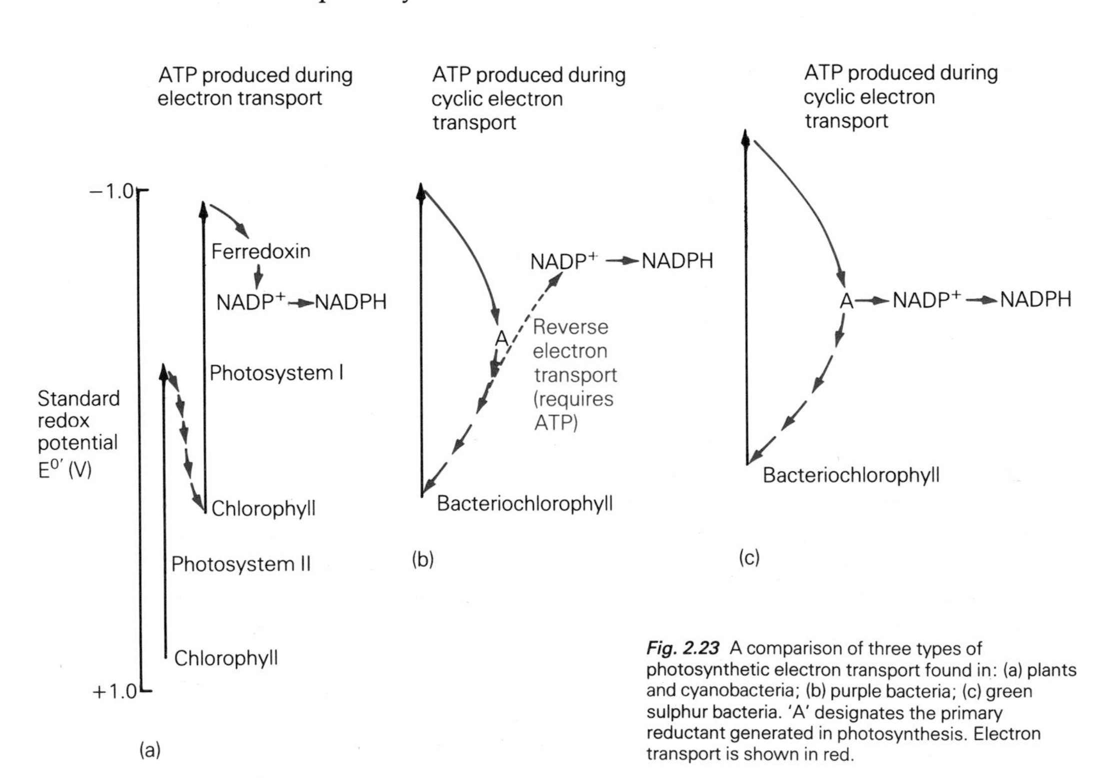

Fig. 2.23 A comparison of three types of photosynthetic electron transport found in: (a) plants and cyanobacteria; (b) purple bacteria; (c) green sulphur bacteria. 'A' designates the primary reductant generated in photosynthesis. Electron transport is shown in red.

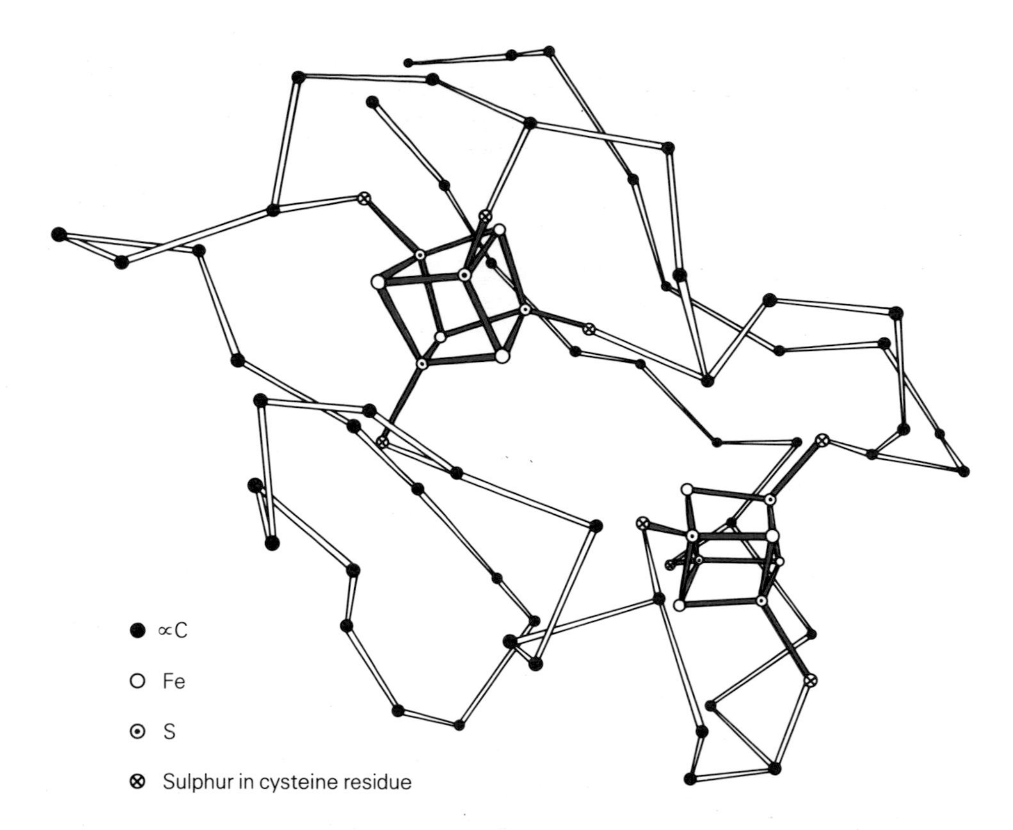

Fig. 2.24 Model of ferredoxin showing the locations of 2 Fe–S clusters (shaded red). Each cluster comprises 4 iron atoms, 4 sulphur atoms and 4 cysteine residue sulphur atoms. Redrawn from G. Zubay (1988), *Biochemistry*, 2nd edn, Macmillan, London, UK.

Aerobic photosynthesizers extract electrons from water using two separate and distinct photosynthetic systems, photosystems I and II, (Fig. 2.23), whereas anaerobic photosynthesizers only have a single system. Photosystem I is similar to the single photosystem of anaerobic photosynthesizers but is based on chlorophyll rather than bacteriochlorophyll. Photosystem II provides a redox couple with a lower reducing potential than water (Table 2.1). Water loses its electrons with the liberation of molecular oxygen (O_2). Thus, the absorption of light in the two photosystems generates a strong reductant from water, which has a reducing potential sufficient to reduce the protein **ferredoxin** (Fig. 2.24), which in turn reduces $NADP^+$ to NADPH. The oxidation–reduction reactions occurring also form an electron transport chain, which generates ATP via a H^+ gradient. Thus, electron transport in aerobic photosynthesizers produces ATP and NADPH.

See *Biosynthesis*, Chapter 2

□ Ferredoxin is a protein containing an iron–sulphur centre, which functions as the primary electron acceptor in oxygenic photosynthesis.

In anoxic photosynthesis in purple sulphur bacteria, light is used primarily to drive the generation of ATP, rather than producing reducing potential. Electron transport is coupled to ATP production but a reductant with a strong enough potential to reduce $NADP^+$ is not formed. Nevertheless, NADPH is required by these organisms and NADPH synthesis is achieved by coupling ATP hydrolysis and $NADP^+$ reduction via **reverse electron transport** (Fig. 2.23b). This ATP-driven synthesis of NADPH is also a feature of non-photosynthetic autotrophs. Other photosynthetic bacteria, such as the green sulphur bacteria possess electron acceptors capable of directly reducing $NADP^+$ to NADPH (Fig. 2.23c).

Some photosynthetic bacteria can utilize organic molecules rather than H_2S as electron donors. The purple non-sulphur bacteria (e.g. *Rhodobacter* sp.) are examples of bacteria in this group. In darkness these bacteria grow aerobically

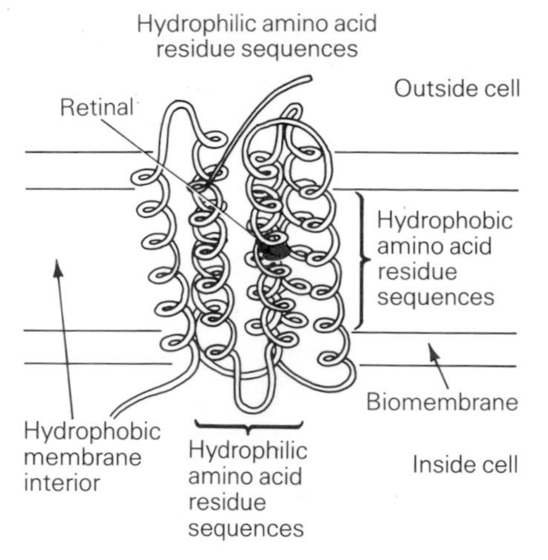

Fig. 2.25 Bacteriorhodopsin has seven hydrophobic sequences of amino acid residues embedded in the core of the membrane. Linking hydrophilic portions are located on each side of the membrane. Retinal attached to the middle of one helix captures light, triggering the movement of H^+ across the membrane. Note the gross similarity between this structure and that of photosystem II (Fig. 2.21). Both have transmembrane portions, which contain a light-harvesting moiety.

Fig. 2.26 Giant tube worms of the family Pognophora from a deep-sea thermal vent in the Galapagos rift. The plumes of the worm (dark) protrude from the sheaths (white) and are rich in blood vessels. They serve to extract oxygen and hydrogen sulphide from seawater. Courtesy of K. Crane, Woods Hole Oceanographic Institution, Public Information Office, Woods Hole, MA, USA.

as heterotrophs. Since they can utilize sunlight as an energy source they are called **photoheterotrophs**. Halobacteria are classified as heterotrophs. However, under conditions of low aeration, *Halobacterium* produces a membrane-bound protein, **bacteriorhodopsin** (Fig. 2.25), which (like the visual pigment of the eye, rhodopsin), changes conformation on absorption of light. This light-stimulated transformation results in the transport of protons to the outside of the bacterial cell membrane. A proton gradient is established, which is sufficient to promote the production of ATP.

Bacteria display a wide diversity in their use of light as a primary source of energy. Compared with green plants, however, bacteria play only a minor role in the fixation of solar energy in the biosphere.

LITHOTROPHS OR CHEMOAUTOTROPHS, in contrast to the photosynthetic autotrophs and photoheterotrophs, derive energy from the chemical oxidation of reduced inorganic compounds in the environment. ATP generation is coupled to the oxidation of an environmental reductant in an electron transport system. The process is similar to the aerobic oxidation of NADH. Environmental reductants are generally weaker than NADPH, so $NADP^+$ cannot accept electrons from environmental donors. NADPH is, therefore, produced in a process similar to reverse electron transport, a process that consumes ATP.

'Colourless sulphur bacteria', so called to distinguish them from the photoautotrophic purple and green bacteria which have bacteriochlorophyll, use reduced sulphur compounds as electron donors. The most commonly used are hydrogen sulphide (H_2S), sulphur (S) and thiosulphate ($S_2O_3^{2-}$). The final product is the much more oxidized sulphate (SO_4^{2-}). The maximum amount of work possible from the transfer of electrons from these sulphur compounds and other environmental reductants to molecular oxygen is shown in Table 2.7.

Electrons from sulphur compounds enter the electron transport chain of the sulphur bacteria at a variety of points (Fig. 2.27). The transfer of electrons to molecular oxygen as electron acceptor leads to the synthesis of ATP.

Until recently, it was thought that chemolithotrophs contributed relatively little to the overall ***primary productivity*** of the biosphere. However, some colourless sulphur bacteria obtain their energy from reduced compounds in the heated mineral waters of thermal vents in the deep rifts of oceans. These organisms solely support a diverse fauna of heterotrophs, including giant clams up to 40 cm in diameter and large pognophoran tube worms up to 2 m in length (Fig. 2.26).

The pognophoran worms exist in a symbiotic relationship with sulphur bacteria. Although these tube worms lack both mouth and anus, they have a highly specialized gastrointestinal tract with an abundant supply of **trophosome** tissue (Fig. 2.28) containing large numbers of bacteria. These bacteria resemble the free-living sulphur bacteria *Thiovulum* sp. The tissue also contains sulphur granules and the enzymes of the Calvin cycle, the anabolic pathway used by most autotrophic organisms to synthesize

Table 2.7 *Standard free-energy change ($\Delta G^{0'}$) on oxidation of environmental electron donors*

Electron donor	Oxidized product	$\Delta G^{0'}$ (kJ mol^{-1})
Hydrogen sulphide (H_2S)	Sulphur (S)	−203
Sulphur (S)	Sulphate (SO_4^{2-})	−588
Ammonia (NH_4^+)	Nitrite (NO_2^-)	−217.5
Nitrite (NO_2^-)	Nitrate (NO_3^-)	−75.7
Ferrous (Fe^{2+})	Ferric (Fe^{3+})	−71

Primary productivity: *the production of biomass by autotrophs.*

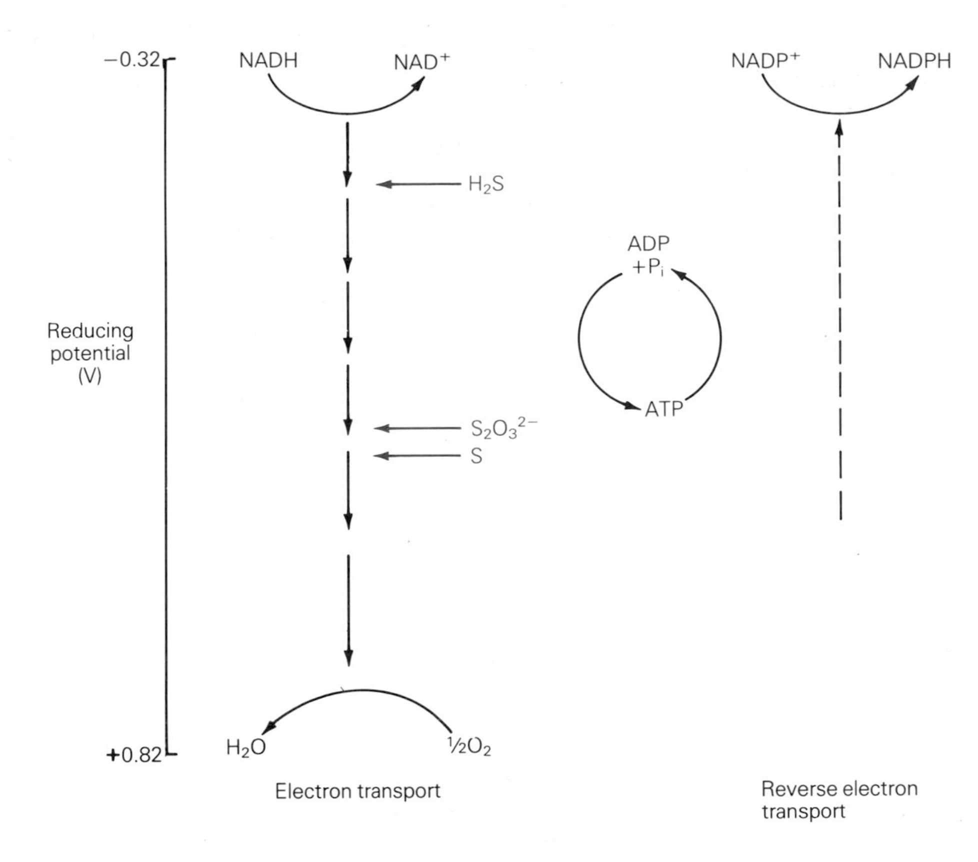

Fig. 2.27 Electron transport in lithotropic sulphur bacteria. The points of entry of electrons from H_2S, $S_2O_3^{2-}$ and S are indicated.

Outer membrane
Plasma membrane
Nucleus
Trophosome cells enclosed in an outer membrane
Spherical lithotrophic sulphur bacteria

Fig. 2.28 A representation of trophosome tissue, indicating the presence of bacteria within the animal cells.

carbohydrate from CO_2. The tube worms are heterotrophs, living off the excretory products and dead cells of their autotrophic symbionts. Colourless sulphur bacteria are therefore an exception to the idea that *all* ecological productivity depends on photosynthesis.

Other examples of organisms that use the oxidation of reduced inorganic compounds to provide energy include:

- *Nitrosomonas* sp., which oxidize ammonia to nitrite in ammonia-rich, alkaline sewage effluent: $NH_3 + 2O_2 \rightarrow NO_2^- + H_2O + H^+$.
- *Alcaligenes eutrophus*, which bind molecular hydrogen to a membrane-bound hydrogenase, before transporting electrons to O_2: $H_2 + O_2 \rightarrow 2H_2O$.
- *Acidophilic Thiobacillus ferrooxidans*, which oxidize ferrous to ferric ions at low pH: $Fe^{2+} + \frac{1}{4}O_2 + H^+ \rightarrow Fe^{3+} + \frac{1}{2}H_2O$.

Heterotrophic nutrition

Heterotrophs show one of two methods of acquiring fuel molecules. One involves consumption followed by digestion and cellular assimilation. Organisms which use this method are called consumers and include the herbivorous (plant-eating), carnivorous (flesh-eating) and detritivorous (detritus-eating) animals. The other means of acquiring organic compounds from the environment involves the extracellular digestive processes carried out by ***decomposer*** organisms such as fungi (Fig. 2.29) and bacteria.

Only a proportion of the food eaten by heterotrophs is assimilated and used to produce energy and precursors for biosynthesis. Figure 2.30 depicts the fate of the energy of food eaten by heterotrophic animals.

☐ There are approximately 400 known species of insectivorous plants (also called carnivorous plants). These plants have specialized devices for trapping and digesting insects.

Venus' flytrap, *Dionaea* sp., an insectivorous plant. Note the gaping jaws of the trap, which actively close around insect prey.

They are commonly found in nutrient-poor environments, such as bogs and fens. The carnivorous habit of these plants enables them to extract nutrients, such as nitrate and phosphate, from their prey. Like all higher plants, insectivorous plants are photoautotrophs synthesizing their organic requirements using the energy of sunlight.

Acidophile: *an organism that can live at low pH.*
Decomposer: *an organism that uses dead organic matter as a source of food. It breaks the food (proteins, polysaccharides, etc.) down extracellularly, absorbing smaller carbon compounds and releasing ions and molecules to the environment. Decomposers are heterotrophs.*

Reference Brock, T.D. and Madigan, M.T. (1988) *Biology of Microorganisms*, 5th edn, Prentice Hall, Englewood Cliffs, NJ. 835 pp. A well-established microbiology text, which includes good comparative accounts of metabolic processes in microorganisms. Chapters 4 and 16 are especially useful.

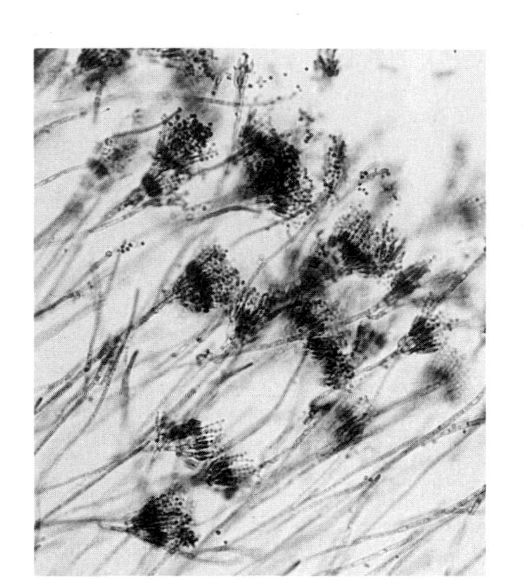

Fig. 2.29 The fungus *Penicillium* sp. (× 90).

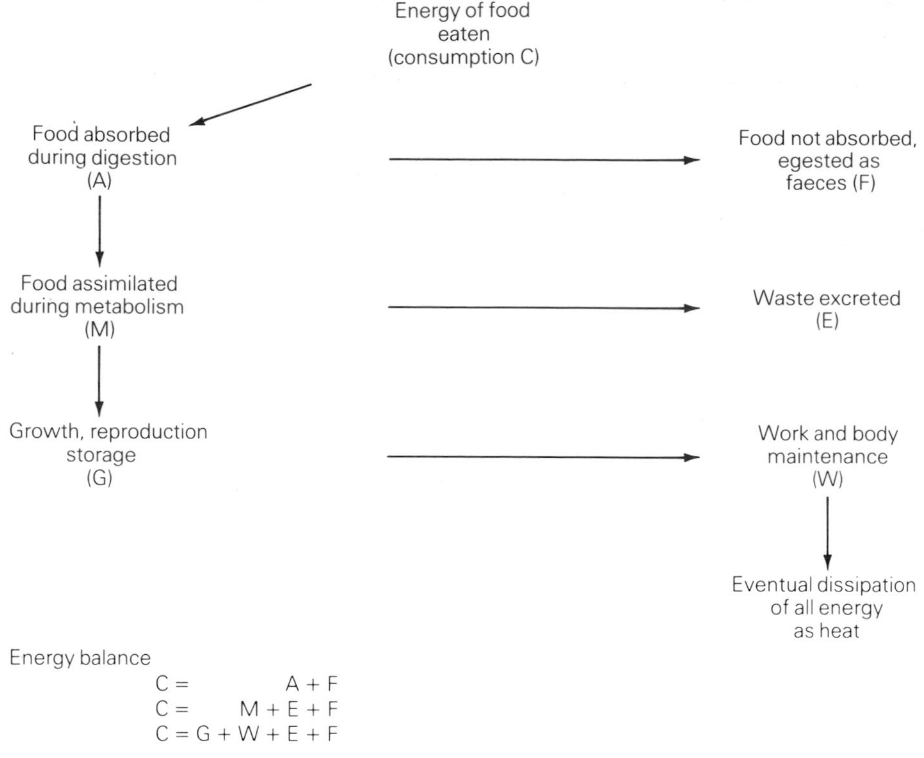

Fig. 2.30 The fate of the energy of food eaten by a heterotrophic herbivore.

Of the food consumed, some energy is lost in faeces and in the hydrolysis of food molecules. Monosaccharides, fatty acids and amino acids, are absorbed by the cells of the digestive tract and then transported around the body of the organism. Some molecules are used directly in the replacement of cell components, growth and reproduction. Other molecules are catabolized, providing energy and intermediates for anabolism. Excess molecules are converted to fuel storage compounds, such as glycogen and lipids. These compounds can be hydrolysed to monomers, which can be oxidized when environmental food supplies are scarce.

See Chapter 7

If the intake of amino acids exceeds that which an animal needs for biosynthesis, excess amino acids are deaminated producing simple nitrogenous compounds and α-oxoacids. The α-oxoacids are catabolized, while the excess nitrogenous compounds are excreted and do not contribute useful energy to the organism.

Exercise 7

An active elephant uses energy at the rate of about 180 000 kJ day^{-1} for growth, work and general body maintenance. On average, elephants consume 24 kg of plant material each day. Assume that the energy lost by excretion is relatively small and can be ignored. The calorific value of both plant material and faeces is 16.5 kJ g^{-1}. (a) Calculate the daily mass of faeces produced by the elephant. (b) Calculate the efficiency of assimilation as a percentage of food eaten.

2.5 *Strategies for generating ATP and NADPH in catabolism*

Autotrophs make ATP and NADPH during the transport of electrons from environmental electron donors. Catabolism of food molecules is the *only* source of electrons in animals, fungi and the majority of prokaryotic organisms. Plants and photoheterotrophs can also catabolize such (stored) fuel compounds in the absence of environmental energy. There are three major modes of catabolism: ***aerobic respiration; anaerobic respiration*** and ***fermentation***.

See Chapters 4 and 5

Aerobic respiration: *involves reactions which bring about the complete oxidation of food molecules, such as glucose. Oxygen is required as a final electron acceptor.*
Anaerobic respiration: *the partial catabolism of organic molecules in the absence of oxygen. Electron transport to acceptors, other than oxygen, as well as substrate level phosphorylation enables ATP to be generated.*
Fermentation: *the partial breakdown of food molecules in the absence of external electron acceptors such as oxygen. Reducing potential is conserved in the process and ATP is produced only by substrate level phosphorylation Chapter 5.*

Box 2.6
The use of absorbable staples in surgery

The non-specific nature of many hydrolytic enzymes has been exploited in the use of self-dissolving surgical staples. These staples can be manufactured from a copolymer of lactic acid and glycolic acid. The following reaction shows how the polymer is formed:

$$\underset{\text{lactic acid}}{\mathrm{HO{-}\underset{\displaystyle CH_3}{\underset{|}{CH}}{-}COOH}} + \underset{\text{glycolic acid}}{\mathrm{HO{-}CH_2{-}COOH}} \longrightarrow$$

$$\mathrm{HO{-}\underset{\displaystyle CH_3}{\underset{|}{CH}}{-}\underset{\displaystyle O}{\underset{\|}{C}}{-}O{+}\left[CH_2{-}\underset{\displaystyle O}{\underset{\|}{C}}{-}O{-}\underset{\displaystyle CH_3}{\underset{|}{CH}}{-}\underset{\displaystyle O}{\underset{\|}{C}}{-}O\right]_n{-}CH_2{-}\underset{\displaystyle O}{\underset{\|}{C}}{-}OH} + n\mathrm{H_2O}$$

The polymer is slow to hydrolyse in the body, taking 6 to 8 weeks to break down. Both the acids released are normal metabolites, so the staples do not release toxic material into the body. The staples retain sufficient strength long enough for healing of the wounds to occur. The advantage of using such staples is the reduced trauma due to faster operating times.

Aerobic respiration

Aerobic respiration is the oxidation of glucose to CO_2 and H_2O. The importance of this process rests on its central role in catabolism. Some of the stages of glucose oxidation are common to the catabolism of fatty acids and amino acids. Glucose degradation takes place in three stages (Fig. 2.31). The first is glycolysis (Table 2.4), in which NADH and ATP are generated. In the presence of oxygen, pyruvate is degraded to a key intermediate, acetyl CoA,

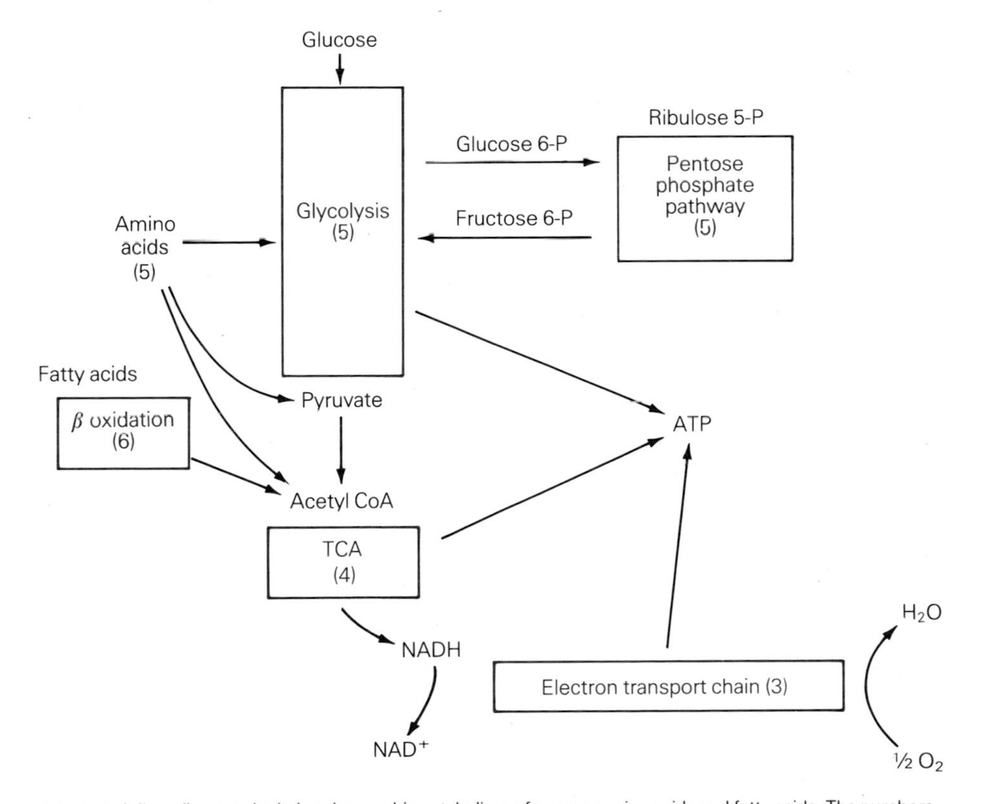

Fig. 2.31 A flow diagram depicting the aerobic catabolism of sugars, amino acids and fatty acids. The numbers (in brackets) indicate various areas of this text where the processes are treated in detail. TCA, tricarboxylic acid cycle; ETC, electron transport chain.

Reference Wiegart, R.G. (ed.) (1976), *Ecological Energetics* (Benchmark papers in ecology 4), Dowden, Hutchinson, Ross, 457 pp. A collection of early papers, which laid down the foundations of the energetics of food and feeding. The early part of the text is most relevant to biochemistry.

☐ The formation of ATP from ADP in chemical reactions involving group transfer is called substrate level phosphorylation.

with the liberation of carbon dioxide. Acetyl CoA enters the second stage of glucose catabolism, the tricarboxylic acid (TCA) cycle. The TCA cycle completely degrades acetyl CoA, generating NADH. NADH is oxidized in an electron transport chain (ETC), generating ATP by oxidative phosphorylation. The three stages are distinct, both chemically and physically.

Glycolysis occurs in the cytosol, the TCA cycle in the matrix of mitochondria, and the ETC on the inner mitochondrial membrane.

See Chapters 3–7

The complete aerobic oxidation of fatty acids (β-oxidation) yields acetyl CoA which can be degraded in the TCA cycle and ETC. Amino acids are degraded by deamination, their carbon-skeletons being catabolized by entering the glycolytic pathway and the TCA cycle (Fig. 2.31).

Pentose phosphate pathway

See *Biosynthesis*, Chapter 3

The pentose phosphate pathway, which is the major supplier of NADPH for biosynthetic activities, is another route for oxidizing glucose. The oxidation of glucose 6-phosphate, an intermediate of glycolysis, to ribulose 5-phosphate and CO_2 is used to produce NADPH. Fructose 6-phosphate, also an intermediate of glycolysis, is generated in the pentose phosphate pathway. The pentose phosphate pathway is thus an alternative to the single glycolytic reaction in which glucose 6-phosphate is converted to fructose 6-phosphate.

Anaerobic respiration

In anaerobic respiration ATP is produced when electrons from NADH are transported to environmental electron acceptors other than oxygen. The acceptor is reduced (Table 2.8). The reduced products are potential sources of energy and nutrients for other organisms. Electron acceptors are useful to anaerobic respirers, only if the difference in the reducing potential between the acceptor and NADH is sufficient to generate ATP.

Fermentation

Fermentation is a catabolic strategy for utilizing an energy source in the absence of an external electron acceptor. Food molecules are degraded, forming two kinds of products. One is a reduced form of carbon, such as ethanol, lactate or succinate. The other is an oxidized form of carbon, usually carbon dioxide. Since reducing potential is not exchanged with the environment the balance of oxidants and reductants must be maintained throughout the process. Figure 2.32 illustrates ethanolic fermentation in yeast cells. In the absence of oxygen, NADH cannot supply electrons to the electron transport chain, and the TCA cycle cannot operate. Glycolysis is maintained because NADH is used to reduce pyruvate to ethanol, with the liberation of carbon dioxide and the regeneration of NAD^+. In this way, glycolysis can take

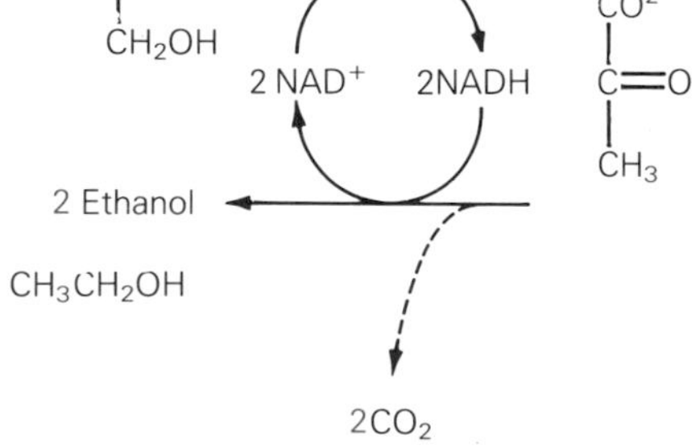

Fig. 2.32 Ethanol fermentation showing the conservation of reducing potential and the generation of ethanol and carbon dioxide.

Table 2.8 *Electron acceptors in anaerobic respiration*

Respiratory type	Electron acceptor	Reduced product
Sulphur respiration	Sulphur (S)	Sulphide (HS^-)
Sulphate respiration	Sulphate (SO_4^{2-})	Sulphide (HS^-)
Carbonate respiration (acetogenic bacteria)	CO_2	Acetate (CH_3COO^-)
Carbonate respiration (methanogenic bacteria)	CO_2	Methane (CH_4)
Nitrate respiration	Nitrate (NO_3^-)	Nitrite (NO_2^-)
Iron respiration	Ferric (Fe^{3+})	Ferrous (Fe^{2+})
Organic compounds	e.g. Fumarate, glycine	e.g. Succinate, acetate

Box 2.7
Methane and the greenhouse effect

Carbon dioxide and methane *may* be responsible for the current warming of the earth by the greenhouse effect. The build up of methane in the atmosphere during the last 300 years has been dramatic. A major proportion of the methane, some 198 million tonnes, is generated each year in natural biochemical processes. Bacterial fermentation in the anaerobic mud of rice paddy fields generates 120 million tonnes per year. Some 78 million tonnes of methane per year are produced by bacteria in the guts of ruminants and released by flatulence.

place in the absence of oxygen. There is a net production of ATP but, unlike aerobic respiration, there is **not** a net production of reducing potential.

Many types of bacteria are capable of fermentation. Some organisms are able to ferment the products of others. Thus, ethanol, lactate or succinate may be reduced to molecules such as acetate or butyrate. The ultimate products of fermentation are methane (CH_4), the most reduced form of carbon, and carbon dioxide (CO_2), the most oxidized form.

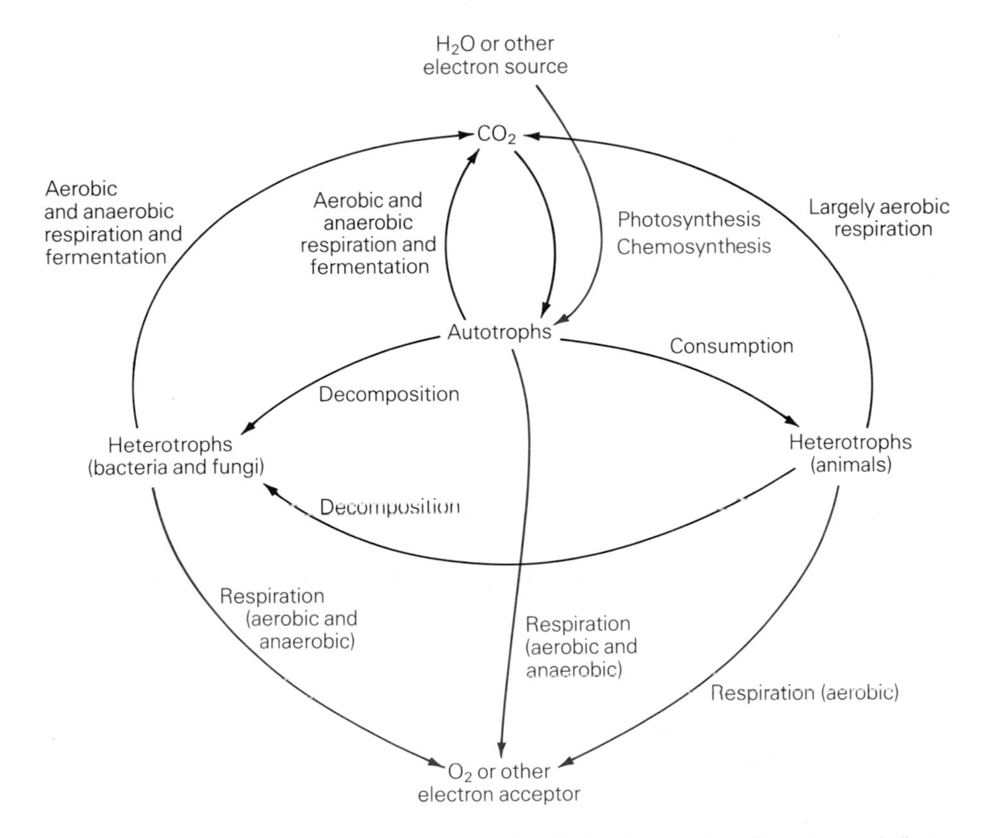

Fig. 2.33 Summary of the modes of nutrition and catabolism displayed by organisms. The red arrows indicate the flow of reducing potential. The dark arrows indicate the path of carbon.

2.6 Overview

Biosynthesis and work are necessary for survival, growth and reproduction. All organisms produce:

- molecular precursors for biosyntheses
- reducing potential in the form of NADPH
- chemical potential in the form of ATP.

Some organisms use the heat generated by metabolic processes as a means of maintaining their body temperature above that of the environment.

Autotrophs derive their molecular and energy needs directly from the environment. They synthesize complex organic food or fuel molecules from more simple inorganic precursors. Heterotrophs, however, derive all their needs by consuming and assimilating ready-made food molecules (Fig. 2.33).

The requirements for both energy and biosynthesis in cells of heterotrophs and autotrophs is met by the balance of catabolic and anabolic pathways.

Catabolism comprises a large number of enzyme-catalysed steps, many of which operate close to equilibrium. Such a strategy enables reactions to proceed efficiently, maintaining pools of metabolites at relatively constant concentrations within cells. Control over whole pathways is largely vested in just a few specific reactions, which are displaced from equilibrium. These reactions, catalysed by regulatory enzymes, control the concentrations of key intermediates in the cell.

Answers to exercises

1.

Question	(a) Redox couple	(b) Reduced form	Oxidized form	(c) Group-transfer molecules
(a)	$NADP^+$/NADPH Phosphoglucono-lactone/ glucose phosphate	NADPH Glucose phosphate	$NADP^+$ Phosphoglucono-lactone	$NADP^+$/NADPH
(b)	FAD/$FADH_2$ Fumarate/ succinate	$FADH_2$ Succinate	FAD Fumarate	FAD/$FADH_2$
(c)	NAD^+/NADH Acetyl CoA/ pyruvate	NADH Pyruvate	NAD^+ Acetyl CoA	NAD^+/NADH CoASH/acetyl CoA

2. Enthalpy changes are *additive*. The total enthalpy changes for the carbohydrate, lipid and protein are 0.38 × 17.6, 0.28 × 39.7 and 0.27 × 18.0 kJ respectively. The sum of these three values is 22.6 kJ. The energy value of the food is thus 22.6 kJ g^{-1}.

3. (a) The ratio of [Y] [Z]/[X] is 10^{-7} mol dm^{-3}. This ratio is 100 times smaller than the equilibrium constant. To restore equilibrium, the reaction would tend to move in the direction X $\rightarrow$ Y + Z. It is worth noting that, even at equilibrium, the concentration of X is much greater than the concentrations of the reactants. (b) If a reaction utilizing Z could not take place, the concentration of Z would tend to increase. This would have the effect of making the ratio [Y] [Z]/[X] larger and closer to equilibrium.

4. (a) Applying the method of calculation given on page 29 the adenylate energy charge is 0.65. (b) The value is less than that quoted for active cells. Possible reasons for this are the previous history of the cultured cells or losses of adenine nucleotides during extraction. You might consider whether or not physicochemical methods such as HPLC could be supplemented by *in vivo* or bioassay methods.

5. Temperature regulation provides a suitable thermal environment for enzyme-driven metabolic reactions . Note, however, that biological work made possible by enzymes does not depend on heat exchange but rather on the maintenance of a suitable cellular environment.

6. The hydrolytic reactions of digestion are not coupled to the production of reducing potential and work processes. The enthalpy is dissipated as heat and is not recovered as free energy. The reaction, therefore, does not contribute to an increase in body weight.

7. Assimilation is energy of catabolism used for growth, work and general body maintenance. Since excretion is taken as zero, the energy absorbed must be 180 000 kJ. The energy of food eaten is mass × calorific value. Therefore, energy of food eaten = 24 000 × 16.5 = 396 000 kJ day^{-1}. The energy of faeces = energy of food eaten − energy of food utilized. Therefore, faecal energy = 396 000 − 180 000 = 216 000 kJ day^{-1}. (a) The mass of faeces produced per day is the energy of faeces produced, divided by the calorific value of faeces. Therefore, faecal production is 13.1 kg day^{-1}. (b) The percentage efficiency of assimilation is given by: assimilation efficiency = energy absorbed/energy eaten × 100. Therefore, assimilation efficiency = 180 000/396 000 × 100 = 45%.

QUESTIONS

FILL IN THE BLANKS

1. Light that drives photosynthesis is absorbed primarily by the green pigment _______ . All green plants have _______ but photosynthetic bacteria have special _______ called _______. Halobacteria have no _______ or _______ but capture light using _______ . Oxygen is evolved in _______ photosynthesis. The oxygen comes from _______ . Purple sulphur bacteria carry out photosynthesis by oxidizing reduced sulphur compounds, such as _______ _______ to _______ or _______ . In eukaryotes photosynthesis occurs in the _______ . During photosynthesis, reducing power in the form of _______ is generated. ATP is produced during photosynthetic _______ _______ , which generates a _______ _______ . This _______ _______ activates an _______ , which catalyses the phosphorylation of _______ to produce ATP.

Choose from the following: ADP, ATPase, aerobic, bacteriochlorophyll (2 occurrences), bacteriorhodopsin, chlorophyll (3 occurrences), chloroplast, electron gradient (2 occurrences), hydrogen, NADPH, proton (2 occurrences), sulphate, sulphide, sulphur, transport, water.

2. Complete the following sentences with the name of a process.
A. Sulphur is an environmental source of electrons for _______ .
B. Sulphate can accept electrons during _______ .
C. Methane is the ultimate reduced state of carbon in _______ .
D. Water donates electrons in _______ .
E. Oxygen is the electron acceptor for _______ .
F. A net reducing potential is not generated in _______ .

3. Complete the following sentences.
A. Fatty acids, hexose sugars and _______ are the principal fuel molecules for heterotrophs.
B. In plants and animals, NADPH is catabolically generated in the _______ pathway.
C. The three principal products generated during the complete catabolism of glucose are _______ , _______ and _______ .

MULTIPLE CHOICE QUESTIONS

4. Which of the following biochemical transformations do not generate (i) NADPH; (ii) NADH; (iii) ATP.

A. glucose $\rightarrow$ pyruvate (in glycolysis)
B. pyruvate $\rightarrow$ carbon dioxide
C. glucose 6-phosphate $\rightarrow$ fructose 6-phosphate (in the pentose phosphate pathway)
D. glucose $\rightarrow$ ethanol
E. NADH $\rightarrow$ NAD^+ (in electron transport)
F. Electrons from water $\rightarrow$ ferredoxin

5. State whether each of the following is True or False; if false give a correct statement.

A. NADH is a form of reducing potential used in reductive biosynthesis.
B. Photosynthesis is the only source of ATP and NADPH in green plants.
C. ATP is generated in catabolism solely by electron transfer-linked phosphorylation of ADP.
D. Methane is a highly reduced form of carbon, so it can act as an environmental source of electrons for autotrophs.
E. All autotrophs generate NADPH during electron transport.

SHORT-ANSWER QUESTIONS

6. Examine the data in the figure in Box 2.5, which shows the ^{31}P-NMR spectra of root tips incubated in air or in nitrogen. Use the data to calculate phosphorylation potentials and ATP/ADP ratios. Discuss the reasons for the differences you observe.

7. List as many examples of biological processes involving work as you can. Classify them in terms of the ways in which the processes are linked to catabolism, as depicted in Figure 2.12.

8. Survey the types of plants, animals, fungi and the prokaryotes and list at least three examples of organisms that can be classified in each of the catagories listed in Tables 2.5 and 2.6.

9. Since organisms deprived of oxygen cannot use it as a terminal electron acceptor, how can they catabolize fuel molecules in its absence?

10. Glucose is a relatively reduced metabolite. Discuss how this property justifies the classification of glucose as a fuel molecule.

11. Explain briefly why heat is sometimes considered to be a waste product of catabolism.

12. Distinguish briefly between the following:

A. digestion and catabolism
B. fermentation and anaerobic respiration
C. photoautotrophy and photoheterotrophy
D. aerobic and anaerobic photosynthesis

ESSAY QUESTION

13. The sun is the primary source of energy for life on earth. Describe the principal biochemical processes by which this assertion is justified.

3

Electron transport

Objectives

After reading this chapter you should be able to:

- ☐ describe and compare the electron transport chains of mitochondria, chloroplasts and microorganisms;
- ☐ explain how electron transport is used to generate proton gradients across membranes.
- ☐ discuss possible mechanisms of ATP generation by membrane-bound ATPases.

3.1 Introduction

Animal, plant and many types of microbial cells obtain energy by oxidizing highly reduced organic compounds such as carbohydrates and fats (and sometimes proteins), which they have either taken up or have stored. In the course of the oxidation, electrons are removed from the compounds to be passed along a chain of carrier molecules referred to as the **electron transport chain**. Molecular oxygen is usually the final electron acceptor, although in some microorganisms the acceptor may be an organic compound such as fumarate or an inorganic compound such as nitrate. Passing electrons along a chain of electron carriers releases energy which is conserved and used to synthesize ATP from ADP and P_i. This is the major way in which organisms other than photosynthetic ones obtain their energy for driving biological processes.

In photosynthesis, electrons are taken from a donor, commonly water, and transferred to a chain of carriers. This is the *photosynthetic* electron transport chain and it also generates ATP. However, the electrons taken from the donor eventually find their way to reduce the coenzyme $NADP^+$. Reduced coenzyme (NADPH) and energy (ATP) generated in this way, are used by the photosynthetic organism to convert CO_2 into highly reduced organic compounds such as carbohydrates, fats and proteins.

Electron transport chains in all organisms show many similarities. The electron carriers, for example, are always arranged in membranes in special orientations. In eukaryotic cells these are the membranes of the mitochondria and chloroplasts. In microorganisms they are the cell membrane and the membranes of the chromatophores. In all the systems the electrons, as they proceed through a chain of carriers, are passing from carriers with a relatively high (negative) oxidation–reduction potential to carriers with relatively low oxidation–reduction potentials. Energy is therefore released, and the cells and membranes are organized to conserve this energy as ATP by a process that is still comparatively poorly understood.

This chapter deals with the components and composition of electron transport chains and the mechanism of the ATP synthesis accompanying electron transport.

3.2 Electron transport

□ The three processes, substrate-level phosphorylation, oxidative phosphorylation and photosynthetic phosphorylation, are the only ones that generate ATP. However, in vertebrate muscle, creatine phosphate is present as a store of phosphate bond energy. When the muscle needs ATP to contract rapidly, the reaction:

$$\text{creatine-P} + \text{ADP} \rightleftharpoons \text{creatine} + \text{ATP}$$

generates this ATP quickly. This is not generation in the same sense, however, as the reverse of this reaction is used to store the creatine-P.

Metabolically useful energy in the form of ATP is generated from ADP and P_i by three processes:

1. ***Substrate-level phosphorylation***, which occurs during glycolysis (see Chapter 5).
2. ***Oxidative phosphorylation*** accompanying electron transport in mitochondria and bacterial membranes. This involves the energy-releasing oxidation of reduced substrates such as fatty acids and sugars in a process by which the coenzymes NAD^+ and FAD are reduced to NADH and $FADH_2$ respectively. The reduced coenzymes then donate their hydrogens and electrons to a series of carriers called the electron transport chain. These carriers include flavoproteins, iron–sulphur proteins, quinones and cytochromes. The electron acceptor at the end of the chain may be molecular oxygen or, in the case of anaerobic bacteria, it may be one of a variety of molecules such as nitrate, nitrite or sulphur compounds.
3. ***Photosynthetic phosphorylation*** which occurs in the chloroplasts of plants and in the chromatophores of photosynthetic bacteria. This is the use of light energy to drive the phosphorylation of ADP to ATP and also involves an electron transport chain. Electrons are obtained from the splitting of water by the energy of light. Bacteria also carry out photosynthesis but can obtain electrons from various molecules, including H_2S and a range of organic molecules.

A common theme of electron transport is that the carriers are arranged on or in the membranes of mitochondria, chloroplasts or bacteria. As the electrons pass from carrier to carrier, protons (H^+) are pumped across the membrane.

The structure of the mitochondrion

See *Cell Biology*, Chapter 5

The mitochondrion is typically a cylindrical organelle approximately $0.5 \times 0.3\ \mu m$, although the shape and size can vary a great deal. It has an outer membrane and an inner membrane, separated by an intermembrane space. The inner membrane is infolded to form cristae, which increase its surface area.

The inside compartment of the mitochondrion is called the ***matrix*** (Fig. 3.1). Each part of the mitochondrion has a characteristic enzyme composition (Table 3.1). All of the enzymes of the tricarboxylic acid cycle are located in the matrix or on the inner surface of the inner membrane, which also contains the components of the electron transport chain.

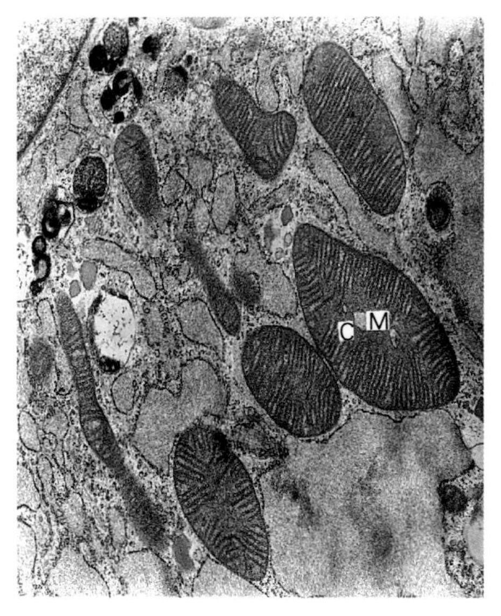

Fig. 3.1 An electron micrograph of a mitochondrion showing the folded membrane system, the cristae. M, matrix, C, cristae of a fibroblast in culture (× 5250), stained with uranyl acetate and lead citrate.) Courtesy of Dr C. Jones, Department of Pathology, University of Manchester.

Table 3.1 *The location of enzymes in the mitochondria*

Outer membrane	Inner membrane space	Inner membrane	Matrix
Monoamine oxidase Fatty acyl CoA synthetase	Adenylate kinase Sulphite oxidase	Cytochrome b, c_1, c, a, a_3 Succinate dehydrogenase Ubiquinone F_1-ATPase Various translocases	Pyruvate dehydrogenase Citrate synthetase Isocitrate dehydrogenase Fumarase- Oxoglutarate dehydrogenase Malate dehydrogenase Enzymes of β-oxidation of fats

Substrate-level phosphorylation: *the generation of ATP from ADP whereby a substrate that is an intermediate in a metabolic pathway, becomes phosphorylated and passes its phosphate group to an ADP.*

Matrix: *the part of the mitochondron enclosed by the inner membrane. From the Latin* matrix, *womb.*

The structural organization of the inner mitochondrial membrane

Electron microscopy shows the folded inner mitochondrial membrane to have an array of small knobs 9–10 nm in diameter. These knobs project into the matrix (Fig. 3.2a) and are the 'headpieces' of an enzyme called the **F_1F_0-ATPase**, which is capable of hydrolysing ATP to ADP and phosphate.

If preparations of the inner membrane are sonicated, the membrane is disrupted and forms **submitochondrial particles** (Fig. 3.2b). Treatment of membrane with detergent yields five fractions, which can be partially separated from one another and catalyse the following reactions:

Fraction I	NADH oxidation–ubiquinone reduction
Fraction II	succinate oxidation–ubiquinone reduction
Fraction III	ubiquinone oxidation–cytochrome *c* reduction
Fraction IV	cytochrome *c* oxidation–oxygen reduction
Fraction V	ATP $\rightarrow$ ADP + P_i

Further analysis of these fractions shows that they contain the components listed in Table 3.2.

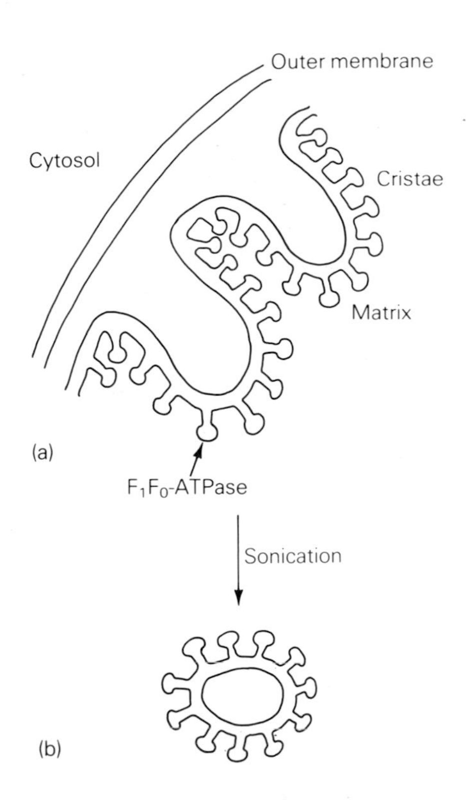

Fig. 3.2 (a) The F_1F_0-ATPase is observed as a series of regular structures on the inner membrane of the mitochondrion. (b) Submitochondrial particles released upon sonication of the mitochondrial membrane are small vesides with ten ATPase 'knobs' pointing outwards.

Table 3.2 *The composition of the respiratory Fractions I–IV*

Complex		$M_r \times 10^3$	Components	Location and role
I	NADH–Q reductase	850	FMN Fe–S (16)*	Protein spans the membrane with NADH site on matrix side, Q site in membrane; translocates H^+
II	Succinate–Q reductase	140	FAD Fe–S	Succinate site on matrix side; Q site; in membrane
III	Cytochrome reductase	250	Haem b_{562} Haem b_{566} Haem c_1 Fe–S	Spans the membrane; cytochrome c_1 on cytosol side; cytochrome *b* in membrane; translocates H^+
	Cytochrome *c*	13		On cytosolic side of membrane
IV	Cytochrome oxidase	160	Haem *a* Haem a_3 Cu	Spans the membrane; cytochrome *c* on cytosolic side; translocates e^- and/or H^+

Fe–S indicates an iron–sulphur centre, Q coenzyme Q (ubiquinone)
* Number of Fe–S centres.
Based on Table 7.1 in Harold, F.M. (1986) *The Vital Force*, Freeman, New York.

Exercise 1

If detergent is necessary to release submitochondrial Fractions I to IV, what information does this provide about their environment in the membrane?

The mitochondrial electron transport chain

The mitochondrial electron transport chain consists of a series of electron carriers arranged in close physical contact on the inner mitochondrial membrane. The carriers are of two types: those that transfer electrons only, and those that carry hydrogens, either in the form of hydride ions or as hydrogen ions *and* electrons. These carriers are reversibly interconverted between oxidized and reduced states. Although some of the carriers deal with hydrogens as well as electrons, it is conventional to speak of the whole system as an electron transport chain. Its main components are NADH-linked dehydrogenases, flavin-linked dehydrogenases, ubiquinone, iron–sulphur proteins and cytochromes. Some of these are enzymes with associated cofactors that catalyse the transfers; others are proteins with prosthetic groups.

DEHYDROGENASES of the electron transport chain catalyse the removal of hydrogens and electrons from a substrate and feed them into the chain. There are two types of dehydrogenase: NAD-linked, in which the acceptor is NAD^+; and flavin-linked, in which FAD is the acceptor. Both FAD and NAD^+ are hydrogen and electron carriers. The electrons carried by NADH enter

Reference Jones, C.W. (1976) *Biological Energy Conservation*, Outline Studies in Biology Conservation Series, Chapman and Hall, London. Although a little old, this gives a comprehensive account of electron transport.

Reference Wittaker, P. and Danks, S. (1978) *Mitochondria: Structure, Function and Assembly*, Longman, London. A good introduction to the structure and functions of mitochondria.

the electron transport chain at NADH–ubiquinone reductase (NADH dehydrogenase). This is a large enzyme (M_r 850 000), which catalyses the transfer of two electrons from NADH to FMN:

$$NADH + H^+ + FMN \rightarrow FMNH_2 + NAD^+$$

Thus, NADH generated in glycolysis, fatty acid β-oxidation pathway, the tricarboxylic acid cycle and other metabolic pathways transfers electrons (and hydrogens) to FMN. In turn, flavin-linked dehydrogenases (Table 3.5) transfer their electrons to ubiquinone.

$FADH_2$ is formed in the tricarboxylic acid cycle by oxidation of succinate to fumarate, in the fatty acid β-oxidation pathway, and by the dehydrogenation of glycerol 3-phosphate by cytosolic glycerol phosphate dehydrogenase. These $FADH_2$ molecules also pass hydrogens and electrons to ubiquinone.

UBIQUINONE also known as coenzyme Q (CoQ or simply 'Q'), is a lipid-soluble electron carrier (Fig. 3.3). The number of isoprene units in the side chain (n) varies from 6 to 10: in humans n is 10. These isoprene groups make Q lipid-soluble and this, along with the fact that it is a small molecule, allow it to move freely in the lipid membrane. Thus, Q can be thought of as a mobile carrier. The quinone part of the molecule may be reduced in two steps by the addition of two hydrogen ions and two electrons (Fig. 3.4) via a semiquinone.

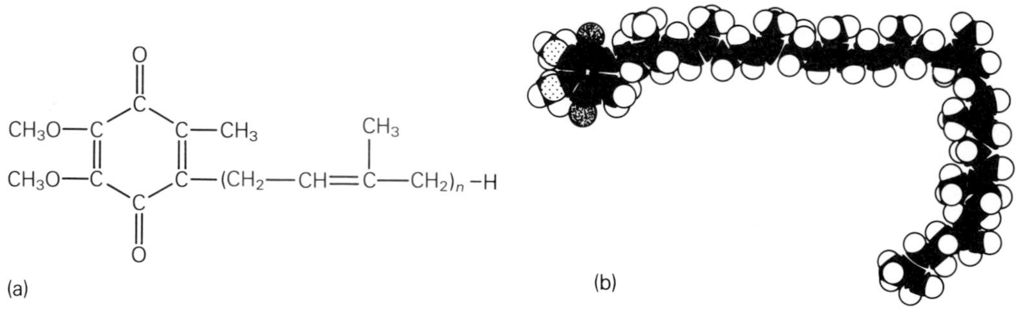

Fig. 3.3 (a) The structure of ubiquinone. The number of isoprene units in the side-chain is variable (n = 10 in humans). (b) Molecular model of ubiquinone.

Box 3.1 Flavin nucleotides, FAD and FMN

Riboflavin, which is synthesized by green plants, fungi and bacteria but not by animals, is the precursor of the coenzymes FMN (flavin mononucleotide or riboflavin 5′ phosphate) and FAD (flavin adenine dinucleotide). The naming system of FAD and FMN is not quite correct in that neither of these compounds is strictly a nucleotide. FMN contains an isoalloxazine (flavin) ring, a sugar derivative (ribitol), and a phosphate. FAD contains in addition AMP linked to FMN by a pyrophosphate bond. Nevertheless, since their structures resemble those of nucleotides they are sometimes called **pseudonucleotides**.

The flavin coenzymes serve as prosthetic groups to a set of enzymes called the flavoproteins that catalyse oxidation–reduction reactions. Flavin-containing enzymes often contain, in addition, a metal atom, either Fe or Mo, and may be associated with iron-sulphur centres.

Flavin coenzymes are usually tightly bound to the apoenzyme. Succinate dehydrogenase, an enzyme of the tricarboxylic acid cycle, contains an FAD molecule covalently linked to a histidyl residue of the protein.

Reference Hatefi, Y. (1985) The mitochondrial electron transport and oxidative phosphorylation system. *Annual Reviews of Biochemistry*, **54**, 1015–69. A detailed review of the structure of complexes I to V.

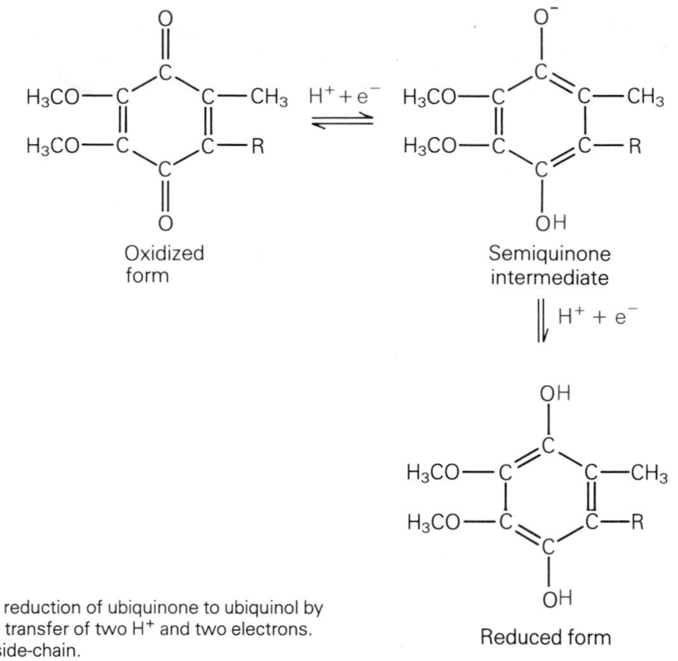

Oxidized form

$\rightleftharpoons$ $H^+ + e^-$

Semiquinone intermediate

$\rightleftharpoons$ $H^+ + e^-$

Reduced form

Fig. 3.4 The reduction of ubiquinone to ubiquinol by the two-step transfer of two H^+ and two electrons. R, isoprene side-chain.

Exercise 2

Draw the structure of an isoprene unit. Why does the addition of isoprene units enhance the lipid solubility of Q?

IRON–SULPHUR PROTEINS which contain iron–sulphur (Fe–S) centres, are components of many of the flavin-linked dehydrogenases in addition to the flavin moiety. The iron in Fe–S proteins can assume a range of oxidation states enabling the Fe–S complex to carry electrons.

Fe–S proteins are compounds of relatively low M_r, consisting of one or more iron atoms linked through the sulphur atoms of cysteine residues to peptides. In addition, non-cysteine sulphur (inorganic sulphur) is usually present. The Fe–S proteins in mitochondria are difficult to extract and little is known of their structure. However, a good deal is known of their electron-carrying behaviour as a result of electron spin resonance studies.

The simplest Fe–S proteins, the **rubredoxins**, contain one Fe–S centre and are found in anaerobic bacteria where they take part in oxidation–reduction reactions. The structure of the iron in rubredoxin (as shown in Fig. 3.5) is tetrahedral.

□ Iron–sulphur proteins contain iron atoms joined to cysteine sulphydryl (–SH) residues in the protein, but in addition many of them also contain elemental sulphur:

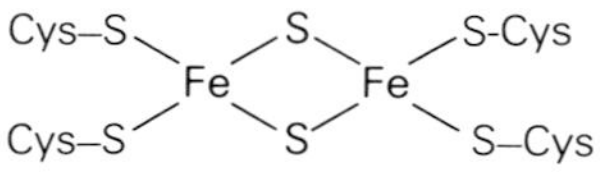

Cys–S, Cys–S — Fe (S, S) Fe — S-Cys, S–Cys

The elemental sulphur may be detected by treating such a protein with acid when H_2S is released.

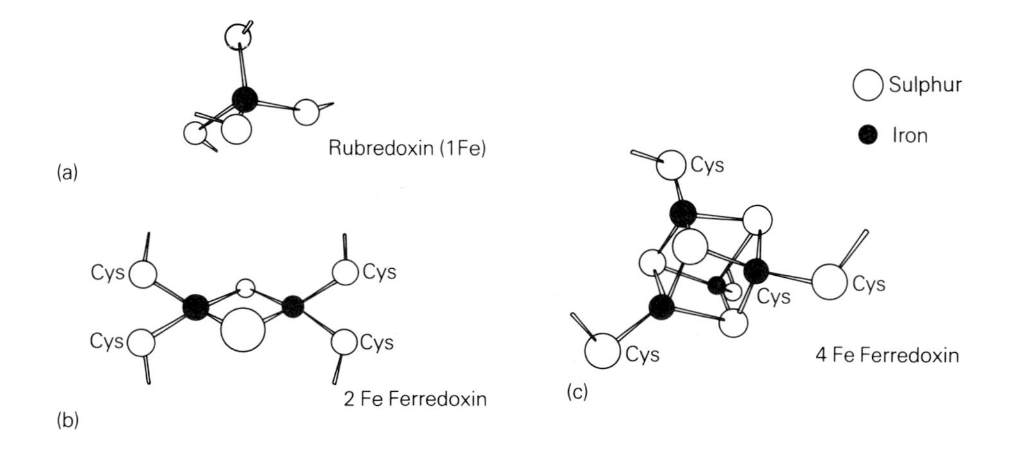

Fig. 3.5 Models of Fe–S complexes containing (a) one, (b) two or (c) four Fe–S centres. Iron atoms are shown in black. Cys, cysteine. Redrawn from Stryer, L. (1988) *Biochemistry*, 3rd edn, W.H. Freeman, New York, p. 403.

Reference Sweeney, W.V. and Rabinowitz J.C. (1980) Proteins containing 4Fe–4S clusters: an overview. *Annual Reviews of Biochemistry*, **49**, 139–61. A comprehensive review of the structure and mechanism of iron–sulphur proteins.

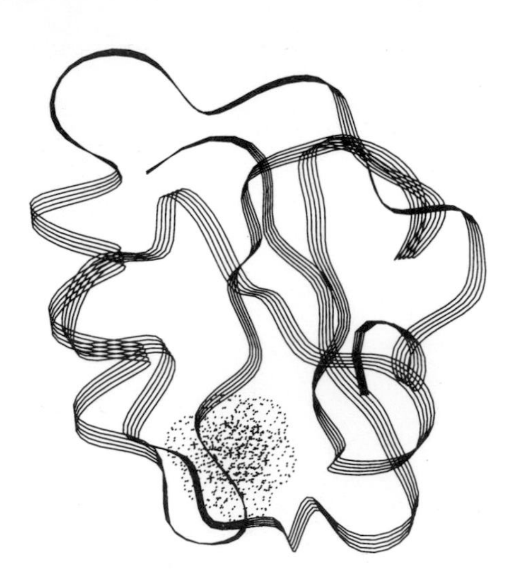

Fig. 3.6 'Ribbon' diagram of *Bacillus thermoproteolyticus* ferredoxin. Courtesy of Dr E.E. Eliopolos, Department of Biology and Molecular Biology, University of Leeds.

Ferredoxins (Fd) contain clusters of two or four iron atoms and have an M_r of 6000–12 000. Two-iron clusters may be represented as $[2Fe–2S]^{n+}$: the normal state is $[2Fe–2S]^{2+}$. On reduction, both iron atoms are involved in accepting an electron (Fig. 3.6).

The **four-iron clusters** $[4Fe–4S]^{n+}$ can have a variety of oxidation states with different numbers of iron atoms reduced at any one time, although each cluster tends to carry only a single electron. For example, in *Clostridial* ferredoxin the state changes from $[2Fe^{3+}, 2Fe^{2+}]$ $(= 10^{+})$ to $[3Fe^{3+}, 1Fe^{2+}]$ $(= 11^{+})$. This type of ferredoxin has an $E^{0'} = -0.42$ V. Other four-item clusters, such as that in succinate dehydrogenase, have the same number of irons but show an $E^{0'}$ of $+0.35$ V. These proteins are called 'HIPIP', or high-potential iron–sulphur proteins.

NADH–Q a reductase contains both 2Fe–2S and 4Fe–4S clusters and accepts electrons from $FMNH_2$ and passes them to Q. Succinate dehydrogenase also contains 2Fe–2S clusters. The protein environment of the Fe–S cluster is hydrophobic and this permits the iron atoms to accept and donate electrons.

CYTOCHROMES are found in all aerobic organisms and are small iron–haem proteins that carry only electrons. ***Haem*** is a porphyrin, that is, it consists of four pyrrole rings linked by methene bridges. The most common type haem in cytochromes is **protoporphyrin IX** (Fig. 3.7) complexed with iron. The iron atom of haem can change oxidation state (Fe^{2+}/Fe^{3+}) and thus carries a single electron in the process.

The cytochromes are classified into types *a*, *b* and *c*, depending on the type of porphyrin they contain; this was originally done on the basis of their absorption spectra. Table 3.3 lists the properties of some mammalian cytochromes. Most cytochromes are difficult to solubilize from membranes without the use of detergents, indicating that they are securely inserted into the membrane. An exception is **cytochrome c**, which is water soluble. In the inner mitochondrial membrane the cytochromes are arranged in an ordered sequence so that electrons can be passed from one to the other. Electron transfer occurs extremely rapidly between cytochromes. One requirement for fast electron transfers is that the geometries of the metal ion in the electron donor and acceptor should be similar.

☐ Because cytochrome *c* is water soluble and easily extractable from mitochondria, a great deal is known about the protein. The three-dimensional structure has been determined by X-ray crystallography. In addition, the amino acid sequences of cytochrome *c* from hundreds of different kinds of organism have been elucidated. It is possible to build up an evolutionary family tree of organisms by looking at the extent to which their cytochrome *c* amino acid sequences differ.

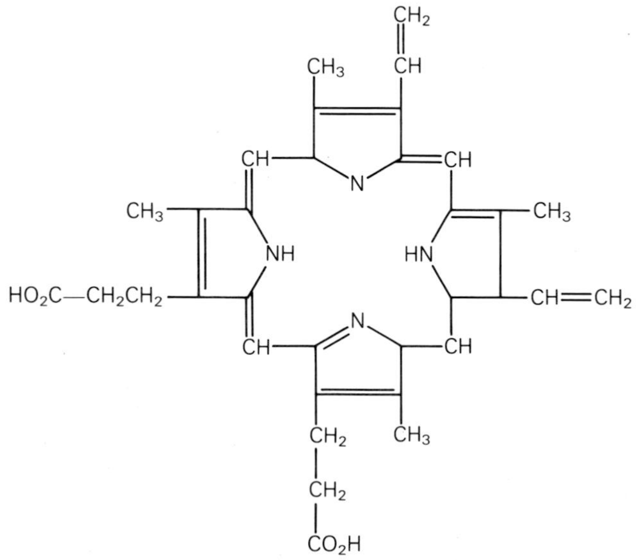

Fig. 3.7 The structure of protoporphyrin IX. Addition of a central Fe atom gives a haem.

Cytochromes: *electron-carrying proteins that contain haem. From the Greek* kutos, *vessel, and* khroma, *colour.*

Haem: *the iron–porphyrin prosthetic group of a number of proteins (haemoglobin, myoglobin and the cytochromes). From the Greek* aima, *blood.*

Table 3.3 Some properties of mammalian cytochromes

Cytochrome	Absorption maxima of reduced form (nm)			E^0 (mV)	M_r
a_3	600	–	445	+200	200 000
a	605	517	414	+340	
c	550	521	416	+260	12 500
c_1	554	523	418	+225	37 000
b_K	562	530	430	+30	25 000
b_T	566	*	*	−30	

* Values not known.

Cytochrome *c* reductase (complex III) catalyses the transfer of electrons from reduced Q to cytochrome *c*. Cytochrome *c* reductase contains two cytochromes, *b* and c_1, Fe–S proteins and several other proteins. **Cytochrome *b*** contains two haem groups in different protein environments and with different electron affinities. The different environments affect their absorption spectra and the haem groups are designated as cytochromes b_{566} and b_{562}, specifying the wavelengths of their absorption maxima.

Exercise 3

Why do CN^-, N_3^- and CO bind irreversibly to iron in the haem group of cytochromes or haemoglobin?

CYTOCHROME C OXIDASE catalyses electron transfer from cytochrome *c* to oxygen, which is reduced to water. This reaction may be written as:

$$4 \text{ cyt } c\,(Fe^{2+}) + 4H^+ + O_2 \rightarrow 4 \text{ cyt } c\,(Fe^{3+}) + 2H_2O$$

From this it can be seen that four electrons are required to be passed to O_2 to reduce it to H_2O. Cytochrome oxidase has eight subunits, two haem groups, called *a* and a_3, and two copper ions. The two haems are in different protein environments but are chemically identical. The two copper ions Cu_A and Cu_B are bound in different areas of the protein. Haem *a* is close to Cu_A and haem a_3 is next to Cu_B. Cytochrome *c* donates an electron to the haem a–Cu_A complex. An electron is next transferred to haem a_3–Cu_B and oxygen is then reduced by a series of intermediates to water. The pathway for the complete reduction of oxygen is shown in Fig. 3.8.

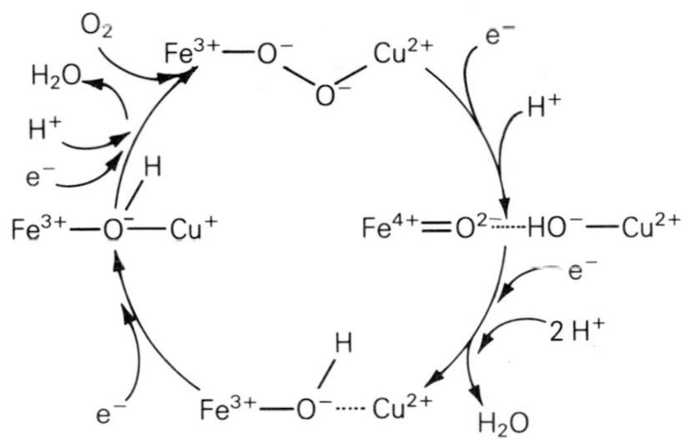

Fig. 3.8 A possible reaction sequence for the complete reduction of O_2 to H_2O by cytochrome oxidase. O_2 is bound to iron and copper and four electrons are required for complete reduction. Redrawn from Stryer, L. (1988) *Biochemistry*, 3rd edn, W.H. Freeman, New York.

The four-electron reduction of O_2 to H_2O presents problems. If the reduction is incomplete superoxide, O_2^-, peroxide, or the hydroxyl radical, may be generated. All of these are highly reactive and potentially dangerous. In order to avoid their production the enzyme does not release partly oxidized intermediates and in some way forces a four-electron transfer safely. Oxygen is bound between Fe^{2+} and Cu^{2+} and electrons are donated in turn to reduce it to H_2O.

Box 3.2
Superoxide dismutases

The enzyme superoxide dismutase was discovered in 1968. In fact, the enzyme activity was found to reside in a copper protein of previously unknown function that had been isolated in 1939 from red blood cells and had been called **erythrocuprein**. In 1969, McCord and Fridovich showed that the protein also contained zinc. Each of the two subunits contained one zinc and one copper atom and the whole protein had an M_r of 32 000. Very quickly this enzyme was demonstrated to be present in the cytoplasm of all eukaryotic cells. Strictly anaerobic bacteria appeared to lack activity. This might explain why they cannot tolerate oxygen, because they have no way of dealing with the highly reactive oxygen species (superoxide, peroxide, hydroxyl radical) that are inevitably produced when oxygen is present.

Soon it was discovered that mitochondria also contained a superoxide dismutase, but that this was a totally different protein from the cytosolic one. It contained manganese instead of copper and zinc, its M_r was 40 000, and it was pink instead of green! Later it was found that this Mn–superoxide dismutase was typical of at least some aerobic bacteria. However, some bacteria such as *Escherichia coli* could, in addition, produce an inducible, iron-containing superoxide dismutase when exposed to oxygen. .

McCord, J.M. and Fridovich, I. (1969) *Journal of Biological Chemistry*, **244**, 6049–95.

Exercise 4

Draw all the intermediate forms when oxygen is reduced to water by accepting four electrons.

The generally accepted sequence of the components of the electron transport chain is shown in Fig. 3.9.

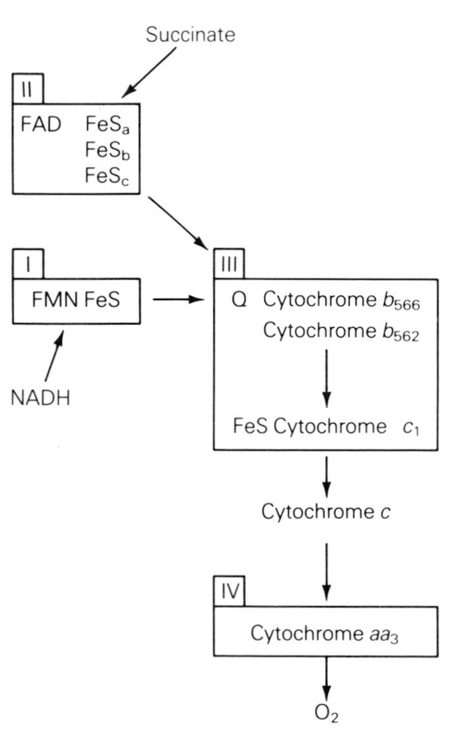

Fig. 3.9 The sequence of the components of the electron transport chain. The components are arranged in four blocks corresponding to Fractions I to IV (see text).

Sequence of carriers in the electron transport chain

A number of experimental approaches have been used to establish this sequence including: (1) comparison of the redox potentials of the carriers, (2) comparison of difference spectra, and (3) the use of inhibitors of electron transport.

REDOX POTENTIALS. Reactions involving electron transfer are oxidation–reduction or redox reactions. The removal of an electron or hydrogen atom is oxidation, the addition of an electron or hydrogen atom is reduction. Two reactions must be linked, one an oxidation, the other a reduction. A/AH_2 and B/BH_2 are called redox couples.

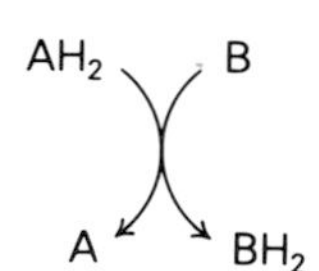

The oxidation–reduction, or redox, potential is a measure of the ability of one compound to oxidize or reduce another. Redox potentials can be measured (Fig. 3.10) but in practice the redox potentials of the components of the electron transport chain cannot easily be measured in the mitochondria under standard conditions. A midpoint potential, $E^{0'}$, is usually recorded at 25°C and pH 7.0 and with equal concentrations of the oxidized and reduced compound. Table 3.3 lists the $E^{0'}$ values of the components of the electron transport chain. The more effective a substance is as a reductant, the more negative its $E^{0'}$. Similarly, oxidants have a positive $E^{0'}$. Knowing the standard redox potential of biochemical reactions allows a prediction of the direction of electron flow to be made when two pairs of redox reactants are linked by an enzyme. For example, from Table 3.4 it can be seen that the $NAD^+/NADH$ pair has a redox potential of −320 mV, and the Q/QH_2 pair a potential of +100 mV. This means electrons will flow from NADH to Q.

The redox potentials of electron transport intermediates would be expected to become more positive on going from NADH to O_2, indicating that this is the direction of electron flow. Therefore, in mitochondria, electrons are

Reference Harold, F.M. (1986) *The Vital Force, a Study of Bioenergetics*, W.H. Freeman, New York. An examination of the general principles of energy production in biology.

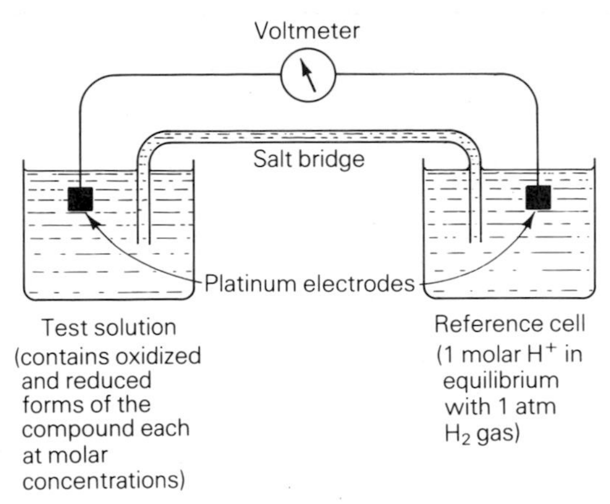

Fig. 3.10 The apparatus used to determine redox potentials. A reference cell containing an H_2/H^+ electrode is used as a standard.

Table 3.4 *Values of some redox couples at pH 7.0 and 25°C*

Reaction	E^0 (mV)
$2H^+ + 2e \rightarrow H_2$	−421
$NAD^+ + 2H^+ \rightarrow NADH + H^+$	−320
fumarate + $2H^+ + 2e^- \rightarrow$ succinate	−31
$2\,cyt\,b_T\,(ox) + 2e^- \rightarrow 2\,cyt\,b_T\,(red)$	−30
$2\,cyt\,b_K\,(ox) + 2e^- \rightarrow 2\,cyt\,b_K\,(red)$	+65
$Q + 2H^+ + 2e^- \rightarrow QH_2$	+100
$2\,cyt\,c_1\,(ox) + 2e^- \rightarrow 2\,cyt\,c_1\,(red)$	+220
$2\,cyt\,c\,(ox) + 2e^- \rightarrow 2\,cyt\,c\,(red)$	+254
$2\,cyt\,a_3\,(ox) + 2e^- \rightarrow 2\,cyt\,a_3\,(red)$	+385
$½O_2 + 2H^+ + 2e^- \rightarrow H_2O$	+816

Ox, oxidized; red, reduced; cyt, cytochrome. See also Tables 1.3 and 2.1.

introduced by NADH or $FADH_2$ and passed along a sequence of carriers to oxygen, equivalent to going *down* a potential energy gradient.

See Sections 1.3 and 2.2.

THE USE OF DIFFERENCE SPECTRA. Difference spectra are recorded using a dual-beam spectrophotometer in which the sample cell contains mitochondria with all the carriers in the fully reduced state and the blank cell contains mitochondria with all the carriers in the fully oxidized state. The resultant spectrum is that of the *reduced-minus-the-oxidized-state* of the electron transport chain. If oxygen is now introduced to the reduced preparation, the components become oxidized starting at cytochrome aa_3 and proceeding sequentially along the chain. This can be followed by the change in absorbance of each of the components as it becomes oxidized. Similarly, if the components in the 'oxidized state' preparation are exposed to a source of electrons (or aerobic conditions) the components become reduced starting at NAD^+, then flavoproteins, Q, etc. (Fig. 3.11).

THE USE OF INHIBITORS. A number of inhibitors are known that block electron flow at specific points along the electron transport chain and thus provide information on the sequence of the carriers. For example, three

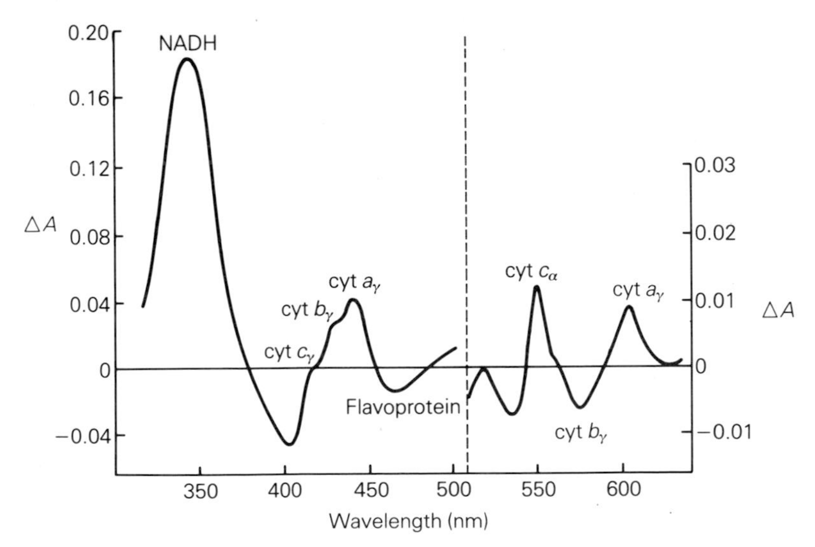

Fig. 3.11 A reduced-*versus*-oxidized difference spectrum of the electron carriers of the respiratory chain. The control is fully oxidized and the test suspension is made anaerobic to ensure that the carriers are reduced. Cyt, cytochrome. Redrawn from Smith, E. *et al.* (1983) *Principles of Biochemistry*, 7th edn, McGraw-Hill, New York.

Box 3.3 Free energy changes in redox reactions

The Nernst equation, which was introduced in Chapter 1, relates the standard redox potential, $E^{0'}$, the measured potential under non-standard conditions (E) and the concentrations of oxidant and reductant as follows:

$$E = E^{0'} + nF \ln [oxidant]/[reductant]$$

where n is the number of electrons transferred, and F is the Faraday, 96 406 J V^{-1}. The free energy released in redox reactions is dependent on the concentration of oxidized and reduced substrate.

The following equation relates free energy ($\Delta G^{0'}$) to $E^{0'}$.

$$\Delta G^{0'} = -nF\Delta E^{0'}.$$

$\Delta E^{0'}$ is the difference in the $E^{0'}$ values of the oxidized and reduced carriers, and n is 2 for components of the electron transport chain.

The energy released when NADH is fully oxidized may be calculated as follows. The $E^{0'}$ of the NADH/NAD$^+$ complex is −0.32 V and for $\frac{1}{2}O_2/H_2O$ the $E^{0'}$ is +0.82 V (Table 3.4).

$$NAD^+ + H^+ + 2e^- \rightarrow NADH \qquad E^{0'} = -0.32\ V \qquad (1)$$
$$\tfrac{1}{2}O_2 + 2H^+ + 2e^- \rightarrow H_2O \qquad E^{0'} = +0.82\ V \qquad (2)$$

subtracting (1) from (2)

$$\tfrac{1}{2}O_2 + NADH + H^+ \rightarrow H_2O + NAD^+ \qquad E^{0'} = 1.14\ V \qquad (3)$$

The free energy of the total oxidation, reaction (3), is given by:

$$\Delta G^{0'} = -nF\Delta E^{0'} = -2 \times 96\,406 \times 1.14$$
$$= -221\ kJ\ mol^{-1}$$

compounds, **rotenone**, **amytal** (a barbiturate), and **piericidin**, bind to NADH dehydrogenase and block electron flow to Q. Antimycin A, an antibiotic, blocks electron flow from cytochrome *b* to cytochrome *c*. The inhibitors **cyanide**, **carbon monoxide** and **azide** combine with cytochrome aa_3 and prevent electron flow to oxygen (Fig. 3.12a).

The site of action of the inhibitors has been determined by observing which intermediates of the chain become reduced and which become oxidized in the presence of the inhibitor and an electron donor. The effect of adding antimycin A to the electron transport chain is shown in Figure 3.12b. All carriers before the block become reduced (electron-rich), and all after the block oxidized (electron-poor) as their electrons drain away to oxygen and are not replaced. Thus, the use of the inhibitors allows the position of intermediates in the chain to be identified.

Exercise 5

Explain what is meant by 'difference spectra'.

3.3 Oxidative phosphorylation

☐ The 'P/O ratio' is also sometimes written 'P : O ratio', or more generally, simply in terms of electrons transferred, 'P : 2e$^-$ ratio'.

The **P/O ratio** is the number of moles of inorganic phosphate taken up to phosphorylate ADP, per atom of oxygen used and is thus a measure of the number of ATP molecules synthesized. Different substrates give different P/O ratios when they donate electrons to the electron transport chain. For NADH-linked substrates P/O is 3; for $FADH_2$-linked substrates it is 2 and for ascorbate it is 1. These three substrates feed in electrons at different points along the electron transport chain. The different P/O ratios indicate that there are three sites of synthesis of ATP.

The relationship between components of the electron transport chain and their redox potentials is shown in Fig. 3.13. At three points (between NADH

Reference Ernster, L. (ed.) (1984) *Bioenergetics*, Elsevier, Amsterdam. An account of electron transport and ATP synthesis in chloroplasts and mitochondria.

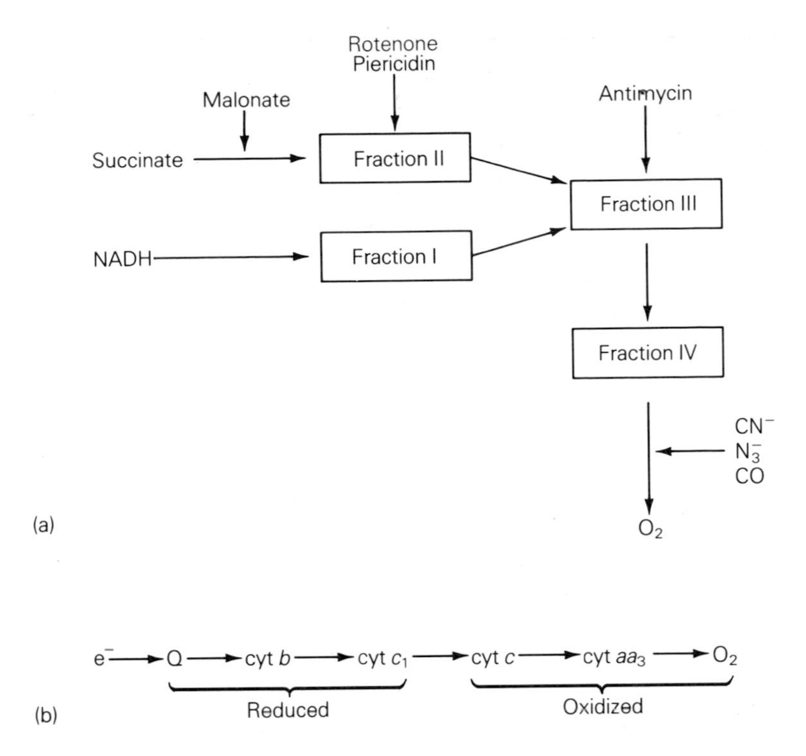

Fig. 3.12 (a) The sites of the action of respiratory inhibitors. (b) If antimycin is added to the chain in the presence of an electron source and oxygen only, components between cytochrome *c* and O_2 become oxidized.

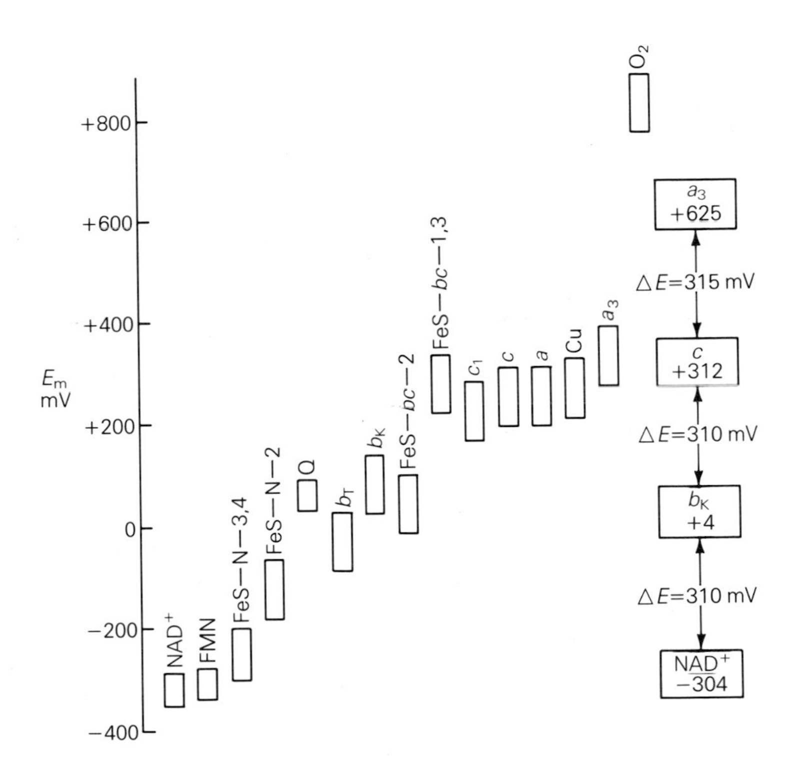

Fig. 3.13 The electromotive potential of the components of the respiratory chain. The box represents the redox potential of the components over a range of 9–91% reduction. The right-hand scale indicates the redox potential of some of the respiratory components and the difference in potential (Δ*E*) where it is large enough to allow ATP synthesis. Redrawn from Smith, E. *et al.* (1983) *Principles of Biochemistry*, 7th edn, McGraw-Hill, New York.

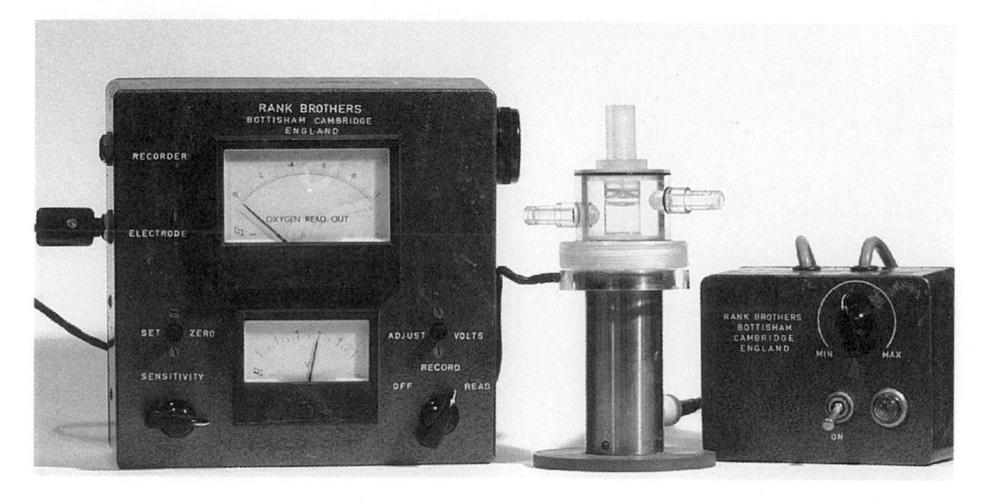

Fig. 3.14 An oxygen electrode. The box on the right drives a magnetic stirrer in the oxygen electrode chamber (centre). The box on the left supplies the polarizing voltage and measures the current (proportional to O_2 concentration). (b) Shows a trace of oxygen consumption against time obtained with respiring mitochondria. The addition of ADP stimulates O_2 consumption. (c) If an uncoupler, such as 2,4-dinitrophenol, is added O_2 consumption occurs without ATP synthesis. If an inhibitor, such as oligomycin, is added O_2 consumption is blocked.

(a)

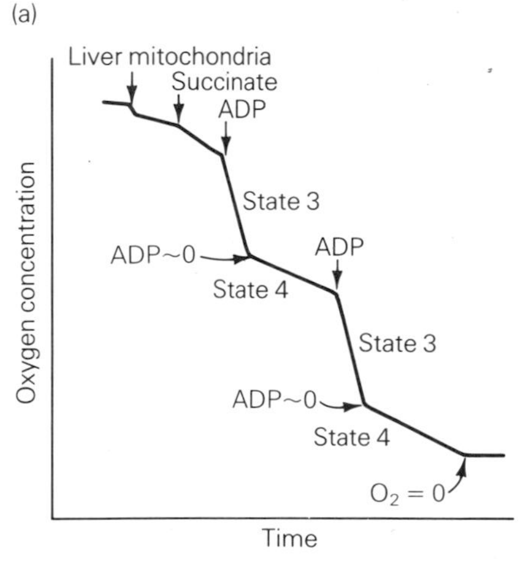

(b)

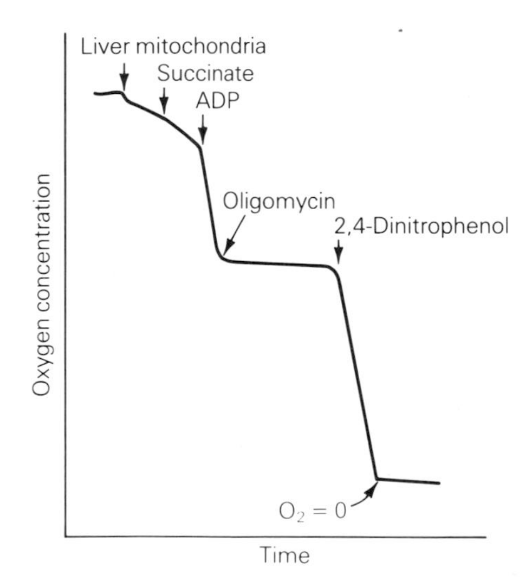

(c)

and cytochrome *b*; cytochrome *b* and cytochrome *c*; and cytochrome *c* and cytochrome *a*; the $\Delta E^{0'}$ is at least 220 mV. This is sufficient to generate 50 kJ, enough energy to synthesize 1 mole of ATP, at least under standard conditions. Since $FADH_2$-linked substrates bypass the first phosphorylation site (NADH dehydrogenase), only two ATP molecules will be produced from the oxidation of $FADH_2$.

The coupling of oxidative phosphorylation to electron transport

If ATP is synthesized during electron flow then the processes of electron transport and phosphorylation of ADP to produce ATP are said to be coupled. The rate of electron transport may be measured as the rate of oxygen uptake by mitochondria. Oxygen uptake may be determined with an oxygen electrode (Fig. 3.14a). In this apparatus, two electrons are required to reduce each oxygen atom.

$$4H^+ + 4e^- + O_2 \rightarrow 2\,H_2O$$

Since each pair of electrons donated by NADH produces 3 ATP, it is possible to calculate a P/O ratio; that is, the amount of ATP generated for each oxygen atom reduced. It is necessary to have some analytical method to measure the amount of ATP produced over the period during which oxygen consumption was measured. ATP is only produced while electrons flow and oxygen is consumed, hence the process is referred to as oxidative phosphorylation.

If an electron-donating substrate and a small amount of ADP are added to mitochondria in the presence of phosphate (P_i), there is a rapid consumption of O_2 (Fig. 3.14b). This active state of ATP synthesis is called 'state 3'. When all the ADP has been converted to ATP, the mitochondria revert to 'state 4', a resting phase where ATP is not being synthesized. The ratio of state 3 to state 4 is the respiratory control ratio and indicates how tightly electron transport is coupled to ATP synthesis. Some chemicals uncouple electron flow from ATP synthesis and are called uncoupling agents. For example, in the presence of 2,4-dinitrophenol, electron flow and O_2 consumption occur *without* ATP synthesis. In contrast, the inhibitor oligomycin prevents both ATP synthesis and electron flow (Fig. 3.14c).

The mechanism of oxidative phosphorylation

The process whereby energy released as electrons flow down a potential gradient in the electron transport chain is coupled to ATP synthesis from ADP

Box 3.4 Ionophores and ion gradients

In addition to the transmembrane proton gradient generated by the electron transport chain, ionic gradients due to unequal distributions of Na^+, K^+ and Ca^+ are also common. The ionic gradient due to the distribution of these ions is important in the function of nerve and muscle.

Antibiotics called **ionophores** have the ability to make membranes permeable to ions, so dissipating ionic gradients. These antibiotics are important experimental tools in the study of the chemiosmotic theory and of ion movement across membranes.

Valinomycin is a cyclic peptide composed of the sequence L-lactate, L-valine, D-hydroxyisovalerate, and D-valine repeated three times. Valinomycin forms a circle 0.8 nm in diameter and binds one K^+ to six carbonyl residues. The non-polar residues of the valinomycin molecule make it lipid-soluble, allowing it to diffuse across the membrane. The K^+ is released at the membrane surface, allowing the K^+ concentration on both sides of the membrane to equilibrate.

Nigericin (a K^+ carrier) and **monensin** (a Na^+ carrier) are other examples of peptide ionophores.

Gramicidin is a polypeptide antibiotic composed of 15 amino acids, which form a helix with a central diameter of 0.4 nm. Two gramicidin molecules, arranged with their amino termini in contact, span the cell membrane and provide a continuous channel. Gramicidin allows K^+, Na^+ and H^+ to pass across the membrane, so dissipating ionic or proton gradients.

Dinitrophenol and other lipid-soluble weak acids bind protons and carry them across membranes, allowing equilibration on either side of this membrane. Dinitrophenol destroys the proton gradient essential for ATP synthesis and is referred to as an **uncoupling agent** because it 'uncouples' oxidation from phosphorylation, that is to say, it allows electron transport in mitochondria to continue but prevents ATP synthesis. Many hundreds of uncoupling agents are known and all are weak acids that are soluble in membrane lipids. Photosynthetic ATP production (photophosphorylation) may also be uncoupled and some herbicides are uncoupling agents.

and P_i and is called oxidative phosphorylation. As electrons flow along the chain of carriers, protons are pumped across the inner mitochondrial membrane. The gradient of protons is somehow used to promote the synthesis of ATP and the generally accepted theory relating the proton flow across the membrane to ATP synthesis is called the **chemiosmotic theory**.

THE CHEMIOSMOTIC THEORY proposes that an **electrochemical gradient** is established across the inner mitochondrial membrane during electron transport. The gradient is due to protons being pumped across the membrane, resulting in a lowered pH and increased positive charge on the cytosolic side of the inner mitochondrial membrane. An analogy can be drawn with a battery where energy is stored as a separation of charge.

The chemiosmotic theory was proposed in 1961 by Mitchell, who received the Nobel prize in 1978. The theory proposes that the components of the electron transport chain are arranged in 'loops' in the inner mitochondrial membrane. The loops connect the matrix and intermembrane space or cytosolic sides of the mitochondrial membrane. The components of the electron transport chain are arranged so that hydrogen and electron carriers alternate. Hydrogens (that is, protons plus electrons) are carried by FMN and Q, while Fe–S centres, cytochrome *b* and cytochrome oxidase carry only electrons (Fig. 3.15a).

Hydrogens entering the respiratory chain from NADH are transferred to FMN and then moved to the cytosolic side of the inner membrane (Fig. 3.15a).

Exercise 6

If oxidative phosphorylation in mitochondria is 'uncoupled' by the addition of 2,4-dinitrophenol, what happens to the energy that would normally be conserved in ATP?

Reference Hinckle, P.C. and McCarty, R.E. (1978) How cells make ATP. *Scientific American*, **238**, 104. A well-written account that unifies this area; splendid diagrams.

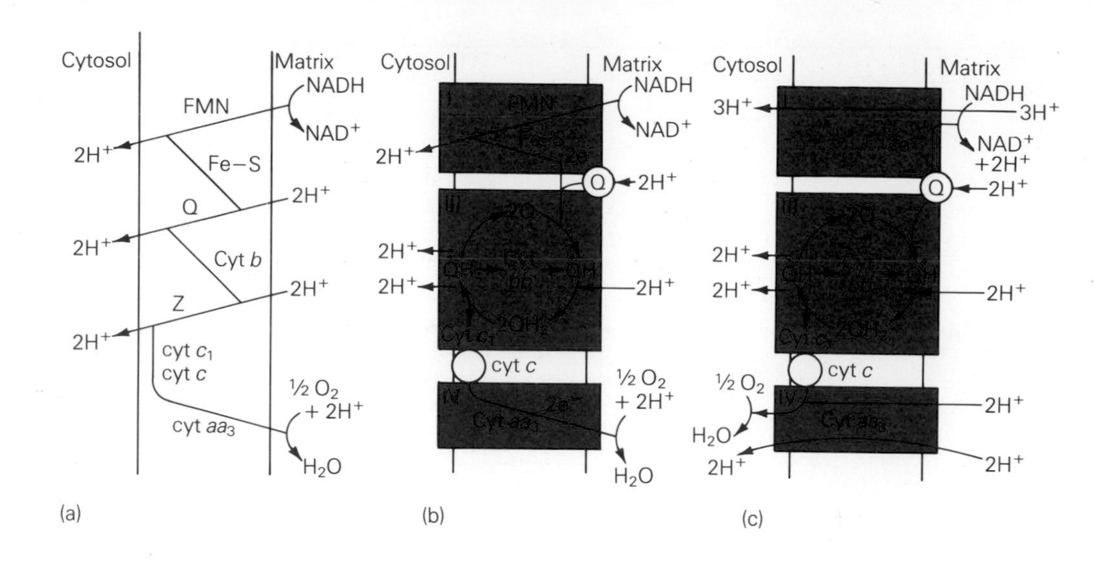

Fig. 3.15 (a) Three redox loops capable of moving protons from the matrix to the cytosolic side of the mitochondrial membrane. The unknown carrier Z is involved in the third loop. (b) An alternative scheme incorporating the Q cycle. (c) A further modification again incorporating the Q cycle. Cytochrome *bb* is an abbreviation for the two distinct forms of cytochrome *b*. Redrawn from Harold, F.M. (1986) *A Study of Bioenergetics*, W.H. Freeman, New York.

Two electrons are passed to Fe–S proteins. At the matrix side of the membrane these two electrons combine with $2H^+$ and are transferred to the cytosolic side via Q. The chain continues with a total of $6H^+$ being moved across the membrane for each two electrons donated from NADH. Substrates feeding in electrons via $FADH_2$ miss the first loop and donate hydrogens and electrons directly to Q. In this case, $4H^+$ move across the membrane.

Evidence for the movement of H^+ across the membrane comes from experiments with mitochondria suspended in an anaerobic, unbuffered saline solution. If a pulse of oxygen is admitted, the medium acidifies, indicating that H^+ have moved out of the mitochondria. However, such experiments are rather difficult to carry out satisfactorily and the interpretation of the data has given rise to much controversy.

The main deficiency of the scheme outlined in Figure 3.15a is the carrier Z, which has not been identified. An alternative scheme (Fig. 3.15b) dispenses with Z and introduces a cycle of reactions involving ubiquinone, called the **Q cycle**. Figure 3.15b shows the three respiratory complexes I, II, IV in series connected by Q and cytochrome *c*. The Q cycle translocates $4H^+$ for every two electrons entering the complex. As with the scheme shown in Figure 3.15a, the cytochrome oxidase complex returns electrons to the matrix where they reduce oxygen. QH is the semiquinone form of Q. The processes of ATP formation from ADP and P_i in mitochondria and in chloroplasts during photosynthesis appear to be very similar and a Q cycle has been detected during electron flow in chloroplasts.

A further modification is shown in Figure 3.15c. This scheme also has three complexes I, III and IV. Complex I extrudes three protons, complex III (a Q cycle) extrudes four protons, and complex IV is envisaged as a proton pump moving $4H^+$ out of the matrix. Two of the protons extruded by complex IV reduce oxygen. It is still uncertain on which side of the membrane oxygen reduction occurs or what the true stoichiometry of the complexes are.

3.4 *Synthesis of ATP*

The chemiosmotic theory proposes that protons flow back across the inner mitochondrial membrane through a membrane-bound ATPase. An ATPase catalyses the hydrolysis of ATP to produce ADP and P_i. However, enzymes merely catalyse the attainment of equilibrium. Therefore, under appropriate conditions an ATPase could catalyse the reaction ADP + $P_i \rightarrow$ ATP.

ATPases have been isolated from a variety of mammalian tissues, bacteria and yeast. The ATPase from mitochondria has an M_r of 480 000–500 000 and Figure 3.16 shows a model of its structure. It consists of a so-called F_1 particle composed of five subunits called α, β, γ, δ and ε. The α and β subunits bind nucleotide (ADP or ATP). The γ subunit allows entry of H^+ and the δ subunit attaches the F_1 particle to the so-called F_0 portion of the ATPase. The ε subunit is believed to have a regulatory function. F_0 is hydrophobic and spans the inner membrane and acts as a proton channel making the membrane specifically permeable to H^+.

The mechanisms of ATP synthesis are very similar in mitochondria and bacteria (which do not possess mitochondria) and in plant and bacterial photosynthesis (see later).

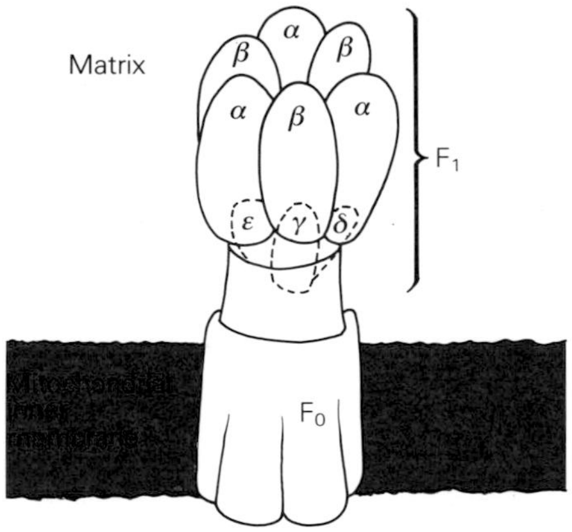

Fig. 3.16 The structure of the mitochondrial ATPase. Note that the enzyme spans the membrane with the headpiece on the matrix side of the mitochondrion.

Bacterial electron transport

Bacteria can use a wide variety of inorganic and organic compounds as electron donors in energy production. The ultimate electron acceptor may be oxygen, or, in anaerobic respiration, a wide variety of molecules such as nitrate, sulphate, carbonate and many organic compounds. Less energy is generated when these electron acceptors are used instead of oxygen but using them allows bacteria to grow in anaerobic conditions.

Box 3.5 Transport of Ca^{2+} into the mitochondria

The transport of cations into the mitochondria is linked to electron transport. The uptake of Ca^{2+} has been most intensively studied, since Ca^{2+} is associated with muscle contraction, neural function and hormone action. It has been estimated that cytostolic Ca^{2+} concentration is only 10^{-8} to 10^{-6} mol dm^{-3}, whereas mitochondrial Ca^{2+} concentration is of the order 10^{-4} mol dm^{-3}.

Ca^{2+} may be taken into the mitochondrion against a concentration gradient, using energy from ATP hydrolysis or from electron transport. If Ca^{2+} uptake is linked to electron transport, then inhibitors of electron flow abolish the process. It is thought that the inner membrane of the mitochondrion contains a uniport inhibited by ruthenium red, lanthanides and other rare-earth elements. Accompanying Ca^{2+} intake there is accumulation of phosphate to equalize charge distribution on either side of the membrane. Deposits of calcium (as $[Ca(PO_4)_2]_3\,Ca\,(OH)_2$) are often visible in mitochondria.

Electron transport cannot support both Ca^{2+} uptake and ATP synthesis, and Ca^{2+} uptake takes precedence. A Ca^{2+}/O ratio can be measured in a similar way to a P/O ratio. For every two electrons passing down the electron chain approximately six Ca^{2+} are accumulated by the mitochondrion. Experiments indicate an uptake of 1.67 Ca^{2+} and 1.0 phosphate ion for each electron pair passing through each coupling site.

The uptake of Ca^{2+}, whether driven by ATP hydrolysis or electron flow, results in H^+ extrusion from the mitochondrion. One H^+ exchanges for one Ca^{2+}. This exchange would be expected to lead to the accumulation of positive charge in the mitochondria, since calcium carries two positive charges. It has been suggested that 2 H^+ exchange with one Ca^{2+} and at the same time a phosphate is taken into the mitochondrion and a OH^- extruded. The OH^- would neutralize the H^+ giving an apparent ratio of 1 H^+ to 1 Ca^{2+}.

Box 3.6 Microsomal electron transport chain

See Box 2.3

Liver mitochondria and liver microsomes contain an electron transport chain including cytochrome P450. Cytochrome P450 absorbs light of wavelength 450 nm when complexed with CO. The electron transport chain containing cytochrome P450 functions in the hydroxylation of steroids, drugs such as phenobarbital and polycyclic aromatic hydrocarbons. The hydroxylation makes the substrates more water-soluble, rendering their excretion easier.

The electron transport chain containing P450 does not generate ATP. The system uses NADPH as a source of electrons, which are donated to a flavin coenzyme of **NADPH–cytochrome P450 reductase**. In the final step of the chain the Fe^{2+} of cytochrome P450 reacts with O_2 to form a cytochrome P450 Fe^{2+}–O_2 complex. One of the oxygen atoms of the complex is added to the substrate as OH^- and the other oxygen atom is reduced to H_2O. If the tissue is exposed to drugs, components of the cytochrome P450 system are induced.

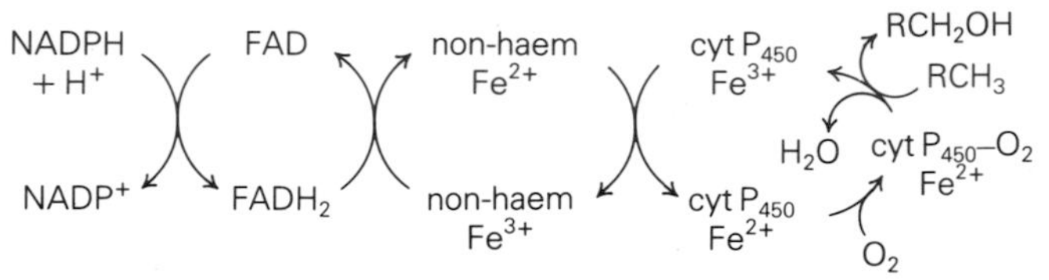

NADPH + H^+ → $NADP^+$; FAD → $FADH_2$; non-haem Fe^{2+} ; non-haem Fe^{3+} ; cyt P_{450} Fe^{3+} ; cyt P_{450} Fe^{2+} ; O_2 ; cyt P_{450}–O_2 Fe^{2+} ; RCH_3 → RCH_2OH ; H_2O

Exercise 7

Cigarette smoke contains a number of carcinogens, including polycyclic aromatic hydrocarbons (PAH). The PAH are activated by the addition of a hydroxyl group. Offer a possible explanation for the fact that not every smoker develops lung cancer.

Lithotrophs have an electron transport chain containing similar components to that in mitochondria and generate a transmembrane proton gradient. However, in this case, the gradient is across the cell membrane because bacteria lack organelles analogous to mitochondria.

The bacterial cell membrane contains dehydrogenases, cytochromes and Fe–S proteins but the respiratory chain of bacteria is generally more complex than that found in mitochondria. Different species of bacteria contain dehydrogenases for succinate, NADH, L-lactate, D-lactate, β-hydroxybutyrate, formate, α-glycerophosphate and many other electron donors. In addition to Q the chain may contain **menaquinone** (e.g. bacterial MK), **dihydromenaquinone** or **2-demethylmenaquinone** (Fig. 3.17). The cytochrome chain may or may not contain cytochrome *c*; it may have more than one type of cytochrome oxidase, e.g. **cytochrome *o***, differing in the haem prosthetic groups. A further complication is that components of the chain may vary with the type of nutrient the bacteria utilize.

Some examples of the electron transport chains found in bacteria are:

Paracoccus denitrificans

$$NADPH \rightarrow FP \rightarrow Fe\text{–}S \rightarrow Q_{10} \rightarrow \text{cyt } b \rightarrow \text{cyt } c \rightarrow \text{cyt oxidase} \rightarrow O_2$$

Escherichia coli

$$NADH \rightarrow FP \rightarrow Fe\text{–}S \rightarrow Q(MK) \rightarrow \text{cyt } b \rightarrow \text{cyt } o \rightarrow O_2$$

Bacillus megatherium

$$NADH \rightarrow FP \rightarrow Fe\text{–}S \rightarrow MK \rightarrow \text{cyt } b \rightarrow \text{cyt oxidase} \rightarrow O_2$$

In some bacterial species, fumarate, NO_3^-, SO_4^{2-}, or CO_3^{2-} is the terminal electron acceptor, rather than oxygen. Some of the components of the bacterial electron transport chain are worth considering in more detail.

DEHYDROGENASES. Table 3.5 lists some of the flavin dehydrogenases of bacteria. These enzymes are membrane-bound and are linked to the electron transport chain. The flavoproteins have FMN or FAD as prosthetic groups and Fe–S centres. However, some dehydrogenases link directly to

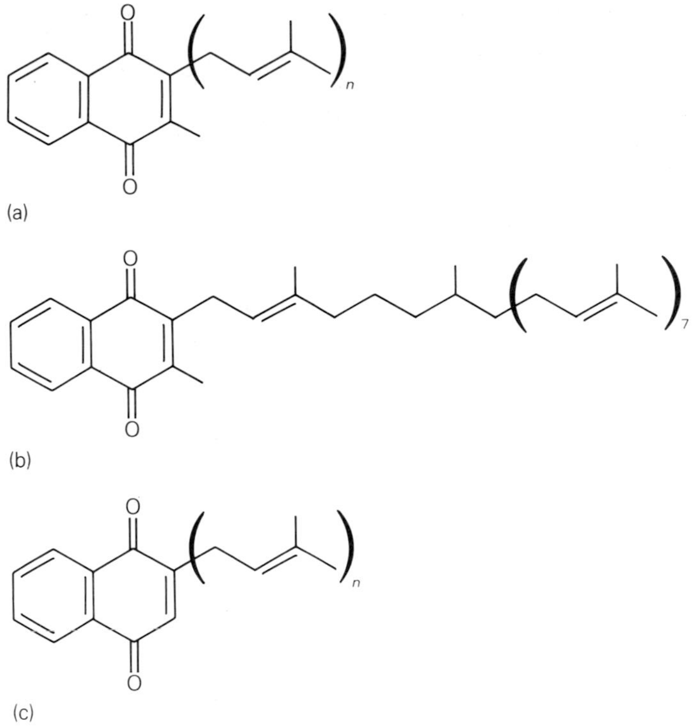

(a)

(b)

(c)

Fig. 3.17 Structures of (a) menaquinone, (b) dihydromenaquinone, and (c) 2-demethylmenaquinone. n = 5–9.

Table 3.5 *Flavoprotein-containing dehydrogenases of bacteria*

Enzyme substrate	Cofactor	Reaction	Organism
NADH	FAD	NADH → NAD^+	*Escherichia coli*
Succinate	FAD	succinate → fumarate	*Micrococcus lactyliticus*
α-Glycerophosphate	FAD	α-glycerophosphate → dihydroxyacetonephosphate	*Streptococcus faecalis*
D-Lactate	FAD	D-lactate → pyruvate	*E. coli*

cytochrome *c* without an intervening quinone. Bacteria may contain membrane-bound, NAD-linked dehydrogenases and soluble cytoplasmic enzymes.

QUINONES. Figure 3.17 and Table 3.6 show the structure of some bacterial quinones. If oxygen is the terminal electron acceptor then ubiquinone is usually found in the chain. If fumarate is the electron acceptor than menaquinone is found in the chain.

CYTOCHROMES. Table 3.6 shows the range of cytochromes found in bacteria. The *b* type cytochromes are found in all cases, but cytochrome *c* may be missing. The cytochrome oxidase component may be the aa_3 system found in mitochondria or cytochromes *o*, *a* or *d*. Bacteria may contain more than one cytochrome oxidase, giving rise to a branched chain. An example found in *Escherichia coli* is:

$$\text{NADH} \rightarrow \text{FP} \rightarrow \text{Q} \begin{array}{l} \nearrow \text{cyt } b \rightarrow \text{cyt } o \rightarrow \text{O}_2 \\ \searrow \text{cyt } b \rightarrow \text{cyt } d \rightarrow \text{O}_2 \end{array}$$

Exercise 8

Generation of ATP by oxidative phosphorylation demands the existence of an intact, impermeable membrane. As bacteria have no mitochondria, how do they generate ATP by oxidative phosphorylation?

Table 3.6 Components of bacterial electron transport chains

Organism	Quinones	*b*-cytochromes oxidases		*c*-cytochromes		Cytochrome oxidases
Escherichia coli	Q8 (MK8)	b_{556}	b_{562}			*o*
Klebsiella pneumoniae	Q8			b_{559}	b_{563}	*o*
Azotobacter vinelandii	Q8			b_{560}	b_{551} c_{555}	*o a, d*
Micrococcus luteus	MK8	b_{557}	b_{562}	c_{549}	c_{552}	a, a_3
Bacillus subtilis	MK7					

MK indicates menaquinone (number indicates isoprene units in the side-chain).
Q is ubiquinone (number indicates isoprene units in the side-chain).

In some bacteria low oxygen conditions result in cytochrome aa_3 being replaced by cytochrome *o*, perhaps indicating that the latter is more efficient in transferring electrons to oxygen.

Anaerobic electron transport

Bacteria may grow under anaerobic conditions using an electron transport chain which has, for example, fumarate or NO_3^- as the terminal electron acceptor. In this case **fumarate reductase** or **nitrate reductase** catalyse the terminal reduction steps. These two enzymes are induced in *E. coli* under appropriate conditions. Figures 3.18a and b shows the organization of the electron transport chains leading to the reduction of NO_3^- to NO_2^- and fumarate to succinate, respectively, in *E. coli*. It can be seen that these electron transport chains can use a wide variety of electron donors including NADH, formate and α-glycerophosphate.

Oxidative phosphorylation in bacteria

Oxidative phosphorylation in bacteria seems to operate by the chemiosmotic mechanism already described for mitochondria but with some differences.

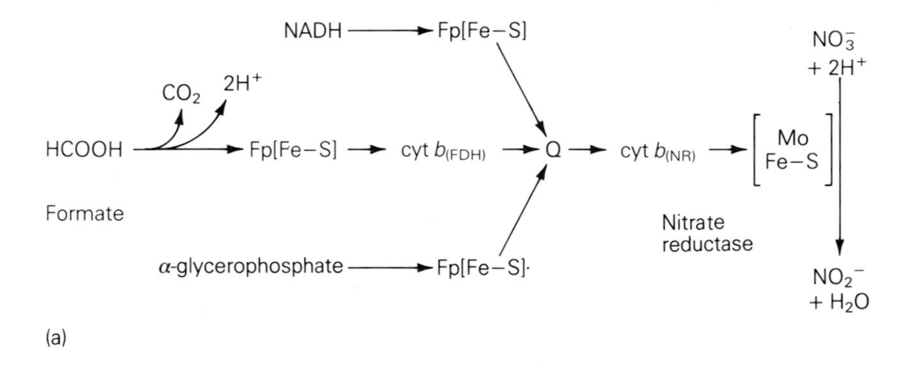

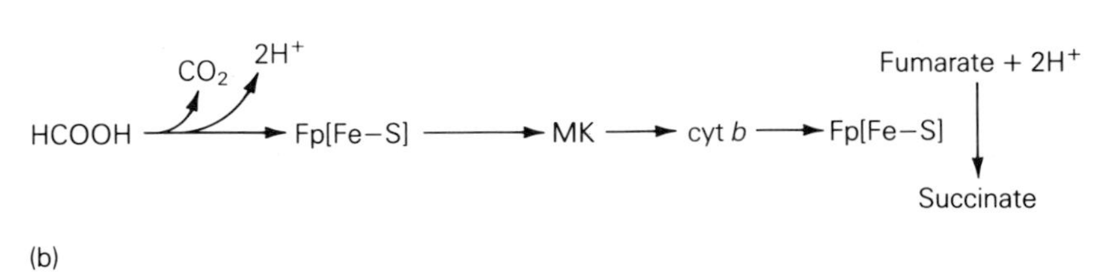

Fig. 3.18 (a) The components of the nitrate reductase electron transport chain of *Escherichia coli*. (b) Fumarate reduction in *E. coli*. Both chains operate under anaerobic conditions.

See *Biosynthesis*, Chapter 2

Reference Silver, S., Nucifora, G., Chu, L. and Misra, T.K. (1989) Bacterial resistance ATPases: primary pumps for exporting toxic cations and anions. *Trends in Biochemical Sciences*, **14**, 76–80. A different role for ATPases.

Bacterial oxidative phosphorylation does not appear to show respiratory control, i.e. it is not regulated by the supply of ADP. Another difference is in the stoichiometry of proton movement accompanying electron transport. In the mitochondrion, H^+/O ratios of about 6 are typical. In bacteria containing a cytochrome *c*, H^+/O ratios of 8 are found; in bacteria with a weak transhydrogenase H^+/O of 6 are the norm and bacteria lacking cytochrome *c* have H^+/O of 4. The H^+/O ratio determines the efficiency of ATP synthesis (see later).

The bacterial electron chain is situated in a plasma membrane, rather than in an organelle, and disruption of the plasma membrane produces respiratory vesicles that can be used in studies of oxidative phosphorylation.

Possible schemes for proton translocation across the *E. coli* plasma membrane for aerobic and anaerobic respiration are shown in Figures 3.19a and b. The mechanism of oxidative phosphorylation is probably analogous to that occurring in mitochondria and **bacterial F_1F_0-ATPase** is closely similar in structure to the mitochondrial enzyme. Thus, bacterial F_1F_0-ATPase contains α, β, γ, δ and ε subunits, although the stoichiometry is uncertain.

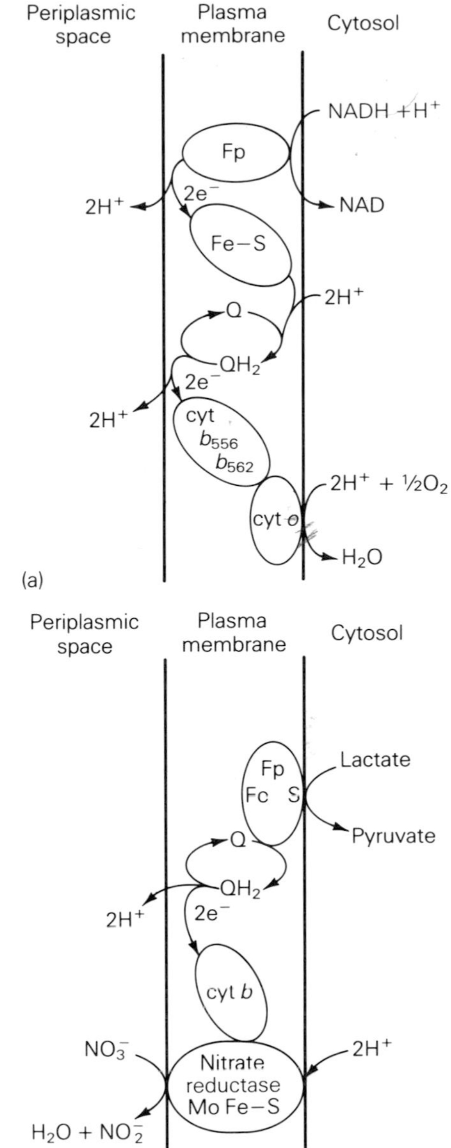

Fig. 3.19 A possible method of proton translocation in *Escherichia coli* during aerobic respiration (a), and anaerobic respiration (b).

3.5 *Photosynthesis*

ATP is generated during photosynthesis in a manner very similar to that in mitochondria, that is to say via a proton gradient. The reactions of photosynthesis are described in detail in *Biosynthesis*, Chapter 2. Here only the electron transport chains associated with photosynthesis will be described. Photosynthesis in green plants occurs in the ***chloroplast*** (Fig. 3.20a). This contains a stack of membranous structures, the thykaloids, which carry the photosynthetic pigments (chlorophylls and carotenoids). The whole structure is enclosed by a double membrane and the central region other than the thylakoids is referred to as stroma (Fig. 3.20b). The thylakoid membrane has an F_1F_0-ATPase of generally similar structure and appearance to the mitochondrial enzyme when viewed with the electron microscope (Fig. 3.21); it is called CF_1CF_0 to distinguish it from the mitochondrial enzyme.

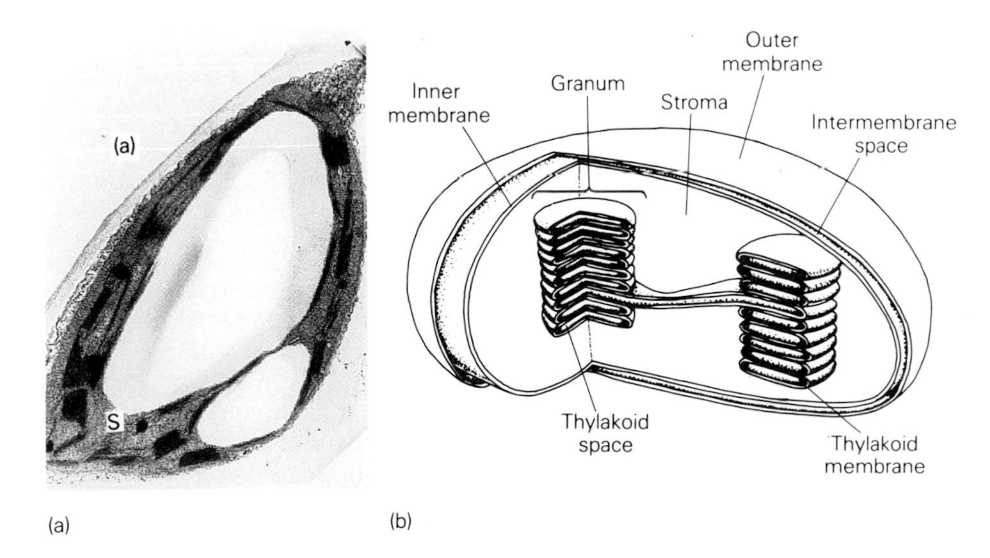

Fig. 3.20 (a) Electron micrograph of a chloroplast from the leaf of tobacco (× 9200). S, starch grain. Courtesy of Dr E. Sheffield, Department of Cell and Structural Biology, University of Manchester. (b) A schematic representation of a cross-section through a chloroplast.

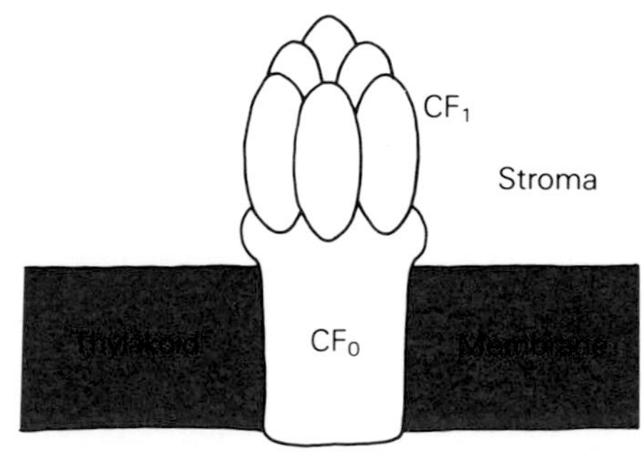

Fig. 3.21 A diagram of the CF_1CF_0-ATPase of chloroplasts. Compare with Fig. 3.6 for similarities to the mitochondrial enzyme.

Reference Shavit, N. (1980) Energy transduction in chloroplasts. *Annual Reviews of Biochemistry*, **49**, 111–38. A full review of the structure and function of the chloroplast ATP complex.

Chloroplast: *from the Greek* chloros, *green, and* plasma, *to mould.*

Exercise 9

Explain why green plants have *both* photosystem I and photosystem II, while photosynthetic bacteria have only photosystem I.

LIGHT ABSORPTION. Photosynthesis may be described as the light-driven reduction of CO_2 to carbohydrate with the concomitant evolution of oxygen:

$$6CO_2 + 6H_2O \xrightarrow{\text{light}} (CH_2O)_6 + 6O_2 \qquad \Delta G^{0'} = +2880 \text{ kJ mol}^{-1}$$

The light is absorbed by specialized chlorophyll molecules, which are arranged in centres called photosystems. **Photosystem I** (PSI) produces reducing power in the form of NADPH, while **Photosystem II** (PSII) transfers electrons from H_2O to an acceptor, liberating oxygen. The reaction centres contain chlorophyll molecules which trap the light energy falling on the photosystem. The chlorophyll molecule that ultimately collects the light energy and passes it on is designated P700 in PSI, and P680 in PSII (the numbers here indicate the wavelength of light absorbed by the chlorophyll).

Isolated chloroplasts may be exposed to light in the presence of an extraneous electron acceptor (shown below as X^{2+}). This process is called the Hill reaction and is characterized by the evolution of oxygen:

$$2H_2O + 4X^{2+} \xrightarrow{\text{light}} O_2 + 4H^+ + 4X^+$$

Respiration involves passing electrons obtained from organic molecules down a respiratory chain to reduce oxygen to water. The light reactions of photosynthesis split water using light energy, evolving oxygen and passing electrons down an electron transport chain. The photosynthetic electron transport chain contains cytochromes, non-haem iron centres and quinones and results in the pumping of protons across a membrane into the thylakoid space. Subsequently, the energy stored in the gradient of protons is used to drive ATP synthesis as in mitochondria.

O
H_3C
H_3C
CH_3
$(CH_2—CH=C—CH_2)_nH$
O

Oxidized, plastoquinone form

(a)

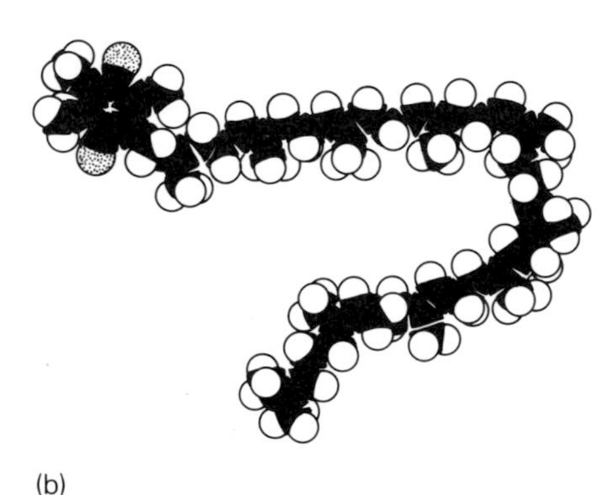

(b)

OH
H_3C
H_3C
CH_3
$(CH_2—CH=C—CH_2)_nH$
OH

Reduced, plastoquinol form

(c)

Fig. 3.22 (a) The structure of plastoquinone with (b) the corresponding molecular model (compare Fig. 3.3) in the oxidized form. The reduced form is shown in (c).

The roles of PSI and PSII

In order to split water using light energy, PSI and PSII interact. PSI generates a strong reductant that reduces $NADP^+$ to NADPH. PSII generates a strong oxidant that liberates oxygen from H_2O.

The absorption of light by PSII excites an electron from the magnesium atom of P680 to a higher energy state (designated P680*) which is transferred to pheophytin. Pheophytin is virtually identical to chlorophyll, except that it lacks Mg. From pheophytin the electron is donated to a protein-bound **plastoquinone** (PQ, Fig. 3.22) which resembles Q of mitochondria. Electrons are subsequently transferred to a second, unbound molecule of PQ (designated PQ_B).

The first phase of photosynthesis can therefore be summarized:

$$2H_2O + 2PQ \xrightarrow{\text{light}} O_2 + 2PQH_2$$

The magnesium atom in P680* is now electron-deficient having transferred an electron to PQ: this may be looked upon as a positively-charged hole. To replace the electron, P680* accepts an electron from water which is split by an enzyme component of PSI. The water-splitting enzyme contains four manganese ions but an unidentified intermediate, Z, is involved in electron transfer between water and P680*. Figure 3.23 summarizes the electron transfer chain of photosynthesis and shows the redox potentials of the components.

□ An electron can absorb energy and jump to a higher energy orbital. The resulting higher energy state is unstable and the electron returns to its original orbital, emitting energy in the process.

THE FLOW OF ELECTRONS FROM PQ TO PSI occurs via a cytochrome called **cytochrome *bf***, which passes electrons from PQ to plastocyanin (PC), a copper-containing protein (M_r about 11 000). During electron transfer

Reference Hunter, C.N., Van Grondelle, R. and Olsen, J.D. (1989) Photosynthetic antenna proteins: 100 ps before photochemistry starts. *Trends in Biochemical Sciences*, **14**, 72–6. An up-to-date account of research on the use of laser spectroscopy to find out about the first fraction of a second of the photosynthetic process.

Reference Govindjee and Coleman, W.J. (1990) How plants make oxygen. *Scientific American*, **262**, 42–51. A good review that emphasizes the light reactions of photosynthesis.

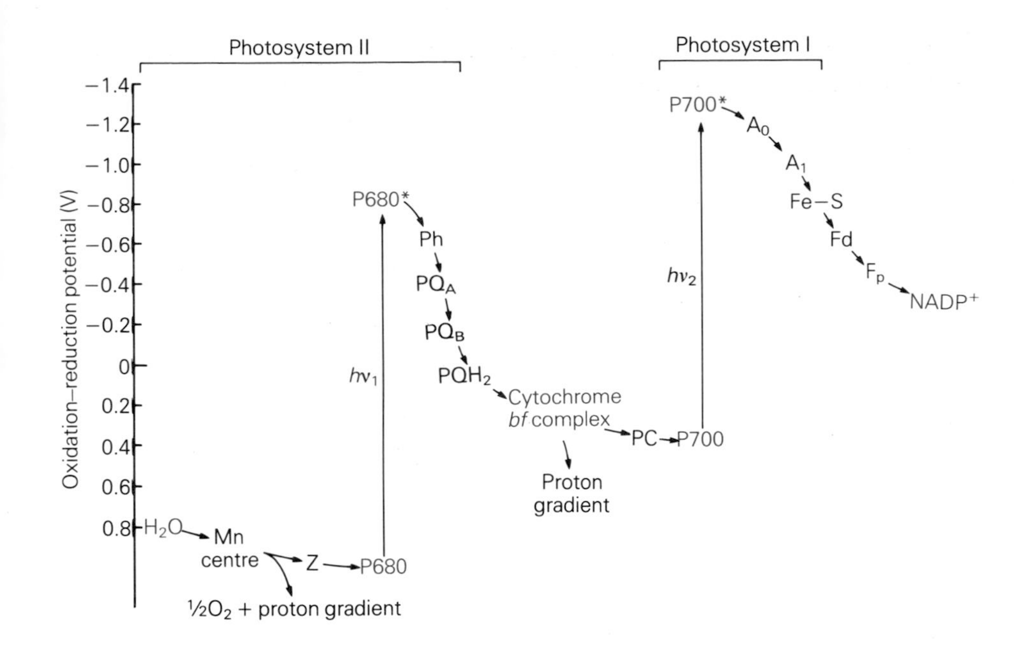

Fig. 3.23 The pathway of electron flow from H_2O to $NADP^+$ during photosynthesis. Two photosystems, PSII (P680) and PSI (P700) are involved. Pc, plastocyanin, Fd, ferredoxin.

between cytochrome *bf* and PC, protons are pumped across the thylakoid membrane into the thylakoid space (inside acidic), generating a transmembrane potential.

Cytochrome *bf* is a complex molecule of four subunits: cytochrome *f* (M_r 34 000), cytochrome b_{563} (M_r 23 000), an Fe–S protein (M_r 20 000), and a further polypeptide (M_r 17 000). The cytochrome *bf* complex closely resembles the cytochrome *c* reductase of mitochondria.

THE PSI ELECTRON TRANSPORT SYSTEM. PSI absorbs light, which excites an electron of Mg in chlorophyll and transfers it via a series of Fe–S centres to **ferredoxin** (Fd). The electron-deficient P700* regains an electron from PC and PSII.

Green plant Fd contains *two* Fe–S centres attached to a polypeptide. Each Fe–S centre may accept an electron and the two electrons eventually reduce $NADP^+$. The reaction is catalysed by ferredoxin–$NADP^+$ reductase which has an FAD prosthetic group.

The combined electron transfer chains of PSI and PSII is called the Z-scheme, because it was originally written in the form of the letter 'Z'. However, it is now more commonly written turned through 90° as a letter 'N' (Fig. 3.23).

THE ARRANGEMENT OF PSI AND PSII IN THE THYLAKOID MEMBRANE is probably as shown in Fig. 3.24. When PSI and PSII absorb light energy, electrons from water reduce $NADP^+$. At the same time, protons enter the thylakoid space, which becomes more acidic approaching a pH of 4.0 during electron transport. The two photosystems lie near the lumenal side of the membrane and the electron chains are arranged so that they span the membrane. Note that while the electron transport chains in mitochondria and bacteria *expel* protons, the thylakoid *accumulates* protons during photosynthesis.

Exercise 10

List some of the electron carriers of mitochondrial oxidative phosphorylation and put against them the corresponding carriers from the photosynthetic electron transport chain.

Reference Winslow, R. and Briggs, A.R. (eds) (1989) *Photosynthesis*, Alan Liss, New York. Keeping up with developments in photosynthesis research is almost impossible; this is a recent book that will give some idea of the sorts of things that are being done and what questions remain to be answered.

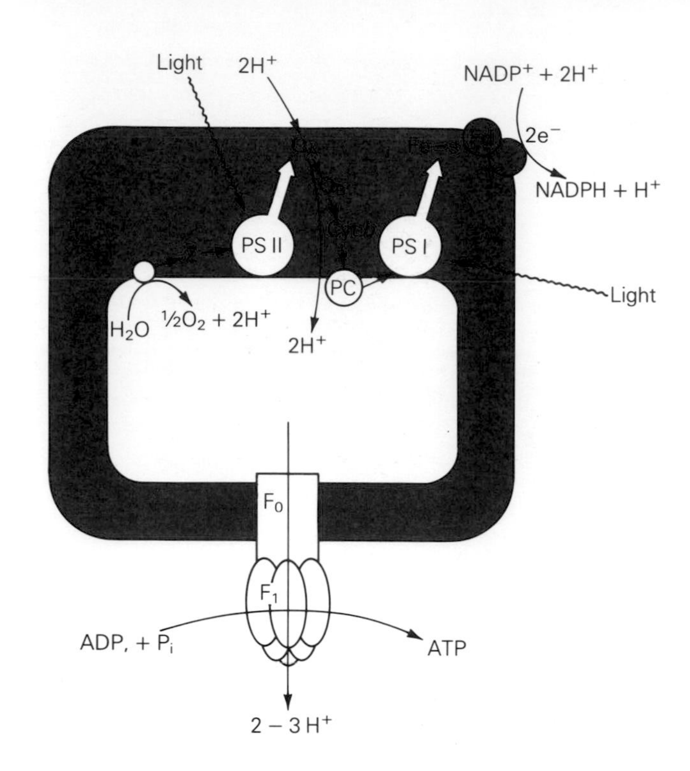

Fig. 3.24 Arrangements of photosystems (PS) I and II in the thylakoid membrane. The electron carriers are looped across the membrane so that for each pair of electrons transported from water four protons appear in the lumen. Redrawn from Harold, F.M. (1986) *A Study of Bioenergetics*, W.H. Freeman, New York.

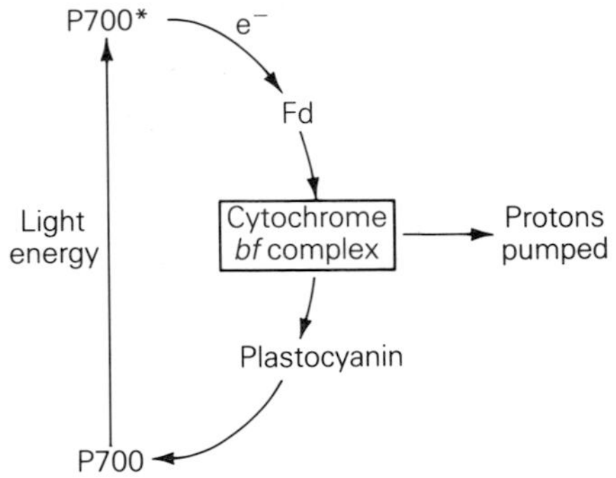

Photosynthetic phosphorylation

The structure of CF_1CF_0 was shown in Figure 3.21. The CF_1 particle consists of the subunits α, β, γ, δ and ε, very similar to the situation in F_1F_0. The stoichiometry of CF_1 is $\alpha_3\beta_3\gamma\delta\varepsilon$. The α and β subunits contain the binding sites for ATP and ADP. The δ subunit binds the CF_1 to CF_0 and the γ subunit regulates proton flow. The ε subunit regulates the overall catalytic activity. The CF_0 particle spans the membrane and allows protons to flow across the thylakoid membrane: as electrons flow, ATP is synthesized.

Fig. 3.25 Electron flow in cyclic photophosphorylation. Electrons from P700 are passed to ferredoxin (Fd), to the cytochrome *bf* complex, then to plastocyanin and back to P700.

CYCLIC PHOTOPHOSPHORYLATION. An alternative pathway of electron flow may occur in PSI. The excited electrons from P700* passes to Fd but instead of reducing $NADP^+$ return to the cytochrome *bf* complex (Fig. 3.25). The electron passing through cytochrome *bf* results in protons being pumped across the membrane which can be used to drive ATP synthesis by a process called **cyclic photophosphorylation**. It is difficult to demonstrate that cyclic photophosphorylation actually occurs *in vivo*, although it is an energetically feasible process.

STOICHIOMETRY OF ATP SYNTHESIS. Three protons flow through the CF_1–CF_0 complex for each ATP synthesized. A flow of three protons corresponds to a $\Delta G^{0'}$ of -61 kJ mol^{-1}, sufficient to synthesize 2 ATP molecules.

PHOTOSYNTHESIS IN CYANOBACTERIA. These unicellular organisms, which used to be called blue-green algae contain chlorophyll a, and carry out oxygen-evolving photosynthesis essentially in the manner described for green plants.

Reference Knaff, D.B. (1988) The photosystem I reaction centre. *Trends in Biochemical Sciences*, **13**, 460–1. One way to try to keep up with the latest developments in photosynthesis research is to read articles in this and similar monthly review journals.

Reference Rutherford, A.W. (1989) Photosystem II, the water-splitting enzyme. *Trends in Biochemical Sciences*, **14**, 227–32. Some of the secrets of this key system are revealed; complete with a cartoon that says 'Zap'.

Bacterial photosynthesis

Many bacteria are capable of carrying out photosynthesis, but bacterial photosynthesis differs from the process described in green plants in a number of ways.

ANAEROBIC PHOTOSYNTHESIS. Except for the cyanobacteria and halobacteria, bacteria carry out anaerobic photosynthesis using substrates such as hydrogen sulphide, sulphur, thiosulphite, hydrogen or a variety of organic compounds, rather than water, as the source of electrons. Examples are the purple sulphur bacteria (Rhodospirillaceae) and green sulphur bacteria (Chromatiaceae) which contain **bacteriochlorophyll** (Fig. 3.26) rather than chlorophylls a and b, and only have one photosystem.

See Section 2.4

Exercise 11

Summarize the main differences between non-cyclic and cyclic photophosphorylation.

As an example, if H_2S is the reductant, the reaction of photosynthesis is:

$$2H_2S + CO_2 \xrightarrow{\text{light}} (CH_2O) + 2S + H_2O$$

but many other reductants can be used in place of H_2S, depending upon the organism.

The arrangement of electron carriers in anaerobic photosynthetic bacteria shows some similarities to that in chloroplasts but the major difference is that there is only one photosystem in these bacteria. The electron transport chain of *Rhodopseudomonas sphaeroides* is shown in Fig. 3.27. The reaction centre, P870, transfers an electron to bacteriophaeophytin and then to a Q–Fe centre. Subsequent electron transfer occurs to Q_B. The electron-deficient P370 accepts an electron from cytochrome c_2, which is linked to Q_B by *b* cytochromes. In bacteria then, the electron flow is cyclic.

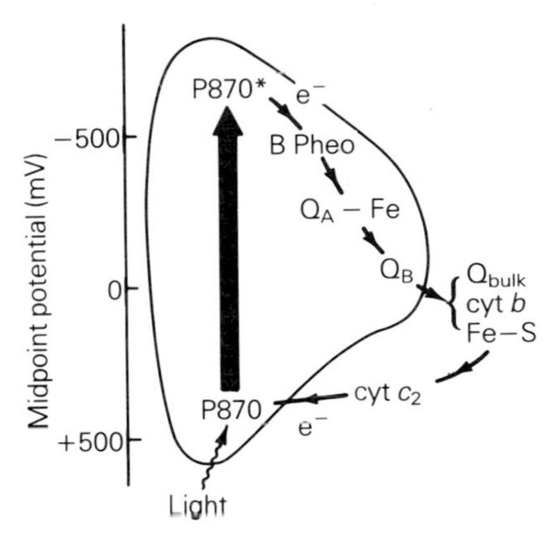

Fig. 3.27 The electron transport chain of *Rhodopseudomonas sphaeroides*. The diagram also shows the redox potentials of the components. Redrawn from Harold, F.M. (1986) *A Study of Bioenergetics*, W.H. Freeman, New York.

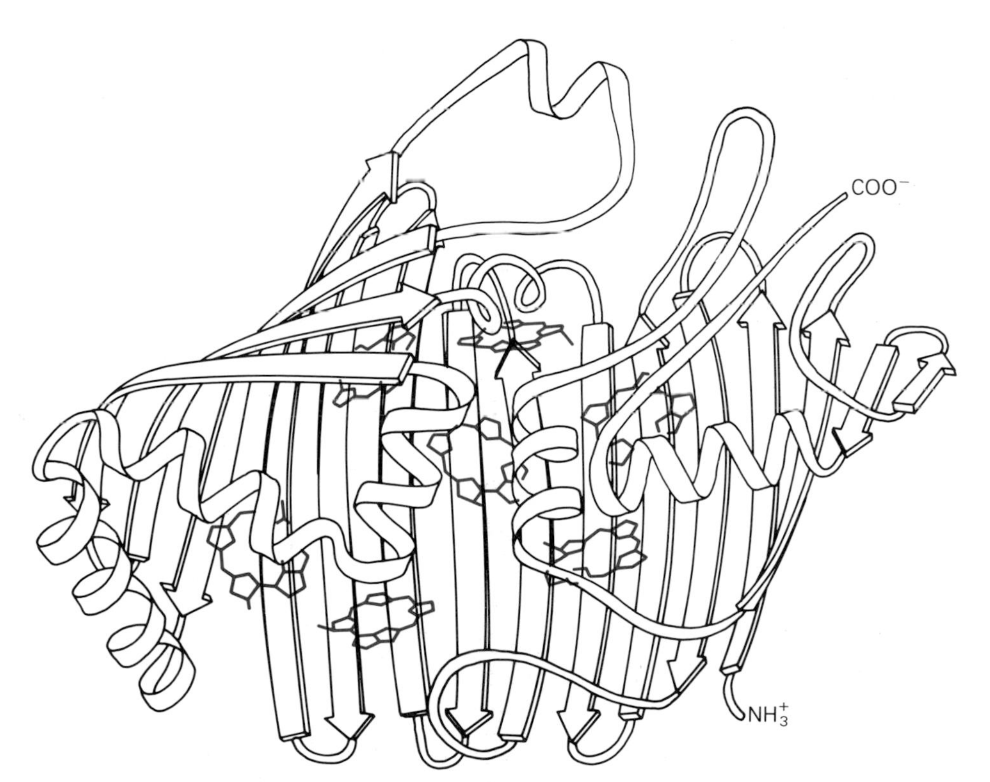

Fig. 3.26 Simplified structure of bacteriochlorophyll in association with its protein, which contains seven bacteriochlorophyll molecules. The phytyl and other side-chains on the porphyrin rings have been omitted. Matthews, B.W. *et al.* (1979) *Journal of Molecular Biology*, **131**, 259–85.

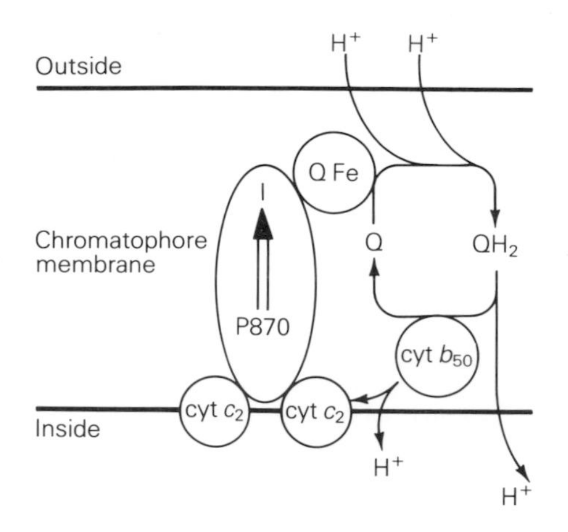

Fig. 3.28 The arrangement of the electron carriers in *Rhodopseudomonas sphaeroides* and the translocation of protons into the bacterium cytosol. Redrawn from Harold, F.M. (1986) *A Study of Bioenergetics*, W.H. Freeman, New York.

HALOBACTERIA contain a light-absorbing protein, **bacteriorhodopsin**, rather than chlorophyll. If halobacteria are grown at low oxygen concentrations and in the light, they generate ATP by photosynthesis. In the presence of light, purple patches containing bacteriorhodopsin appear in the cell membrane. Like rhodopsin, the pigment of the cells of the human eye, bacteriorhodopsin consists of a protein linked to vitamin A aldehyde or retinal. The purple membrane acts as a proton pump. An electrochemical gradient is generated, which can be used for ATP synthesis.

PROTON TRANSLOCATION. Cyclic electron flow in *Rhodospiralles* generates a proton gradient across the bacterial membrane. As shown in Fig. 3.28, the components of the electron transport chain are distributed asymmetrically and protons are pumped *into* the bacteria.

3.6 The mechanism of ATP synthesis

Mitochondria, chloroplasts and bacteria contain very similar membrane-bound F_1F_0-ATPases. Electron transport results in protons being pumped across a membrane and the ATPase catalyses the formation of ATP:

$$ADP^{3-} + P_i^{2-} + H^+ \rightarrow ATP^{4-} + H_2O$$

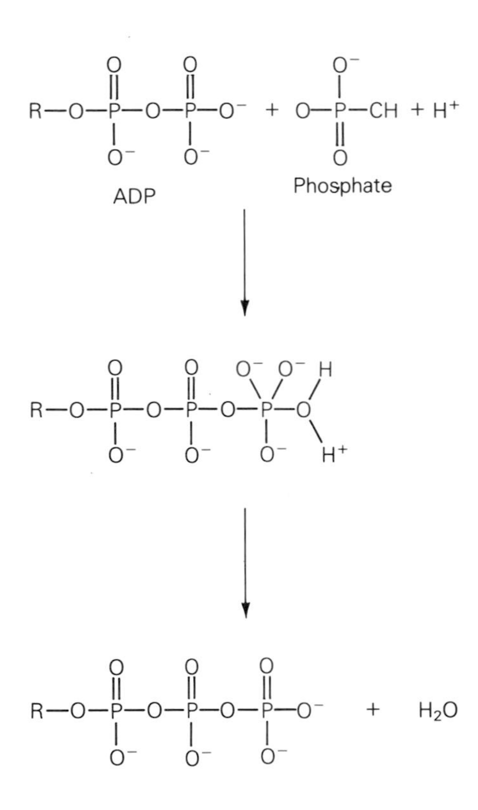

Fig. 3.29 The synthesis of ATP generates an unstable intermediate in which phosphorus has a valency of five. This decomposes to produce water and ATP.

ADP and ATP are bound to Mg^{2+} as complexes. The mode of phosphate linkage to ADP is shown in Figure 3.29 and involves an unstable intermediate in which a phosphorus atom with a valency of five is generated. This then loses water to generate ATP. How does proton flow through the F_1F_0-ATPase cause the generation of ATP? Normally an ATPase would be thought of as catalysing the opposite reaction, namely the hydrolysis of ATP rather than its synthesis. However, it has to be remembered that enzymes catalyse the attainment of equilibrium under the conditions prevailing. In fact it is unclear how ATP is synthesized but a number of models have been proposed. Two of them, named after their proposers, are Mitchell's and Boyer's models.

MITCHELL'S MODEL. In Mitchell's view, the protons flowing across the ATPase participate directly in the synthesis of ATP. Figure 3.30a depicts a possible mechanism, assuming that three protons move across the membrane per ATP generated. The F_1 particle admits ADP and phosphate in their deprotonated forms designated $ARPP^-$ and P_i^{2-}, respectively. Thus, three H^+ will be released into the matrix from the ionization of ADP and phosphate. It

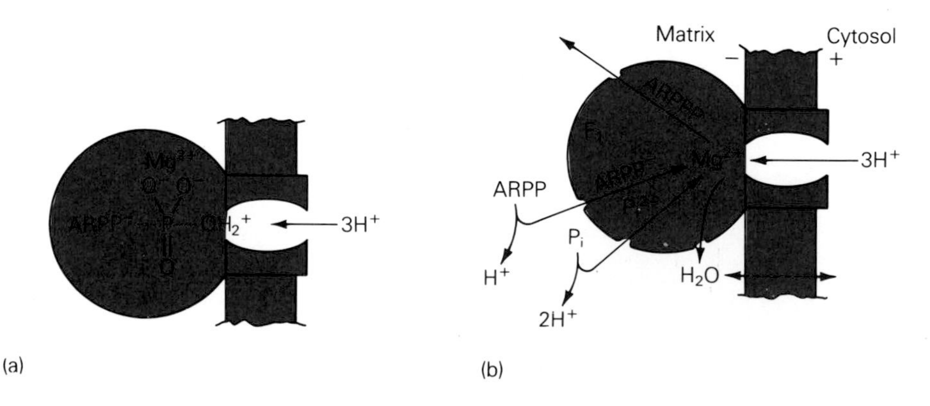

Fig. 3.30 (a) Mitchell's model for ATP synthesis during proton flow through the ATPase. $ARPP^-$ and P_i^{2-} enter the catalytic site probably complexed to Mg^{2+}. Ionization of these compounds leaves three protons in the matrix ($ARPP^-$, ionized form of ADP; ARPPP, ATP). (b) Shows the arrangement of ADP and P_i at the catalytic site. Redrawn from Harold, F.M. (1986) *A Study of Bioenergetics*, W.H. Freeman, New York.

Reference Racker, E. (1981) *Energy Transducting Mechanisms* (MTP International Review of Science), Vol 12, Butterworths, London. A wide-ranging, readable review of energy production in living systems.

is assumed that $ARPP^-$ and P_i^{2-} will be stabilized as salts and by some of the proteins of the F_1 particle. The protons entering the F_0 particle from the mitochondrial cytosol attack one of the oxygen atoms of P_i^{2-} forming water and leaving a chemically reactive form of phosphate which attacks $ARPP^-$ forming ATP (Fig. 3.30b). During the synthesis of one molecule of ATP, three protons are used from the cytosol and three are released into the matrix.

BOYER'S MECHANISM. In the mechanism proposed by Boyer the protons entering the F_1F_0 complex are not directly involved in ATP synthesis. The protons are assumed to initiate a sequence of conformational changes in the F_1 particle, which alter the affinity of the headpiece for ATP and P_i (Fig. 3.31). Initially the ADP and P_i are loosely bound to the catalytic site on the F_1. As the F_0 accepts protons from the cytosol and becomes protonated, the affinity of the catalytic site for ATP increases and tightly bound ATP is synthesized. The conformational change of the ATPase moves the protons to the matrix side where they are released. As the protons are released the original conformation of the enzyme is assumed and ATP is released. The energy-requiring step is not ATP synthesis but ATP *release* from the enzyme.

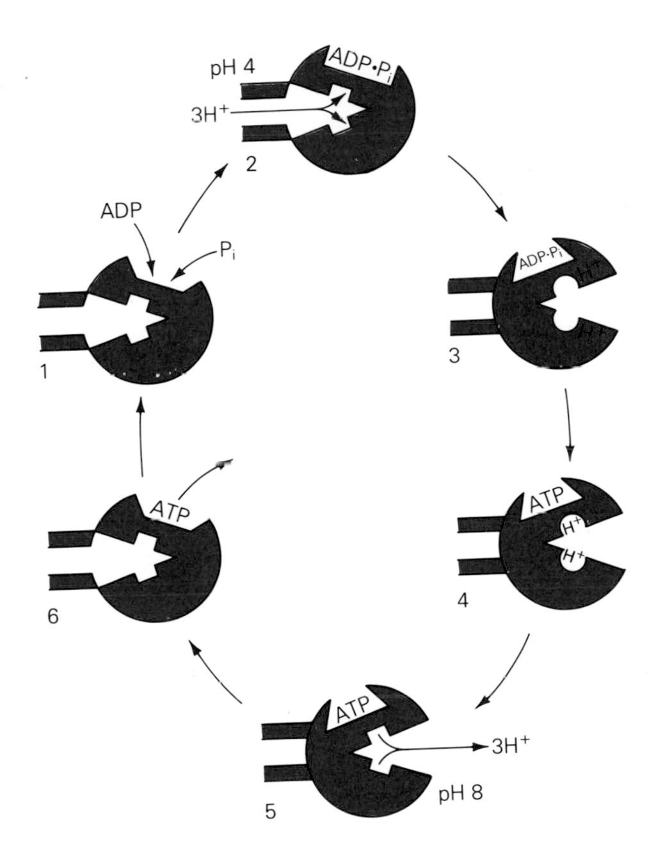

Fig. 3.31 Boyer's mechanism in which protons bring about a conformational change in the headpiece of the ATPase, altering the affinity of the catalytic site for ATP. Redrawn from Harold, F.M. (1986) *A Study of Bioenergetics*, W.H. Freeman, New York.

Reference Prebble, J.N. (1981) *Mitochondria, Chloroplasts and Bacterial membranes*, Longman, London. A detailed study of the function of these structures.

3.7 *Overview*

The major way in which ATP is generated from ADP and inorganic phosphate (P_i) is oxidative phosphorylation in the mitochondria of both animals and plants and in the chloroplasts of green plants. Bacteria, although lacking these organelles, can carry out similar processes. All involve the passage of electrons through a series of membrane-located carrier molecules, with the simultaneous release of energy.

In the mitochondrial electron transport chain electrons donated by NADH and $FADH_2$ are passed down a series of carriers to a final electron acceptor, oxygen, which is reduced to water. Accompanying electron transport there is a movement of protons from the mitochondrial matrix to the cytosol. The F_1F_0-ATPase on the inner mitochondrial membrane utilizes this transmembrane proton potential to drive the synthesis of ATP from ADP and inorganic phosphate.

Bacteria also have electron transport chains that are more diverse than those found in mitochondria. However, these too synthesize ATP using a F_1F_0-ATPase and a transmembrane proton gradient.

In photosynthetic organisms, light energy is used to generate a proton gradient and in green plant chloroplasts a CF_1CF_0-ATPase of similar structure to the F_1F_0-ATPase found in mitochondria and bacteria is required for ATP formation. Green plants effectively use water as the reductant for photosynthesis, that is, the electrons come originally from H_2O. Bacteria use more diverse reductants for photosynthesis, including sulphur compounds and a variety of organic molecules. A number of mechanisms for ATP synthesis accompanying proton flow through the F_1F_0-ATPase have been proposed, but many uncertainties remain as to exactly how ATP synthesis is achieved.

Answers to exercises

1. The requirement for a detergent to release Fractions I to IV indicates they are embedded in the lipid layer of the membrane.
2. Isoprene has the structure:

$$H_2C{=}\underset{}{\overset{CH_3}{\overset{|}{C}}}{-}CH{=}CH_2$$

Isoprene units are hydrophobic and dissolve readily in lipid containing membranes.
3. The oxygen molecule is larger than CN^-, N_3^- or CO. The space above the porphyrin ring is restricted and the smaller molecules will form a stronger bond with the iron atom.
4. The intermediate forms are:

a. $O_2 + e^- \rightarrow O_2^{\cdot -}$
b. $O_2^{\cdot -} + e^- + 2H^+ \rightarrow H_2O_2$
c. $H_2O_2 + e^- + H^+ \rightarrow HO^{\cdot} + H_2O$
d. $HO^{\cdot} + e^- + H^+ \rightarrow H_2O$

5. Difference spectra are obtained by substracting the spectrum of a compound in one of its forms (for example, the oxidized form) from the spectrum of a different form (for example, the reduced form).
6. An uncoupling agent allows electrons to flow through the electron transport chain, but no ATP is generated; the energy that would have been conserved appears as heat .
7. The enzymes that hydroxylate PAH using the cytochrome P450 system will be present in differing amounts in tissues and individuals. The amount and distribution of the enzymes will be genetically determined. If the activation of carcinogen is a requirement for the development of cancer then the genetic constitution and distribution of enzymes are going to be very important.
8. Bacteria use the cell membrane and pump protons out into the periplasmic space.
9. Plants split water to obtain electrons for photosynthesis, whereas photosynthetic bacteria use relatively more powerful reducing agents such as H_2S. Consequently, more light energy is needed by green plant photosynthesis and this is captured by having two not one photosystems.
10. Many of the carriers are similar, differing only in chemical detail.

Mitochondria	*Chloroplasts*
cytochromes *a*, *b*, *c*	cytochrome *bf*
ubiquinone	plastoquinone
iron–sulphur proteins	iron–sulphur proteins
cytochrome oxidase (Cu)	plastocyanin

11. In non-cyclic photo-phosphorylation the electrons obtained from water ultimately end up in NADPH and some ATP is generated as the electrons pass through a photosynthetic electron transport chain as part of the Z-scheme. In cyclic photophosphorylation, the electrons are raised to a high potential using light energy, and drop down through an electron transport chain generating ATP, to return to the chlorophyll molecule from which they originally came. No NADPH is generated. It is difficult to demonstrate the occurrence of cyclic photophosphorylation because only ATP is generated: this is, of course, generated by the simultaneous non-cyclic photo-phosphorylation that is going on.

QUESTIONS

FILL IN THE BLANKS

1. The synthesis of ATP requires the establishment of a _______ gradient across a biomembrane. In _______, the _______ are pumped across the inner membrane making the intermembrane space more _______. _______ flow back across the membrane through the _______. This enzyme catalyses the synthesis of ATP. It has a quaternary _______ and consists of five subunits _______, _______, _______, _______, and _______. The _______ and _______ subunits bind nucleotides (_______ and _______). The _______ subunit allows the movement of _______ across the membrane, while the _______ subunit attaches the _______ particle to the _______ portion embedded in the membrane. The _______ subunit has a regulatory role.

According to _______ chemiosmotic hypothesis, _______ _______ move across the membrane for each _______ molecule phosphorylated. An alternative hypothesis, proposed by Boyer, suggests that _______ are involved in initiating a _______ change in the _______ particle. This alters the _______ of the protein for _______ and _______.

Choose from: α, β, γ, δ, ε (2 occurrences of each), ADP (2 occurrences), ATP (2 occurrences), F_1 (2 occurrences), F_0, F_1F_0-ATPase, P_i, acidic, affinity, conformational, Mitchell's, mitochondria, proton, protons (4 occurrences), structure, three protons.

MULTIPLE-CHOICE QUESTIONS

2. State if the following are true or false:

A. ATP is synthesized by oxidative phosphorylation during glycolysis in the cytosol.
B. The formation of ATP from ADP and P_i linked to electron transport is called oxidative phosphorylation.
C. Oxidases catalyse the removal of hydrogen from substrates and transfer it to the electron transport chain.
D. A negative value for $E^{0'}$ indicates a compound is an effective oxidant.
E. An intact biomembrane is required to produce ATP.
F. An enzyme which catalyses the phosphorylation of ADP is the F_1F_0-ATPase.
G. The chemiosmotic theory, proposed by Boyer, states that an electrochemical gradient is coupled to the production of ATP.
H. Sonication of the inner mitochondrial membrane separates five distinct electron-transporting fractions.

3. In the oxidation of succinate to fumarate:

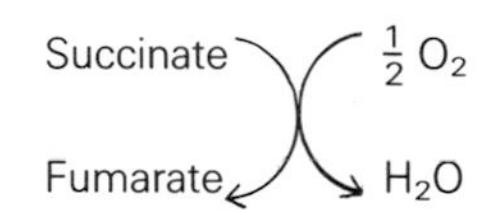

$\Delta G^{0'}$ is 151.2 kJ mol^{-1}. How much greater is the E^0 of the O_2/H_2O redox couple than the $E^{0'}$ of the fumarate/succinate redox couple? (Assume pH 7 and 25°C.)

4. Lactate dehydrogenase catalyses the conversion of pyruvate to lactate:

$$\text{pyruvate} + \text{NADH} + H^+ \rightarrow \text{lactate} + \text{NAD}^+$$

Determine the value of $\Delta G^{0'}$ at pH 7 using the following information:

$E^{0'}$ for the pyruvate/lactate couple is −0.185 V
$E^{0'}$ for the NAD^+/NADH couple is −0.32 V

5. Pyruvate can be reduced to lactate in the presence of NADH:

$$H^+ + \text{pyruvate} + \text{NADH} \rightarrow \text{lactate} + \text{NAD}^+$$

Calculate the equilibrium constant at pH 7 and 25°C for the reduction of pyruvate. Use the following values:

E^0 of the NAD^+/NADH + H^+ couple is −0.32 V
E^0 of the pyruvate/lactate couple is −0.185 V

6. List the following compounds in order of increasing ability to donate electrons:

H_2, O_2, lactate, cytochrome *c* reduced, NADH

7. As electrons pass down the respiratory chain to oxygen there is a decline in $E^{0'}$ The difference in E_0' between NADH and Q is 0.27 V; cytochrome *b* and *c* 0.22 V; and cytochrome *a* and O_2 0.53 V. Calculate the energy available to synthesize ATP at each of these points.

8. List the numbers of electrons and protons that can be carried by: (a) NAD^+, (b) FAD, (c) coenzyme A, (d) ferredoxin, (e) cytochrome *b*.

9. Uncoupling agents are small hydrophobic molecules that carry protons. Use this information to explain why uncouplers prevent ATP synthesis.

4

The tricarboxylic acid cycle

Objectives

After reading this chapter you should be able to:

- ☐ explain the concepts and outline the methods that were involved in elucidating the tricarboxylic acid cycle;
- ☐ describe the individual reactions of the tricarboxylic acid cycle;
- ☐ outline the relationships between the TCA cycle and other major metabolic pathways.

4.1 *Introduction*

From the dawn of biochemistry as a distinct discipline, that is, from the beginning of the twentieth century, two major problems occupied a large part of the research effort: how do yeasts ferment sugars to produce ethanol, and how do mammalian skeletal muscles obtain the energy needed for contraction? At the end of the 1920s, Meyerhof and his colleagues in Heidelberg began the studies that culminated, in 1937, in the complete

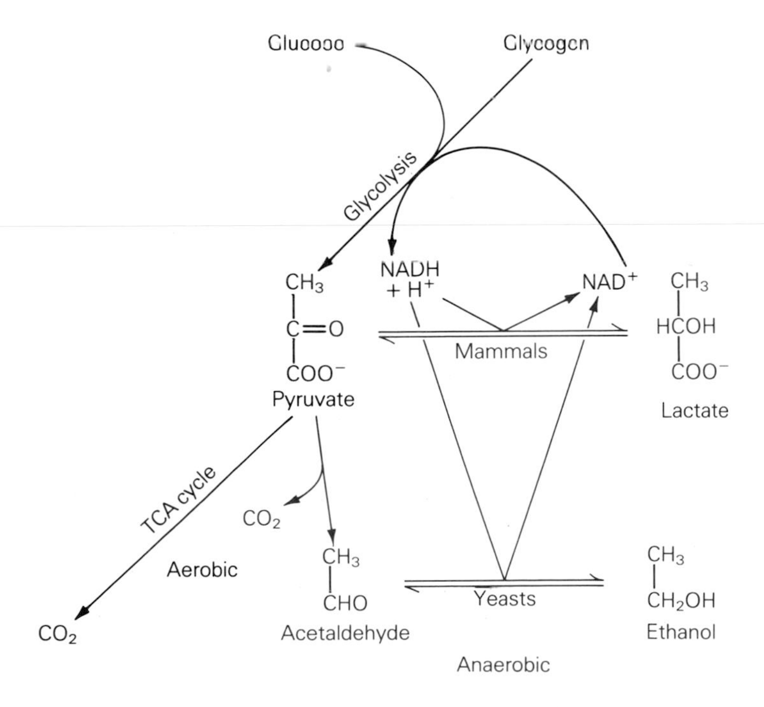

Fig. 4.1 Glycolysis and aerobic catabolism.

elucidation of ***glycolysis:*** the conversion of glucose or glycogen to pyruvate. Under anaerobic conditions the pyruvate is reduced to lactate in muscles or is converted to alcohol in yeasts. This metabolic pathway is common to yeast and muscle cells: indeed, it is ubiquitous. The two big problems of early biochemistry both had the same answer. However, under aerobic conditions, yeasts and muscle cells (and other cell types) do not convert glucose to ethanol or lactate; they oxidize it completely to CO_2 by means of a combination of glycolysis followed by the tricarboxylic acid (TCA) cycle, as it is now called (Fig. 4.1). It was discovered by Sir Hans Krebs, who called it the citric acid cycle, but it is also frequently referred to as the Krebs cycle in his honour.

When the TCA cycle was originally described in the late 1930s, biochemists regarded it as a pathway for the aerobic catabolism of glucose and glycogen but the cycle is now known to be necessary for complete catabolism of lipids and amino acids as well as carbohydrates (Fig. 4.2). It is a mitochondrial process in eukaryotic cells, and provides most of the reduced coenzymes for electron transport. It is, therefore, responsible for the bulk of ATP production in most aerobic cells. Many intermediates of the pathway also provide the starting-points for anabolic (biosynthetic) processes, including the biosyntheses of carbohydrates, porphyrins and many amino acids. Thus, the TCA cycle occupies a central place in metabolism.

See *Biosynthesis*

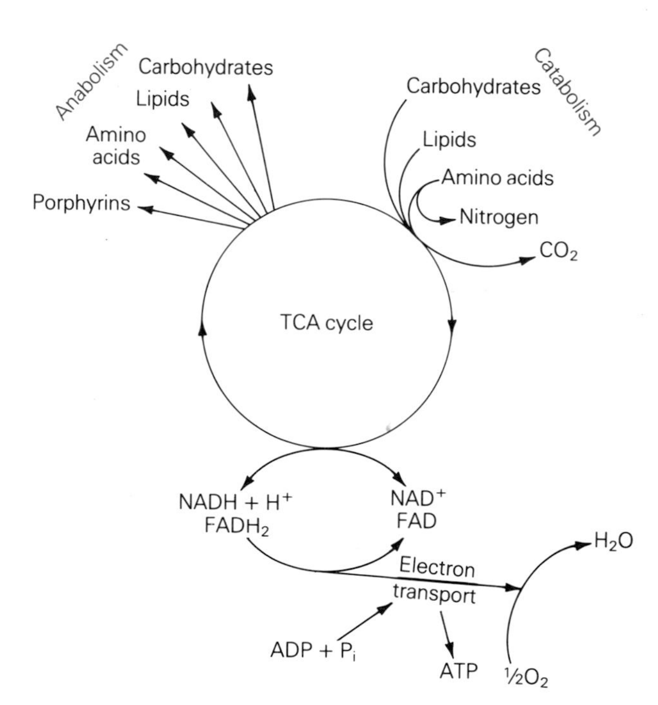

Fig. 4.2 The central position of the TCA cycle in intermediary metabolism.

4.2 *How the cycle was elucidated*

Advances in biochemistry, like those in any other science, generally start from a clear definition of a problem. Over a period of time, through the generation of new ideas and the application of novel techniques, the problem is clarified and solutions emerge. Sometimes, misleading results or incorrect interpretations halt progress; sometimes, apparently irrelevant discoveries prove crucial. The elucidation of the TCA cycle illustrates all these characteristics of scientific research, and this makes it worthwhile to survey, briefly, this part of the history of biochemistry.

Glycolysis: *glucose from the Greek* glykys, *sweet, and* lysis, *dissolution.*

Reference Baldwin, E. (1947, 1949) *Dynamic Aspects of Biochemistry*, Cambridge University Press, Cambridge, UK. A simple account is given of the historical development of knowledge of biochemistry; these first two editions were written before current views were fully established.

The Szent-Györgyi model

In 1933, when Meyerhof's work on glycolysis was no more than half-complete, Szent-Györgyi in Budapest began to address this question in a novel way. Using slices of pigeon breast muscle, he measured the rate of oxygen consumption (Fig. 4.3). He found that within a few minutes of the start of his measurements, the rate decreased markedly, although little of the muscle glycogen and little of the available oxygen had been used up. Why did this happen? Accumulation of toxins that might inhibit respiration could be ruled out experimentally; so presumably, some substance necessary for the process was being lost.

Some years earlier, Warburg, the director of the Kaiser Wilhelm Institute where Meyerhof then worked, had demonstrated the reaction catalysed by succinate dehydrogenase:

$$\underset{\text{succinate}}{\begin{matrix}COO^- \\ | \\ CH_2 \\ | \\ CH_2 \\ | \\ COO^-\end{matrix}} + FAD \underset{}{\overset{\text{succinate dehydrogenase}}{\rightleftharpoons}} \underset{\text{fumarate}}{\begin{matrix}COO^- \\ | \\ CH \\ \| \\ CH \\ | \\ COO^-\end{matrix}} + FADH_2$$

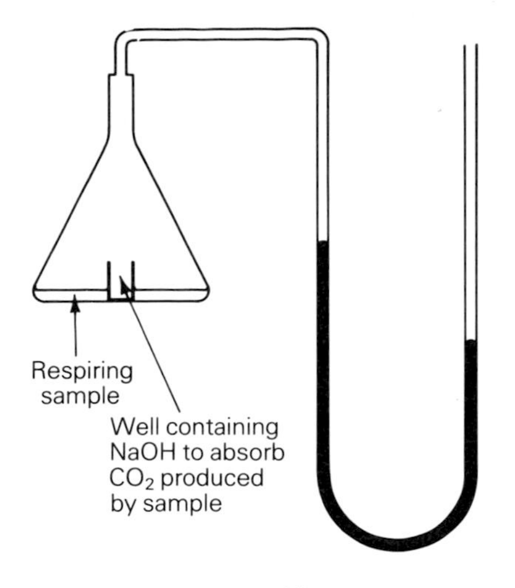

Fig. 4.3 Schematic representation of the Szent-Györgyi experiment for measuring oxygen consumption.

Szent-Györgyi found that by adding tiny amounts of either succinate or fumarate to the pigeon breast muscle slices, he could restore the initial rate of O_2 consumption (Fig. 4.4). An important point is that the effective quantities of these dicarboxylic acids were very much less than the quantities of oxygen consumed or carbon dioxide produced; therefore, they were not the primary source of the carbon appearing as CO_2. Szent-Györgyi related this finding to what was already known about glycolysis and electron transport by the following model shown in Fig. 4.5 'XH_2' represents the hypothetical end-product of glycolysis, which was not known at that time.

This model explained how carbon dioxide production from the unknown end-product of glycolysis could be linked via succinate dehydrogenase and electron transport to the reduction of oxygen. It also explained why tiny amounts of succinate or fumarate restored the original activity. It did not explain why the O_2 utilization rate had decreased in the first place. However, if either succinate or fumarate is interconvertible with a labile chemical, one that spontaneously and (in effect) irreversibly turns into something else, then

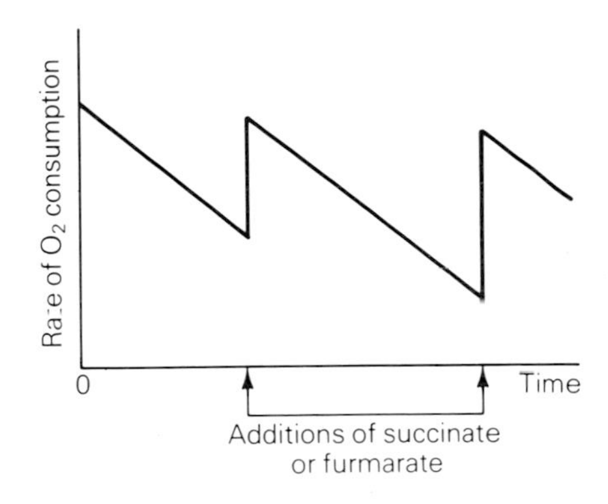

Fig. 4.4 Results obtained in the Szent-Györgyi experiment: effect of succinate or fumarate on oxygen consumption.

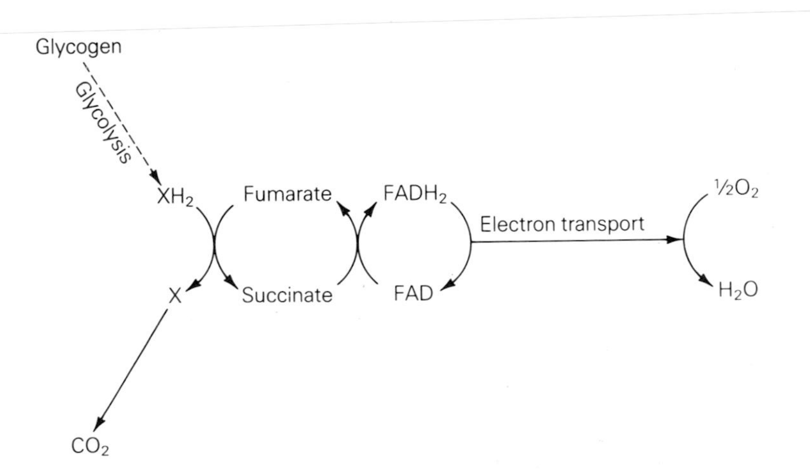

Fig. 4.5 Szent-Györgyi's model for aerobic catabolism.

Exercise 1

Design an experiment that would demonstrate, using the methods outlined in the text, that the decrease in oxygen consumption rate was not due to the accumulation of toxins in the muscle slices.

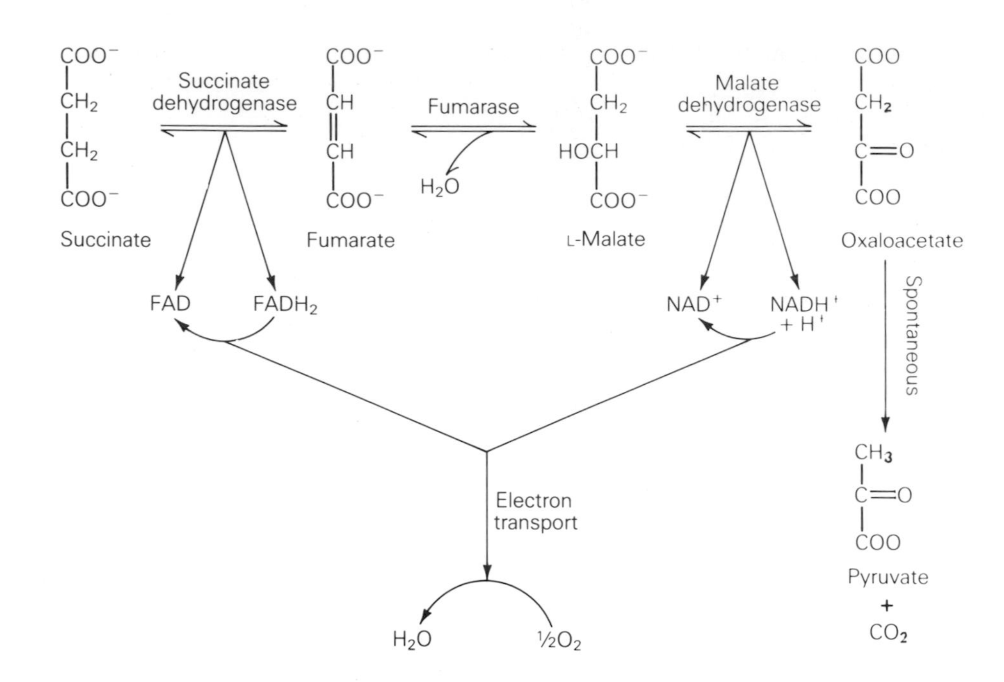

Fig. 4.6 Interconversions of the C_4 dicarboxylic acids.

this particular problem is solved. Experiments with other dicarboxylic acids duly provided the solution. Fumarate may be converted to the L isomer of malate, and malate in turn can be oxidized, with concomitant reduction of NAD^+, to oxaloacetate. Oxaloacetate is converted spontaneously, and quite quickly, into pyruvate and carbon dioxide (Fig. 4.6).

The NADH generated in the malate dehydrogenase reaction is reoxidized to NAD^+ by the electron transport chain. Further support for Szent-Györgyi's model came from the finding that malate and oxaloacetate, once again in tiny amounts, restored the original rate of CO_2 production (O_2 utilization) in pigeon breast muscle slices, as illustrated in Fig. 4.4.

The Krebs model

Krebs, a young researcher in Warburg's department, performed experiments that led to further elaboration of Szent-Györgyi's model. He found that CO_2 production could be restored not only by the C_4 dicarboxylic acids (succinate, fumarate, malate and oxaloacetate) but also by the C_5 compound, α-oxo-glutarate (formerly called α-ketoglutarate), and by the C_6 tricarboxylic acids, citrate, aconitate and isocitrate. The inter-conversions of these compounds were characterized by Martius and Knoop. Krebs found that other 4, 5 and 6-carbon acids were ineffective. Moreover, enzymes present in the muscle could catalyse the conversion of citrate to aconitate and isocitrate, isocitrate (by an NAD^+-linked oxidation coupled with decarboxylation) to α-oxo-glutarate, and α-oxoglutarate, apparently by a similar oxidative decarboxyl-ation, to succinate (Fig. 4.7). Because of these interconversions, all these compounds could be incorporated into Szent-Györgyi's model, but at the cost of making it rather untidy (Fig. 4.8).

Because of the political situation in Germany in the 1930s, Krebs lost his post in Berlin. Through the influence of Szent-Györgyi, he went to Britain and in due course secured a post at Cambridge. It was after Krebs's forced emigration that he made his most salient contribution to this field, which was to win him a Nobel prize and associate his name firmly with the central

Reference Krebs, H.A. and Martin, A. (1981) *Reminiscences and Reflections*, Clarendon Press, Oxford, UK, and Krebs, H.A. (1970) The history of the tricarboxylic acid cycle. *Perspectives in Biology and Medicine*, **14**, 154–70. These provide detailed historical information.

pathway of aerobic catabolism. This contribution was not a particular further discovery or the introduction of a new technique; it was a revised model that made better sense of the existing experimental data than the model formulated by Szent-Györgyi.

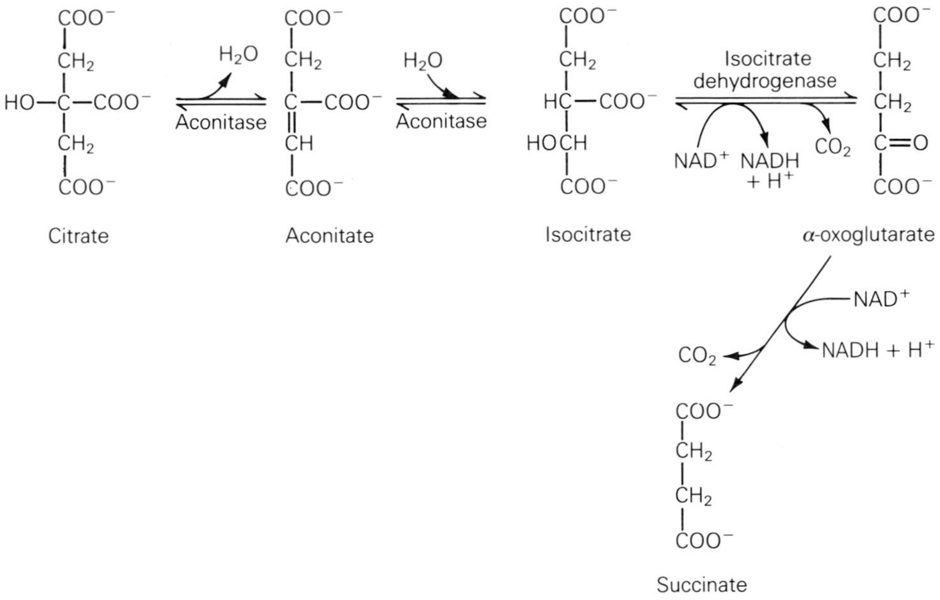

Citrate ⇌ (Aconitase, H_2O) Aconitate ⇌ (Aconitase, H_2O) Isocitrate ⇌ (Isocitrate dehydrogenase; NAD^+ → $NADH + H^+$; CO_2) α-oxoglutarate → (NAD^+ → $NADH + H^+$; CO_2) Succinate

Fig. 4.7 Formation of succinate from C_6 tricarboxylic acids.

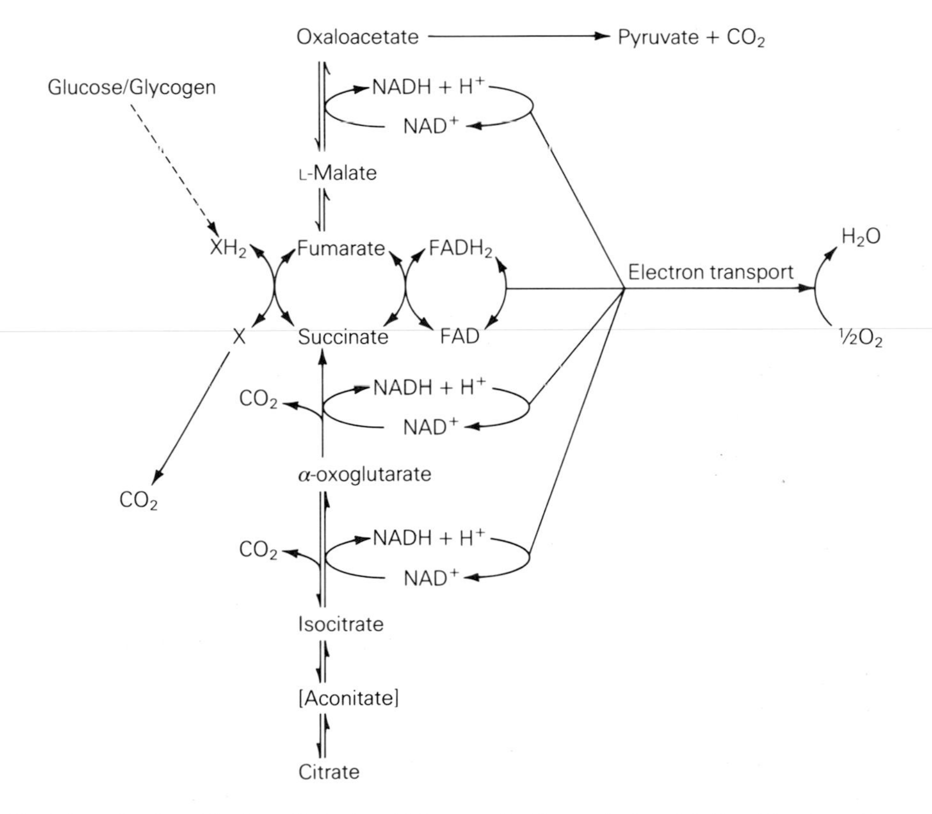

Fig. 4.8 Modification of the Szent-Györgyi model, incorporating the metabolism of 4-, 5- and 6-carbon acids.

4.1 Krebs' proof of the cycle

Knowing that a set of metabolic reactions *may* take place in a cell is not proof that they *do* take place. All the necessary enzymes and coenzymes may be present, but yet there may be reasons why they do not form a 'metabolic pathway'. If a hypothesis is proposed that a particular set of reactions do operate as a pathway, it is necessary to prove it. Here is how Krebs proved that a cycle actually operated. (This was published in a rather obscure journal: Krebs, H.S. and Johnson, W.A. (1937) *Enzymologia*, **4,** 148, after a well-known journal had refused to accept the paper.)

Krebs argued as follows. The reaction catalysed by succinate dehydrogenase was well known and was reversible:

$$\text{succinate} + \text{FAD} \rightleftharpoons \text{fumarate} + \text{FADH}_2$$

In Krebs' proposed cycle there were two possible pathways between fumarate and succinate, one was by a reversal of the succinate dehydrogenase reaction, and the other by going all round the cycle as proposed:

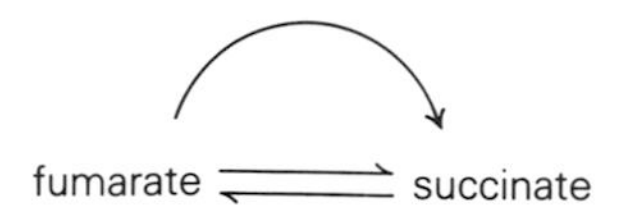

It was also known that malonate was an inhibitor of succinate dehydrogenase. Therefore, in the presence of malonate, the pathway between fumarate and succinate catalysed by succinate dehydrogenase will be blocked. Krebs argued that if *in the presence of malonate* addition of fumarate led to the accumulation of succinate, then this was because a cycle was operating.

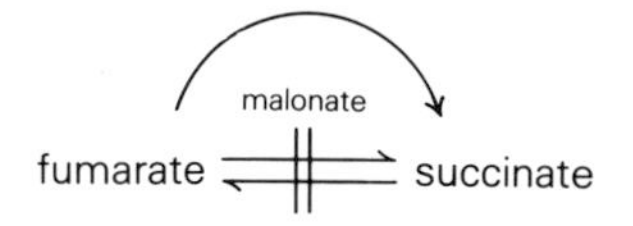

These experiments were carried out in 1936–37. Radioactive isotopes were not available at the time and the test was simply to see if succinate accumulated. This was observed, and provided the proof that Krebs' proposed cycle did indeed operate in the cell.

A revised model was necessitated partly by the increasing complexity of the Szent-Györgyi scheme, but mostly by Meyerhof's finding that the end-product of glycolysis was pyruvate. Pyruvate simply cannot be the 'XH_2' of the original model. The removal of two hydrogens from pyruvate is not chemically possible, at least, not in aqueous solution or under anything like physiological conditions; and even if two hydrogens were removed, how could the product be converted to CO_2? Krebs dealt with this problem by dropping 'X' from the model altogether, and suggesting instead that pyruvate from glycolysis reacted with oxaloacetate to form a C_7 compound (Fig. 4.9). This hypothetical substance was decarboxylated to form the C_6 citrate, which was then reconverted via isocitrate, α-oxoglutarate, succinate, fumarate and malate back to oxaloacetate. During this conversion of the C_7 compound to the C_4 oxaloacetate, three carbons, the equivalent of those added by the pyruvate, were lost as CO_2, and the hydrogens were removed via NAD^+ or FAD and the electron transport chain, ultimately as water. Krebs demonstrated directly that oxygen consumption paralleled pyruvate oxidation in experiments with muscle slices or homogenates, a piece of experimental evidence that clearly supported his model.

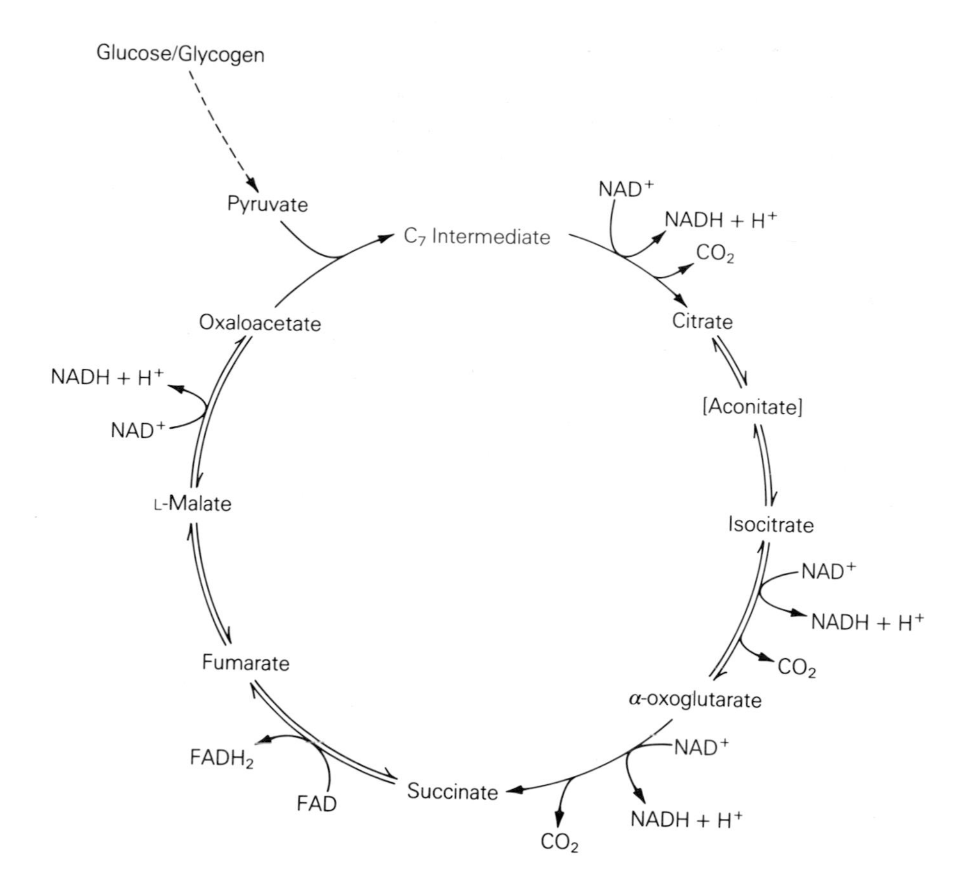

Fig. 4.9 Krebs' *original* model of a cyclic pathway for pyruvate oxidation (now discounted).

Exercise 2

Invent a possible structure for Krebs' hypothetical C_7 compound, and suggest single-enzyme reactions by which it could be (a) synthesized from pyruvate and oxaloacetate, and (b) decarboxylated to citrate.

Modifications to the Krebs model

The model proposed by Krebs retained all the advantages of that proposed by Szent-Györgyi, but it was much neater and fitted the new information more readily. The only obvious remaining difficulty was the precise nature of the C_7 intermediate. The Second World War interrupted further study but, after the war ended, the search for the C_7 compound was resumed. However, no candidate for the C_7 intermediate has ever been identified.

Box 4.2 Coenzyme A

In animals CoA is synthesized from pantothenic acid. Because CoA contains a terminal sulphydryl group, the coenzyme is sometimes abbreviated CoASH. Pantothenic acid is synthesized by green plants and most microorganisms but not by animals.

Acetyl CoA the product of the oxidative decarboxylation of pyruvate is also produced from the metabolism of fats, carbohydrates and amino acids. It may be produced from free acetate by acetyl CoA synthetase as follows:

$$\text{ATP} + \text{coenzyme A} + \text{acetate} \rightarrow \text{AMP} + \text{acetyl CoA} + \text{PP}_i$$

The acetyl group is attached to CoASH through its carbonyl carbon to the sulphur atom of CoASH . The carbonyl carbon is therefore linked to two electronegative, electron-withdrawing atoms (S and O), and this induces a partial positive charge on the carbonyl carbon. The effect is to make one of the methyl hydrogens acidic and the methyl carbon chemically more reactive.

Box 4.3 An analogy between the TCA cycle and the light reactions of photosynthesis

The photosynthetic apparatus in the chloroplast uses light energy to split water in order to reduce hydrogen-carrying coenzymes. The TCA cycle apparatus in the mitochondrion uses the chemical energy of pyruvate to split water in order to reduce hydrogen-carrying coenzymes. This is an unconventional, but perfectly legitimate, perspective on the TCA cycle.

Around 1950, Lipmann purified and characterized two novel compounds: coenzyme A (CoA) and its acetyl derivative, acetyl CoA. It then became apparent that pyruvate does not react directly with oxaloacetate. Rather, it is oxidatively decarboxylated to give acetyl CoA, and the acetyl residue is then combined with oxaloacetate to produce citrate. (These processes will be discussed in more detail later in this chapter.) Similarly, α-oxoglutarate is not directly converted to succinate, as in Krebs' scheme; it is oxidatively decarboxylated to succinyl CoA, which is then converted to succinate. Lipmann's discovery meant that the search for the C_7 compound could be abandoned (Fig. 4.10). Krebs' model had to be modified in a couple of details, but the spirit of it was retained.

A beautifully conceived experiment using the novel technique of radioisotope labelling appeared to cast doubt on Krebs' model. (Although the war had interrupted research, it had made this technique available because of military backing for research in nuclear physics.) This experiment is worth describing in detail, partly because it illustrates how a well-designed and executed experiment may sometimes give misleading results, and partly because the explanation for the results, which came only some ten years after the experiment was first performed, added an important new concept to biochemistry.

Figure 4.11 illustrates the principle of the experiment. Suppose acetyl CoA is labelled on the carbonyl group with ^{14}C and mixed with an active mitochondrial preparation, or muscle slices. When the acetyl residue is added

Box 4.4 Some molecular details of the TCA cycle

Details of the TCA cycle are given in Figure 4.10 and Table 4.1. Citrate synthetase utilizes the high free-energy of hydrolysis of the thioester bond to transfer the acetyl residue from the coenzyme A derivative to carbon-2 of oxaloacetate. Citrate is a tertiary alcohol and is therefore not a substrate for a dehydrogenase reaction. However, it is isomerized to the secondary alcohol isocitrate by the enzyme aconitase, and the isocitrate is then oxidized by isocitrate dehydrogenase, with concomitant loss of a carboxylate group as CO_2. (This decarboxylation is spontaneous: the 2-oxocitrate formed as an intermediate in the reaction is a β-ketoacid and is very unstable.) The resulting α-oxoglutarate is a 2-ketoacid, homologous to pyruvate, and it undergoes an analogous dehydrogenation/decarboxylation reaction. The product of this α-oxoglutarate dehydrogenase reaction is succinyl coenzyme A, just as the product of pyruvate dehydrogenase is acetyl coenzyme A. In the case of succinyl coenzyme A, however, the high free-energy of the thioester bond is utilized not in bond formation (as in citrate synthetase), but in manufacturing a GTP molecule from GDP and phosphate; GDP is a cofactor for succinate thiokinase. The resulting C_4 dicarboxylic acid is reconverted to oxaloacetate by two dehydrogenation reactions (succinate dehydrogenase and malate dehydrogenase) with an intervening hydration (fumarase). Both the acetyl carbons are recovered in the oxaloacetate.

Reference Baldwin, J.E. and Krebs, H.A. (1981) The evolution of metabolic cycles. *Nature*, **291**, 381–2. This compact essay shows what scientific thinking really involves.

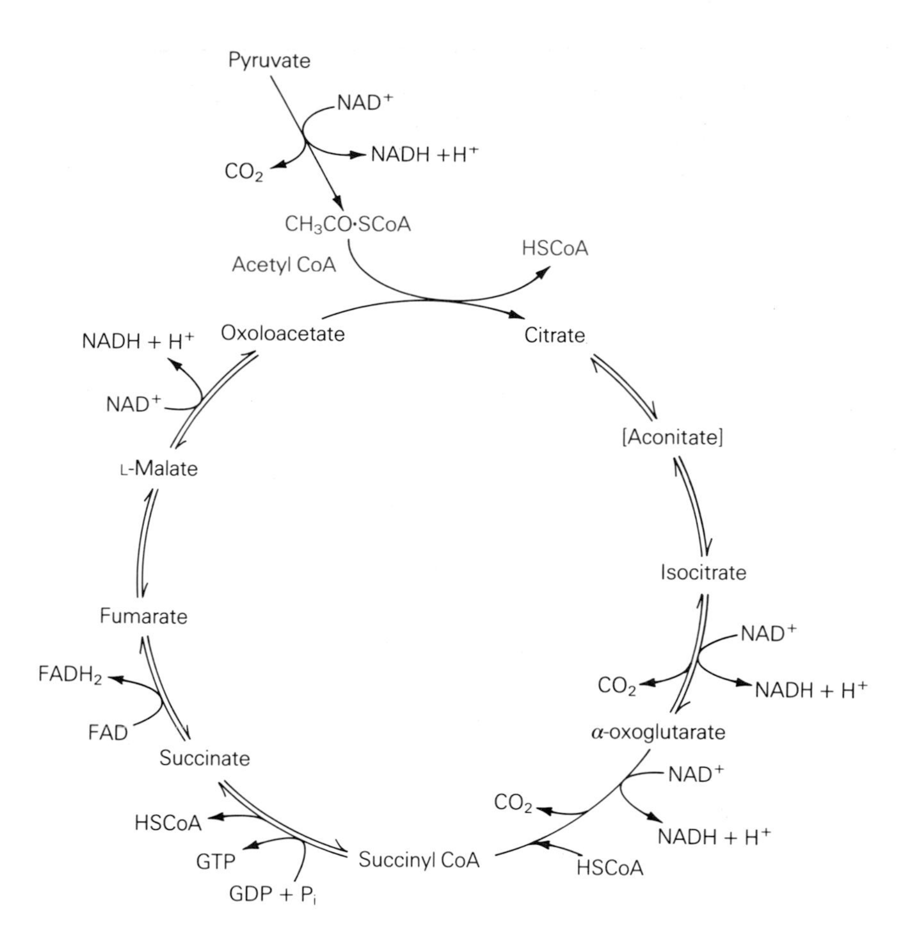

Fig. 4.10 The modern version of the TCA cycle (compare Fig. 4.9).

Exercise 3

Try rewriting Figure 4.10 in structures instead of names. (All the relevant structures are given in this chapter.) To which class of enzymes (oxido-reductase, hydrolase, etc.) does each enzyme of the cycle belong?

to oxaloacetate, the radiolabel is incorporated into the resulting citrate, as shown. Citrate does not have a chiral carbon. Therefore, when it is converted to isocitrate, there should be a 50% chance that the hydroxyl group will move one carbon nearer to the ^{14}C label, and a 50% chance that it will move one carbon further away. Those isocitrate molecules with the hydroxyl nearer to the label will be converted to *unlabelled* succinate; the label will appear in the CO_2 lost at the α-oxoglutarate dehydrogenase step. Those isocitrate molecules with the hydroxyl further away from the label will be converted to *labelled* succinate. The TCA cycle can be blocked at the succinate dehydrogenase step, for example, by adding malonate ($^-OOC–CH_2–COO^-$), which is a competitive inhibitor of succinate dehydrogenase. The prediction that half the added label will be recovered in the succinate was tested experimentally, but it was not fulfilled. *All* the added label was recovered in the succinate; *none* was lost in the CO_2.

This result led to the widespread belief that there was something seriously amiss with Krebs' model. At best, it was argued, citrate could not be an intermediate in the pathway; it must lie on a side-branch to the main cyclic process. However, in all other respects, the model, duly modified to accommodate the discovery of coenzyme A, continued to fit the experimental data. This situation persisted until 1956, when Ogston provided an explanation.

According to this explanation, citrate is transferred directly from the citrate synthetase to aconitase, and it binds to the active site of the enzyme at three

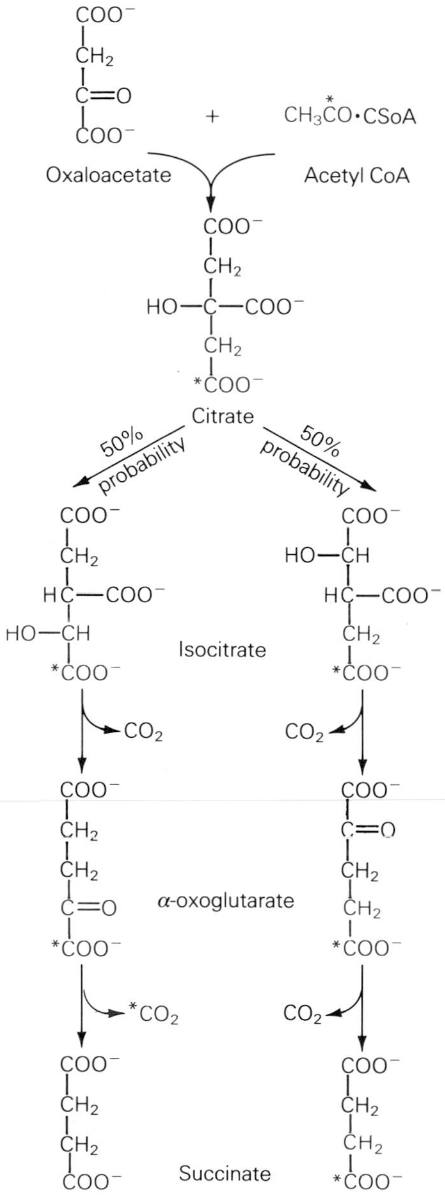

Fig. 4.11 Predicted outcome of the labelling experiment described in the text. The radiolabelled carbon atom is indicated by the asterisk.

Reference Stryer, L. (1988) *Biochemistry*, 3rd edn, Freeman, New York, USA. An excellent overview of the TCA cycle with clear and informative diagrams in Chapter 16, pages 373–96.

Exercise 4

In which reactions in the TCA cycle (Fig. 4.10) does water participate? Can you account for the three molecules of water needed to balance the overall equation for the complete oxidation of pyruvate to CO_2?

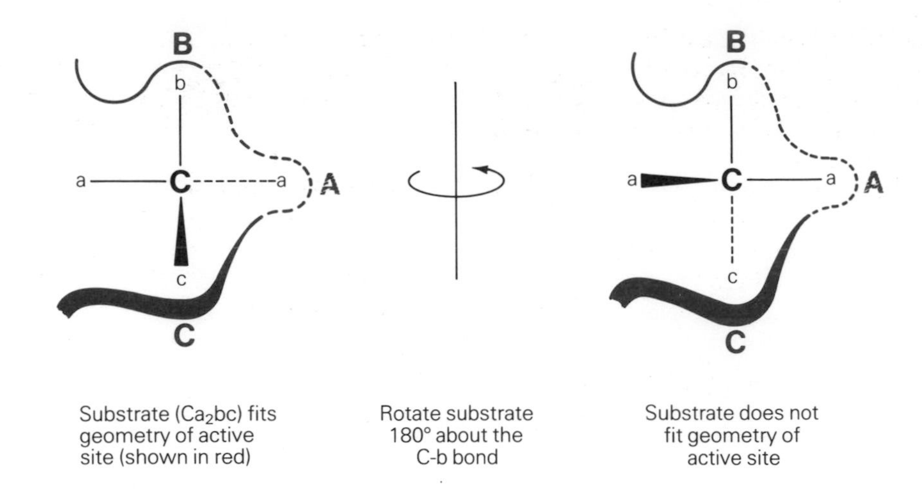

Fig. 4.12 The Ogston concept, explaining the 'asymmetric' behaviour of citrate.

points. This mode of binding means that the citrate can only fit in one way to the active site. This mode of enzyme binding confers asymmetry on a non-chiral (symmetrical) substrate (Fig. 4.12). This concept, the idea that a non-chiral compound can behave as if it were chiral by virtue of its mode of binding to an enzyme, has proved valuable in explaining other metabolic processes. Examples are provided by the pathway of cholesterol biosynthesis. In the case of the TCA cycle, the concept saved Krebs' scheme. The hypothetical route that involves movement of the hydroxyl group to the carbon nearer the label, resulting in the formation of unlabelled succinate, is not followed.

Ogston's contribution can be taken to mark the final establishment of our accepted modern version of the TCA cycle. It was the final chapter in a story that began with the elegant studies performed by Szent-Györgyi in the early 1930s.

4.3 Some biochemical details

The TCA cycle is central to present understanding of how intermediary metabolism works in aerobes; that is, how aerobic cells obtain an adequate energy supply and how they produce many of the raw materials for biosyntheses.

See Chapter 3

The cycle is ubiquitous in aerobes: it is found throughout the animal and plant kingdoms and in aerobic prokaryotes. The utilization of the reduced cofactors in ATP production has been described. In Section 4.4 this will be illustrated by indicating the connections between the TCA cycle and other metabolic pathways and by showing how these connections explain a number of biological observations that would otherwise be difficult to understand. A few details about some of the reactions will facilitate this discussion.

Exercise 5

Using Figure 4.11 and the related part of the text as a guide, determine which carbon in succinate will become labelled if the acetyl coenzyme A is ^{14}C-labelled on the methyl group of the acetyl residue.

A sledgehammer to crack a nut?

Why do organisms use a complicated cycle of eight enzyme-catalysed reactions to oxidize something as simple as an acetyl residue to CO_2? A reasonable answer to this might be that the methyl group of the acetyl residue is resistant to oxidation. The process could only be carried out directly under vigorous, non-physiological conditions. The attachment of the methyl group to oxaloacetate, generating citrate, has enabled organisms to overcome this

Reference Gottschalk, G. (1986) *Bacterial Metabolism*, 2nd edn, Springer-Verlag, Heidelberg, Germany. Includes a compendious survey of the ways in which TCA cycle activities are adapted to the needs of different bacterial species.

problem. The remainder of the cycle is a remarkably direct and efficient way of ensuring complete oxidation, by reshuffling bonds to facilitate an appropriate sequence of dehydrogenations and decarboxylations. It is 'efficient' in the sense that the generation of reduced cofactors, and hence of ATP, is maximized.

The conversion of citrate to isocitrate exemplifies this point. Citrate is a tertiary alcohol, and as such is not readily amenable to dehydrogenation. Isocitrate is a secondary alcohol, and can readily be dehydrogenated. L-Malate, another secondary alcohol, undergoes a similar dehydrogenation. In both cases, the product is a reactive compound, an α-oxoacid.

Acyl coenzyme A derivatives

Two acyl CoA derivatives are shown in Fig. 4.10: the acetyl and succinyl thioesters. More thioesters of the same general chemical kind will appear later.

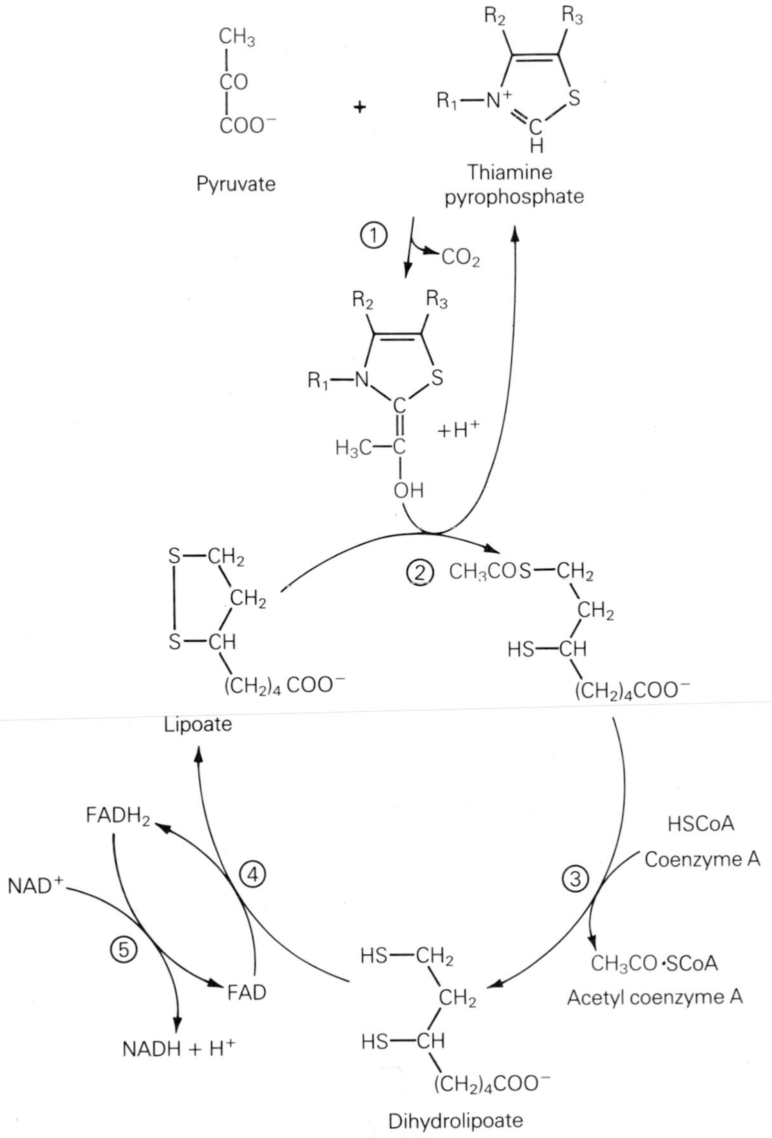

☐ Unlike α-oxo acids, which are stable, β-oxo acids tend to decarboxylate spontaneously. Oxaloacetate is one example. Another example appears in Figure 4.25. The reaction mechanism shown here helps to explain the instability.

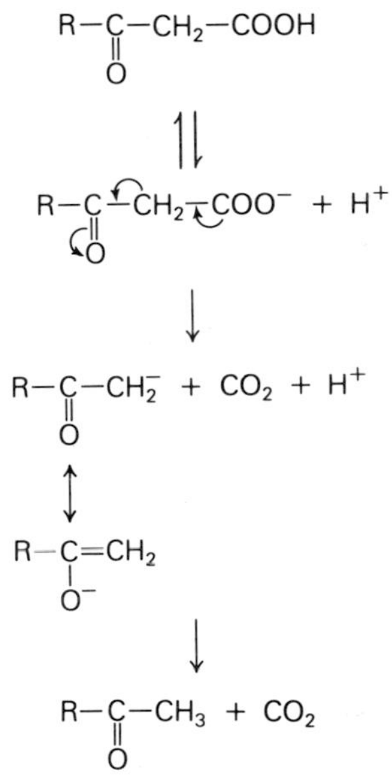

Fig. 4.13 The reactions catalysed by the pyruvate dehydrogenase complex.

See *Biological Molecules*, Chapter 2

OXIDATIVE DECARBOXYLATIONS that give rise to acetyl and succinyl CoA are catalysed by **multienzyme complexes**. In both cases, the substrate is an α-oxoacid: pyruvate or α-oxoglutarate. Pyruvate dehydrogenase is a multienzyme complex catalysing the overall reaction:

$$\text{pyruvate} + NAD^+ + \text{CoA} \rightarrow \text{acetyl CoA} + NADH + H^+ + CO_2$$

Details of the component reactions are shown in Fig. 4.13. The first reaction involves thiamin pyrophosphate, the biologically active form (coenzyme) of the vitamin thiamin. Thiamin contains a five-membered **thiazole** ring, which is the key to the mechanism of decarboxylation of the α-oxo acid. In the second reaction, the newly formed acetyl residue is transferred from the thiamin pyrophosphate to lipoic acid (lipoate), and in the third it is transferred from the lipoate to coenzyme A. The resulting dihydrolipoate is reoxidized by an FAD-containing enzyme, which then, in the final reaction, transfers the hydrogens from the $FADH_2$ to NAD^+.

□ In the thiazole ring, the carbon between the nitrogen and the sulphur is electron-deficient, because the cationic nitrogen withdraws electrons from the N–C bond and electrons cannot be withdrawn from the C–S bond. The hydrogen attached to this carbon is therefore dissociable, a fact that can be demonstrated by nuclear magnetic studies of thiazoles dissolved in ^3H_2O. The resulting conjugate base, a structure known as an **ylid**, is a potent nucleophile that attacks the electron-deficient carbonyl carbon of the α-keto acid. The resulting unstable intermediate rapidly decarboxylates.

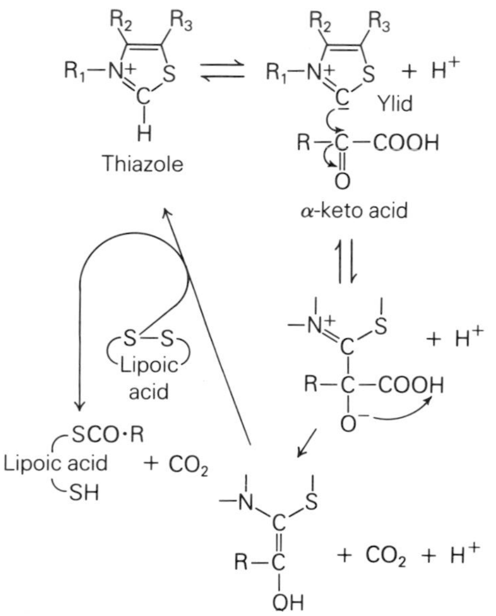

Thioesters have two important chemical properties that distinguish them from ordinary esters resulting from the fact that electrons are less readily accommodated in π-orbitals in carbon–sulphur bonds than in carbon–oxygen bonds. This means that the thioester bond is relatively unstable, and therefore has a higher free energy of hydrolysis than most 'ordinary' ester bonds. This is exploited biologically in the conversion of succinyl CoA to succinate in the TCA cycle. The energy released in this process is sufficient for the synthesis of a pyrophosphate bond, and this reaction is coupled to the synthesis of a molecule of GTP from GDP and phosphate. It also means that the electron-deficient carbonyl carbon tends to pull electrons from the α-carbon of the acyl group, making this group nominally acidic: that is, it can be pictured as dissociating to yield a proton and a carbanion (Fig. 4.14a).

THE CITRATE SYNTHETASE reaction becomes more easily understandable when this second property of thioesters is taken into account (Fig. 4.14b). The formation of citrate depends on the attack by the α-carbon of the acetyl group in the thioester on the electron-deficient carbon of the oxo group in oxaloacetate.

Reaction mechanisms of this kind involving the α-carbon of an acyl coenzyme A derivative are quite widespread, for example in the catabolism of the rare odd-chain-length fatty acids and in fatty acid biosynthesis.

Replenishing the supply of oxaloacetate

As pointed out earlier (Fig. 4.6), oxaloacetate is unstable and degrades spontaneously to form pyruvate and carbon dioxide. Oxaloacetate also participates in a number of other metabolic interconversions, such as gluconeogenesis. The concentration of oxaloacetate always tends to fall, therefore, because it is continuously being removed by these various processes. Consequently, the TCA cycle and metabolism in general (and respiration in particular) will slow down unless there is some means by which oxaloacetate can be regenerated continuously by the organism.

□ Lipoic acid is 6,8-dithiooctanoic acid and was identified in 1949 as a growth factor for certain microorganisms. The isolation procedure started with about 10 tons of liver and yielded 30 mg lipoic acid. In the cell this compound is not found free but is covalently combined with the amino group of a lysyl side-chain of the enzyme protein. For this reason the cofactor is often referred to as lipoamide.

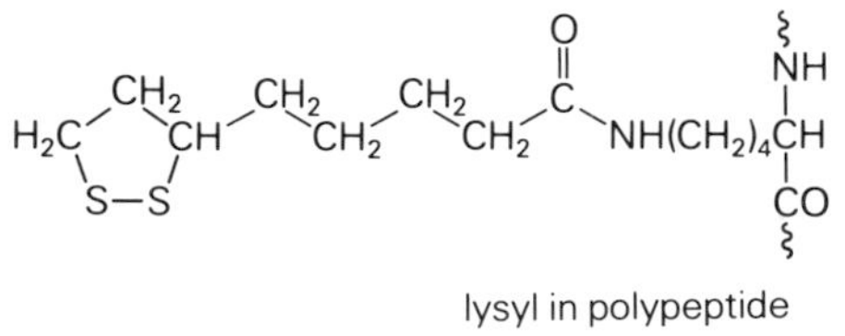

lysyl in polypeptide

Oxaloacetate can be formed directly from the amino acid aspartate. This reaction is an example of a transamination. Many other amino acids give a net production of one or more TCA cycle intermediates during catabolism, and therefore (since one TCA cycle intermediate is converted to another) replenish the oxaloacetate. However, a major source of oxaloacetate is pyruvate and two mechanisms are known for making oxaloacetate from pyruvate. One is a carboxylation, requiring ATP and biotin, catalysed by pyruvate carboxylase. The other is a simultaneous carboxylation (again biotin-dependent) and NADPH-dependent reduction, catalysed by the so-

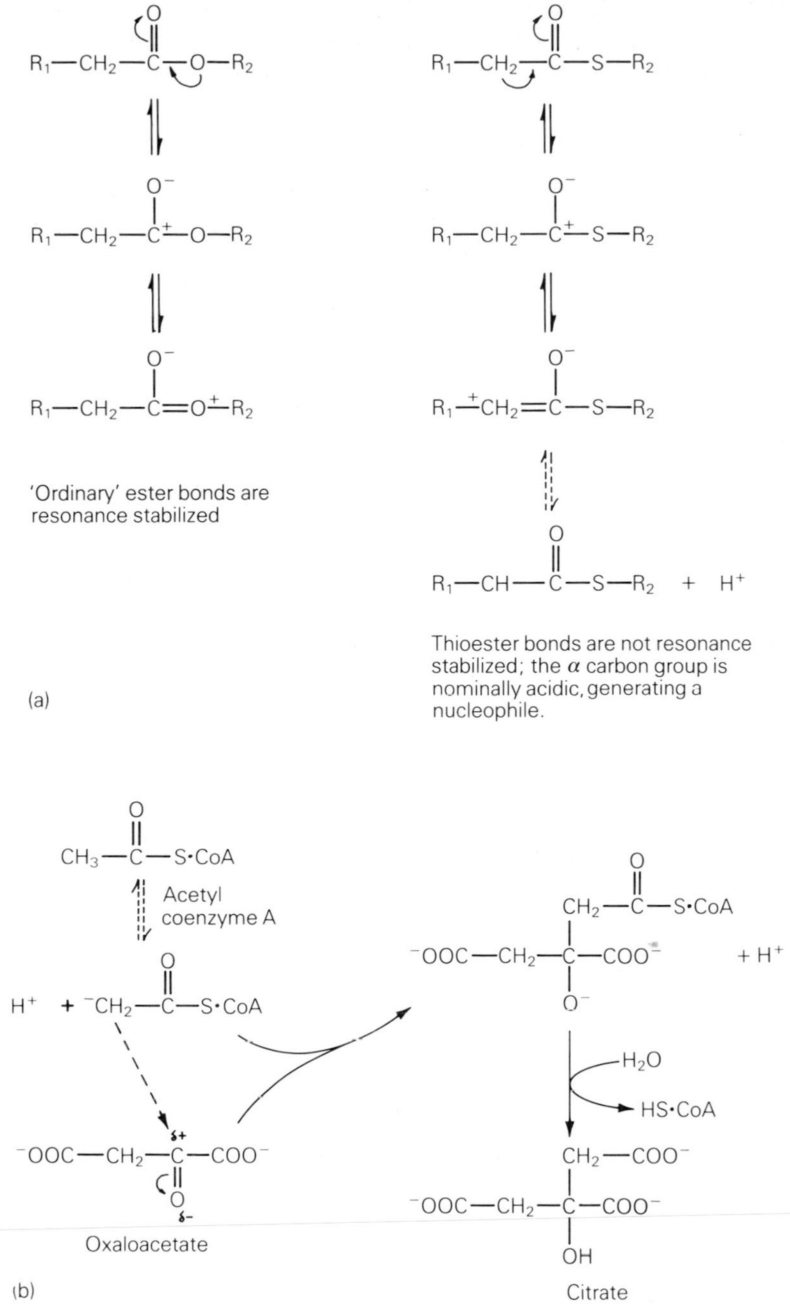

'Ordinary' ester bonds are resonance stabilized

Thioester bonds are not resonance stabilized; the α carbon group is nominally acidic, generating a nucleophile.

(a)

Acetyl coenzyme A

Oxaloacetate

Citrate

(b)

Fig. 4.14 Chemistry of thioesters. (a) Contrast between esters and thioesters; (b) reaction between acetyl CoA and oxaloacetate.

☐ Like thioesters, the esters of α-amino acids also have high free-energies of hydrolysis. This is because the amino group, which is positively charged at physiological pH, withdraws electrons from the ester linkage and decreases the amount of resonance stabilization. This fact is relevant to the synthesis of peptide bonds during protein biosynthesis: the bond between the amino acid and its tRNA is an ester.

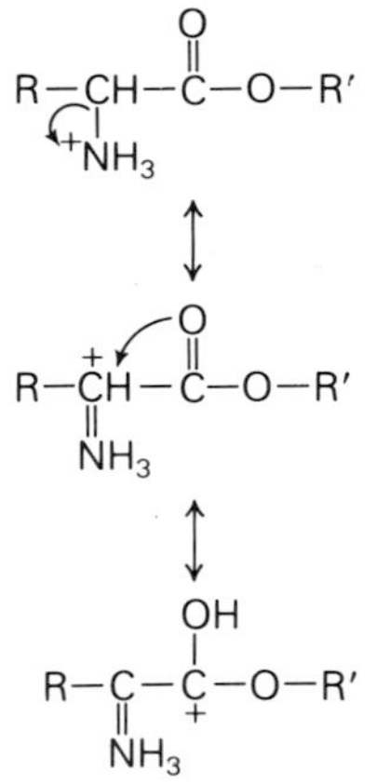

called malic enzyme (Fig. 4.15). Reactions of this kind, which serve to 'top up' the continually diminishing supply of oxaloacetate and other TCA cycle intermediates, are known as ***anaplerotic* reactions**.

Control of the TCA cycle

The relevance of the TCA cycle to bioenergetics is obvious from its relationship to the electron transport chain and therefore to oxidative

Anaplerotic: *from the Greek* ana, *up, and* pleroma, *full. It refers to reactions that 'fill up' the supply of intermediates that tend to become depleted during metabolism.*

Exercise 6

The equilibrium of the malate dehydrogenase reaction favours malate formation; that of citrate synthetase favours citrate formation. Given this information, would you expect the Michaelis constants of enzymes for which oxaloacetate is a substrate to be large or small?

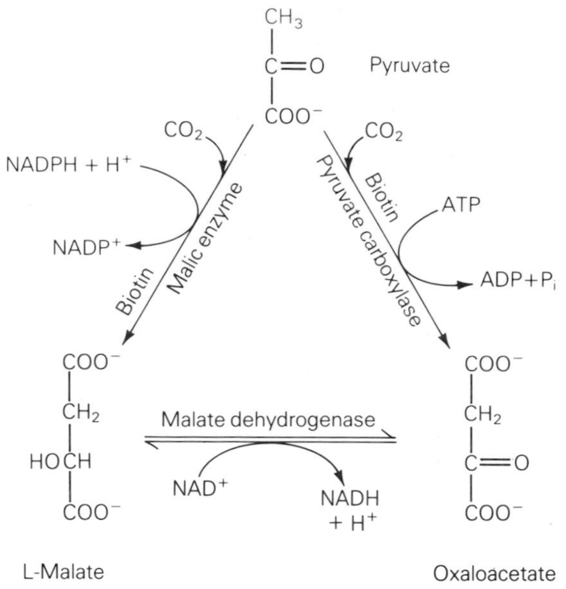

Fig. 4.15 Formation of malate and oxaloacetate from pyruvate: the anaplerotic reactions.

☐ Negative feedback is a widespread mechanism in metabolic control. Typically, the end-product of a metabolic pathway is an allosteric inhibitor of the key regulatory enzyme in that pathway. The pathways of biosynthesis of fatty acids and of cholesterol provide good examples. See *Biosynthesis*.

phosphorylation. The cycle is directly involved in producing the bulk of the ATP in aerobic cells, hence control of the cycle is crucial. For this reason, it makes good biological sense to use the current ATP concentration within the cell as the main regulator of the rate of TCA cycle activity (Fig. 4.16).

ISOCITRATE DEHYDROGENASE is the key regulatory enzyme of the cycle. In many types of cell, the reaction catalysed by this enzyme proceeds more slowly than the others (Table 4.1), that is to say, it is rate-determining. Isocitrate dehydrogenase is an allosteric enzyme, and the main allosteric inhibitor is ATP. Therefore, an increase in the concentration of ATP inhibits isocitrate dehydrogenase, retards the TCA cycle and, in turn, decreases ATP synthesis from oxidative phosphorylation. Conversely, if the ATP concentration falls, the inhibition of isocitrate dehydrogenase is relieved, the turnover of the TCA cycle is accelerated, and oxidative phosphorylation is increased. ADP and AMP also serve as allosteric activators.

Table 4.1 *Some general properties of the enzymes of the TCA cycle*

Step	Enzyme	M_r ($\times 10^{-3}$)	$\Delta G^{0'}$ (kJ mol) of the reaction catalysed	Cofactors	Subunits
1	Citrate synthetase	98	−31.4	–	2
2	Aconitase	65	+6.6	Fe–S	1
3	Isocitrate dehydrogenase	380	−8.7	Lipoate, Mg^{2+}, NAD^+, FAD	8
4	α-Oxoglutarate dehydrogenase	*c.* 4500	−30.1	TPP, lipoate, NAD^+, FAD	*c.* 60
5	Succinyl thiokinase	56	−3.3	GDP, P_i	1
6	Succinate dehydrogenase	97	0	FAD, Fe–S	2
7	Fumarase	200	−3.8	–	4
8	Malate dehydrogenase	130	+29.6	NAD^+	2

Reference Williamson, J.R. and Cooper, R.H. (1980) Regulation of the citric acid cycle in mammalian systems. *FEBS Letters*, **117** (suppl.), K73–85. An excellent survey of the control of the TCA cycle.

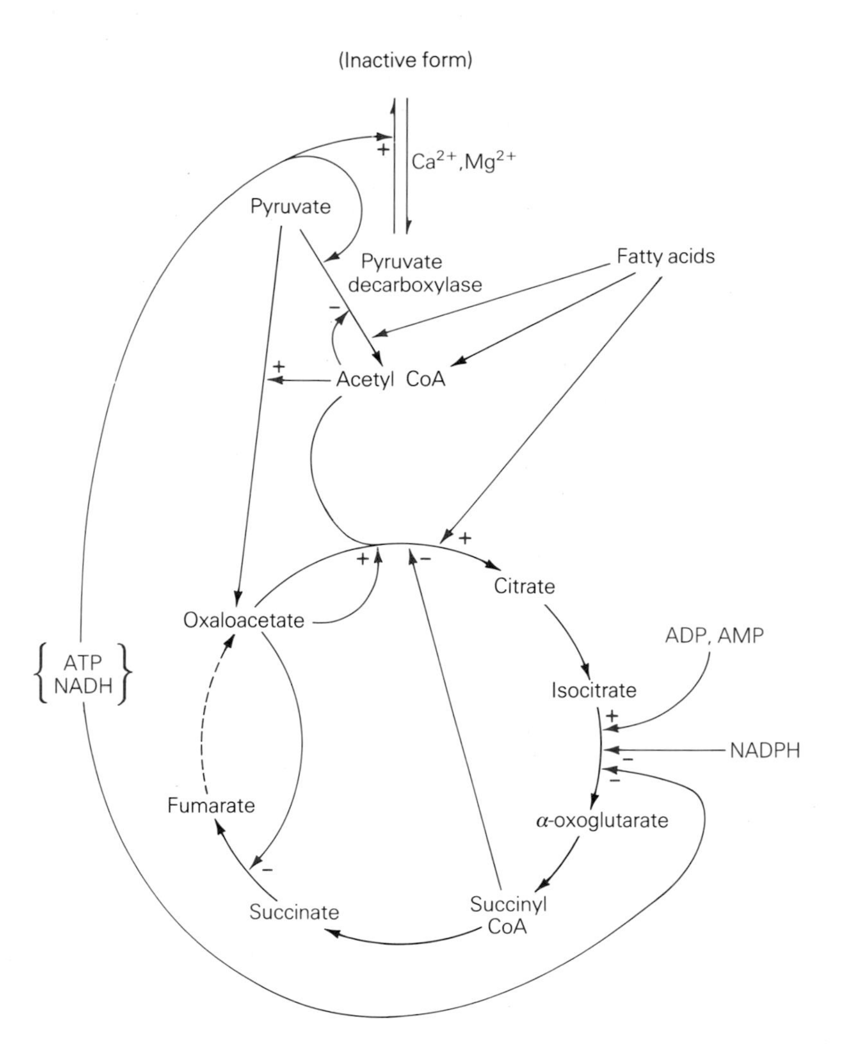

Fig. 4.16 Summary of major mechanisms of control of the TCA cycle.

Exercise 7

In humans, the symptoms of thiamin deficiency (beriberi) generally include weakness, muscular soreness, fatigue after even slight exertion, loss of appetite and insomnia. Can you relate any or all of these symptoms to the biochemical role of pyruvate as described in Figure 4.13 and the related text?

PYRUVATE CARBOXYLASE is another allosteric enzyme implicated in control of the TCA cycle. Acetyl CoA functions as an allosteric activator. It is produced from pyruvate and from other sources such as fatty acid and amino acid degradation, but complete oxidation of the acetyl moiety to CO_2 can only take place through the TCA cycle, and therefore requires oxaloacetate. The previous discussion emphasized the importance of pyruvate carboxylase in producing oxaloacetate. Because of the allosteric control of pyruvate carboxylase, acetyl CoA itself regulates the supply of oxaloacetate needed for complete oxidation of the acetyl residue (always assuming, of course, that there is an adequate supply of pyruvate as substrate for the enzyme).

PYRUVATE DEHYDROGENASE, the very large multienzyme complex that generates acetyl CoA is allosterically inhibited by NADH, ATP and acetyl CoA. It is also inhibited by ATP-dependent phosphorylation and stimulated by dephosphorylation. This complex will therefore tend to be more phosphorylated, hence inhibited, when the ATP concentration inside the mitochondrion is high. This makes the same good biological sense as the allosteric control of isocitrate dehydrogenase.

These and other aspects of the control of the TCA cycle are summarized in Table 4.2 and Fig. 4.16.

Table 4.2 *Control of the TCA cycle, indicating which enzymes are stimulated (+) or inhibited (−) by which effectors. The biological significance of some aspects of this regulation is discussed in the text.*

Enzyme	Effector	+/−	Mechanism
Pyruvate dehydrogenase	ATP + NADH	−	Phosphorylation
	Ca^{2+}, Mg^{2+}	+	Dephosphorylation
	ATP	−	Allosteric
	NADH	−	Allosteric
	Acetyl CoA	−	Allosteric
	Long-chain fatty acids	−	Allosteric
Pyruvate carboxylase	Acetyl CoA	+	Allosteric
Citrate synthetase	Succinyl CoA	−	Competitive
	Oxaloacetate	+	Allosteric (?)
	Fatty acids	+	Allosteric
Isocitrate dehydrogenase	ADP	+	Allosteric
	ATP	−	Allosteric
	NADH, NADPH	−	Allosteric
Succinate dehydrogenase	Oxaloacetate	−	Competitive

A short-cut through the cycle

Plants and some microorganisms contain isocitrate lyase, which cleaves isocitrate to succinate and glyoxylate. The glyoxylate condenses with acetyl CoA in a reaction catalysed by malate synthetase to produce L-malate (Fig. 4.17). In conjunction with the TCA cycle enzymes, these reactions makes up the so-called 'glyoxylate cycle', which results in a net synthesis of malate, and

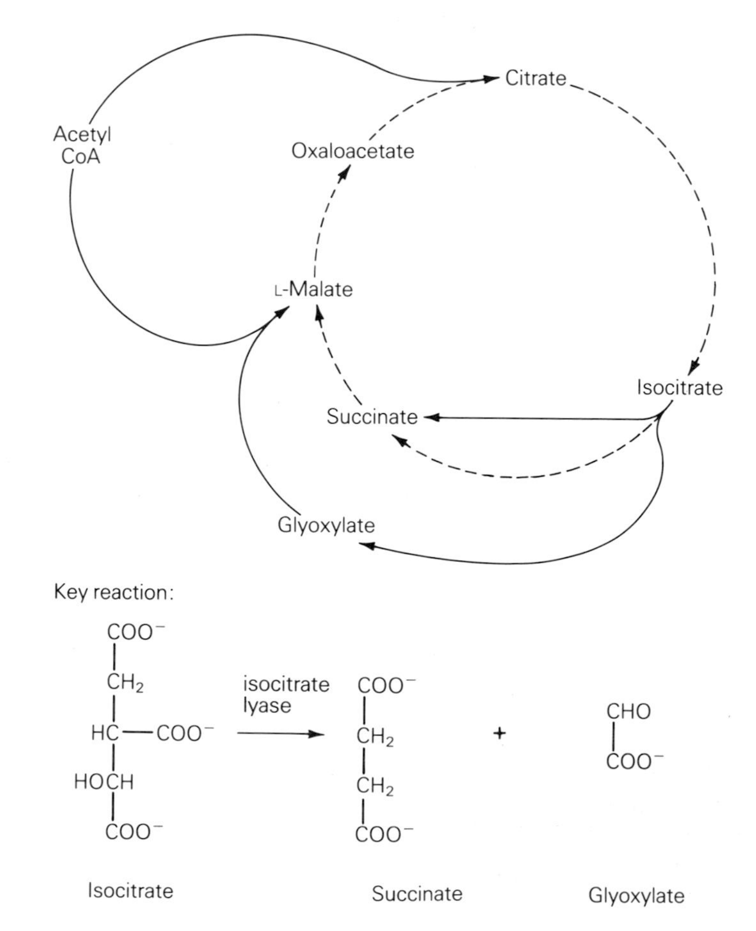

Fig. 4.17 The glyoxylate cycle.

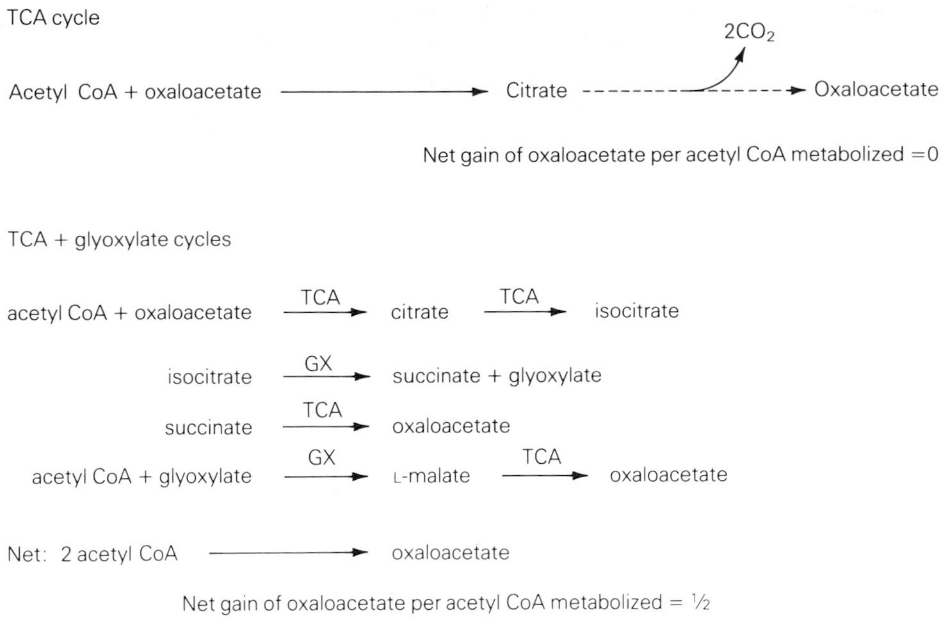

TCA cycle

Acetyl CoA + oxaloacetate ⟶ Citrate - - - -($2CO_2$)- - - → Oxaloacetate

Net gain of oxaloacetate per acetyl CoA metabolized = 0

TCA + glyoxylate cycles

acetyl CoA + oxaloacetate —TCA→ citrate —TCA→ isocitrate

isocitrate —GX→ succinate + glyoxylate

succinate —TCA→ oxaloacetate

acetyl CoA + glyoxylate —GX→ L-malate —TCA→ oxaloacetate

Net: 2 acetyl CoA ⟶ oxaloacetate

Net gain of oxaloacetate per acetyl CoA metabolized = ½

Fig. 4.18 Why the glyoxylate cycle, but not the TCA cycle, gives a net synthesis of oxaloacetate from acetyl coenzyme A. Key: TCA, tricarboxylic acid cycle reactions; GX, glyoxylate cycle reactions.

hence of oxaloacetate, from acetyl CoA. This contrasts with the situation in animal cells, which lack a glyoxylate cycle; here there can be no net synthesis of oxaloacetate from acetyl CoA. Each turn of the cycle generates an oxaloacetate molecule and uses an acetyl CoA molecule; but an oxaloacetate molecule is also used up. Thus, there is not a net production of oxaloacetate (Fig. 4.18).

4.4 *The TCA cycle in relation to other cellular processes*

The modern version of the cycle (Fig. 4.10) enables sense to be made of many physiological observations, of which the process of aerobic glycogen catabolism is just one example. Figure 4.2 shows the interrelationships between the TCA cycle and parts of carbohydrate, lipid and amino acid metabolism. These interrelationships reveal some examples of those physiological observations that the cycle helps explain.

The TCA cycle enzymes are mitochondrial

The TCA cycle reactions occur in the mitochondria in eukaryotic cells. This fact is fundamental to the interrelationships between the cycle and other metabolic processes. Except for succinate dehydrogenase, which is an integral part of the inner membrane, the enzymes of the cycle are located in the mitochondrial matrix. The NADH and $FADH_2$ generated during the cycle are therefore supplied directly to the electron transport chain components, which are also in the inner membrane. That the TCA cycle is a mitochondrial process has the following implications. If a pathway of carbohydrate, fat or amino acid metabolism is wholly or partially cytosolic, then its products must cross the mitochondrial membranes to feed into the TCA cycle. Also, any extramitochondrial anabolic pathway that needs TCA cycle intermediates as

□ In aerobic prokaryotes, TCA cycle enzymes are located on the inner faces of the surface membranes.

its starting-point is only viable if these intermediates can be transferred across the mitochondrial membranes. As a general rule, it is the *inner* membrane of the mitochondrion, not the outer one, that constitutes a significant permeability barrier to ions and small metabolites. Therefore, a key issue in the integration of the TCA cycle with other metabolic processes is the crossing of the inner mitochondrial membrane by the relevant metabolic intermediates.

Crossing the inner mitochondrial membrane

THE PRODUCT OF GLYCOLYSIS, PYRUVATE, can cross the membrane. Within the mitochondrion it may be converted to oxaloacetate or to acetyl CoA. However, glycolysis under aerobic conditions involves not only pyruvate production but also the reduction of two moles of NAD^+ per mole of glucose oxidized, and neither NAD^+ nor NADH can cross the inner mitochondrial membrane. So how does the NADH produced in glycolysis become reoxidized by the electron transport chain?

Figure 4.19 offers a solution to this problem and to the integration of the reactions involved with the TCA cycle. The key point is that although oxaloacetate cannot cross the membrane, L-malate can. Furthermore, there are malate dehydrogenases in the cytosol as well as in the mitochondrial matrix. There are alternative mechanisms for taking reducing equivalents across the membrane (for transporting NADH in spirit, as it were, if not in body) but the mechanism described in Fig. 4.19 is one of the most important quantitatively.

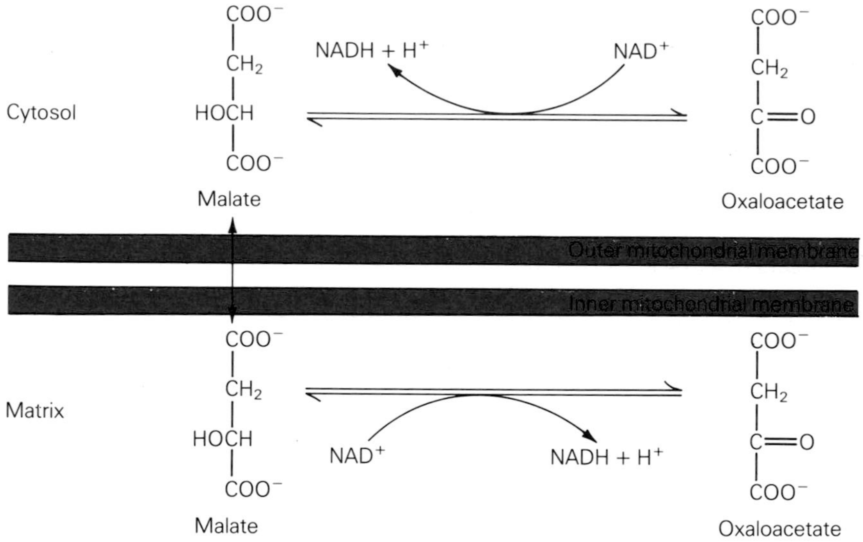

Cytosol: Malate (COO^-—CH_2—HOCH—COO^-) ⇌ Oxaloacetate (COO^-—CH_2—C=O—COO^-), NADH + H^+ / NAD^+

Outer mitochondrial membrane

Inner mitochondrial membrane

Matrix: Malate (COO^-—CH_2—HOCH—COO^-) ⇌ Oxaloacetate (COO^-—CH_2—C=O—COO^-), NAD^+ / NADH + H^+

Fig. 4.19 The malate shuttle, taking reducing equivalents across the mitochondrial membranes.

$(CH_3)_3N^+$—CH_2—CH(OH)—CH_2—COO^- Carnitine

\+

R—CO—S·CoA Fatty acyl coenzyme A

⇌

$(CH_3)_3N^+$—CH_2—CH(O—CO—R)—CH_2—COO^- + HSCoA

Acyl carnitine

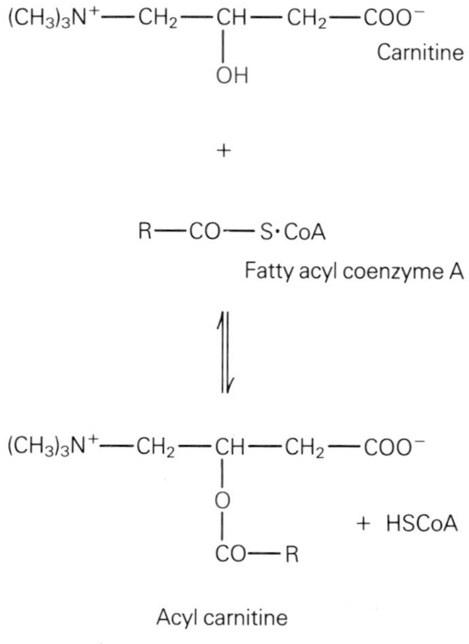

Fig. 4.20 The formation of acyl carnitine.

FATTY ACID BREAKDOWN. Free (non-esterified) fatty acids are oxidized to acetyl CoA within the mitochondrion. This process (β-oxidation) generates considerable quantities of reduced coenzymes. For this oxidation to occur, the free fatty acids have to be transported across the inner membrane by covalent linkage to a 'carrier' molecule, carnitine, which is a component of that membrane (Fig. 4.20). This mode of transport (use of a carrier to take a substrate across a membrane barrier), known as **facilitated diffusion**, will be discussed in more detail later.

See Chapter 6

FATTY ACID BIOSYNTHESIS also involves carriers of a different sort in a minor but indispensible role. Fatty acid biosynthesis is a cytosolic process (usually the enzymes are associated with smooth endoplasmic reticulum) requiring a supply of acetyl CoA. However, acetyl CoA is produced in the mitochondria but cannot cross the mitochondrial inner membrane and therefore cannot be delivered directly to the cytosolic compartment. However, citrate *can* be transported out of the mitochondrion by facilitated diffusion, once again involving a carrier, this time a transport protein or channel, and in the cytosol the citrate can be cleaved in a reaction involving ATP hydrolysis to generate cytosolic acetyl CoA. The functions of these carriers, and the relationships between the processes involved and the TCA cycle, are summarized in Fig. 4.21.

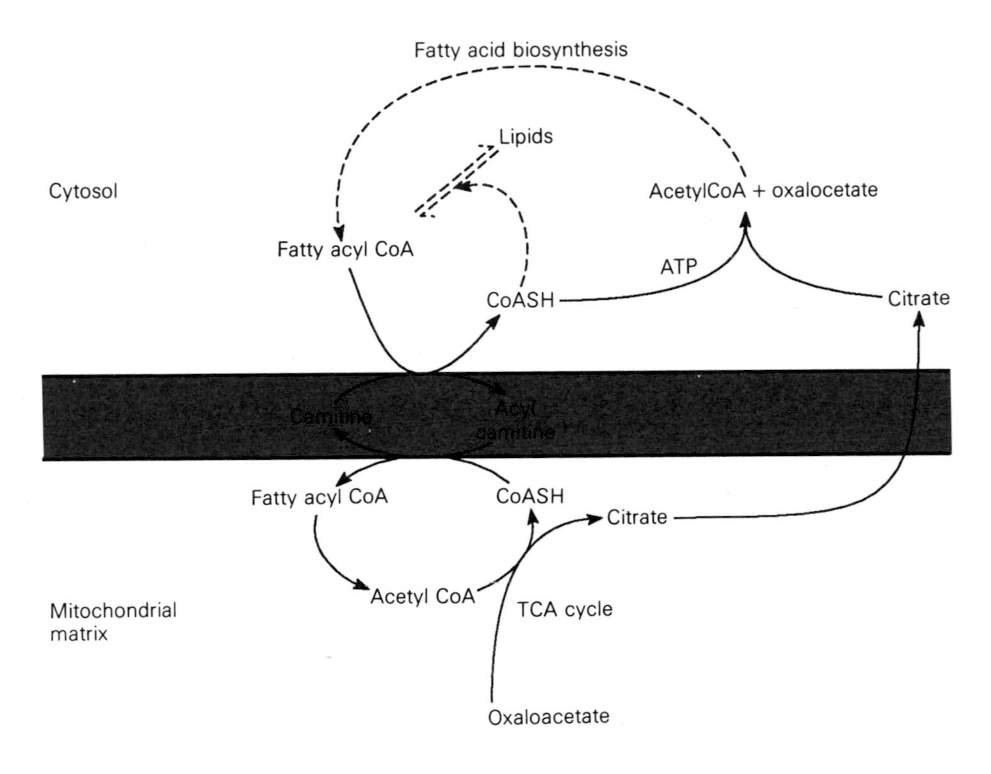

Fig. 4.21 Carnitine, the citrate carrier, fat metabolism and the TCA cycle.

Other interrelationships

CARBOHYDRATE BIOSYNTHESIS from non-carbohydrate precursors entails a number of processes and a rather roundabout route. The non-carbohydrate precursors are transported into the mitochondrial matrix, or converted into something that can be transported there, and catabolized into a TCA cycle intermediate (Fig. 4.22). This produces malate, which can leave the mitochondrion, enter the cytosol, and undergo oxidation to oxaloacetate. Oxaloacetate is the starting material for the process of ***gluconeogenesis***, the biosynthesis of glucose and glycogen from metabolic intermediates. The problem is that the pyruvate kinase reaction is not reversible under the conditions existing in the cell, and therefore phosphoenolpyruvate needs to be produced some other way for gluconeogenesis. This is done by phosphoenolpyruvate carboxykinase for which the substrate is oxaloacetate. However, oxaloacetate cannot leave the mitochondrion, but malate can, as described above.

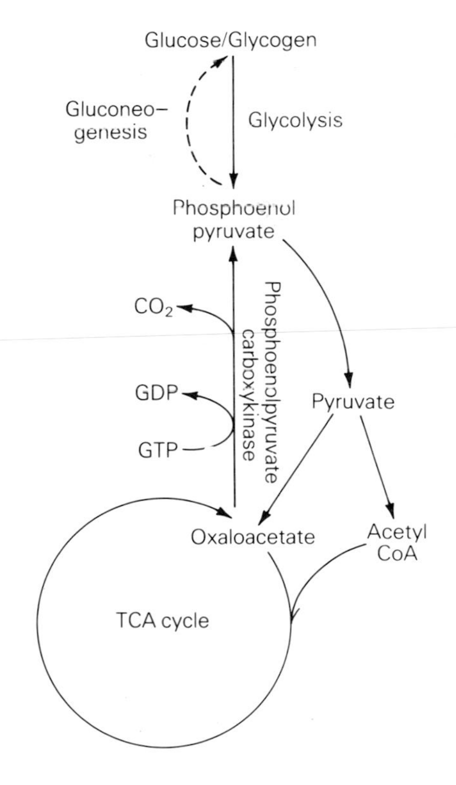

Fig. 4.22 The TCA cycle, glycolysis and gluconeogenesis.

Gluconeogenesis: *from glucose, and the Greek* neos, *new* genis, *produce.*

AMINO ACID CATABOLISM involves the removal of the nitrogen from an amino acid and the conversion of the rest of the molecule, the 'carbon skeleton', to a metabolic intermediate, usually acetyl CoA or a TCA cycle intermediate. In some cases these conversions are simple and direct; in other cases they involve more complex pathways. The TCA cycle provides a means whereby these carbon skeletons can be oxidized to carbon dioxide, with concomitant production of ATP (Fig. 4.23). Alternatively, it provides a means whereby the carbon skeletons can be converted, via malate and oxaloacetate, to carbohydrate. The reactions by which amino acids are fed into the cycle are also reversible. This means that, given a supply of nitrogen in the appropriate form, carbon skeletons may be taken from the cycle and used for amino acid biosynthesis.

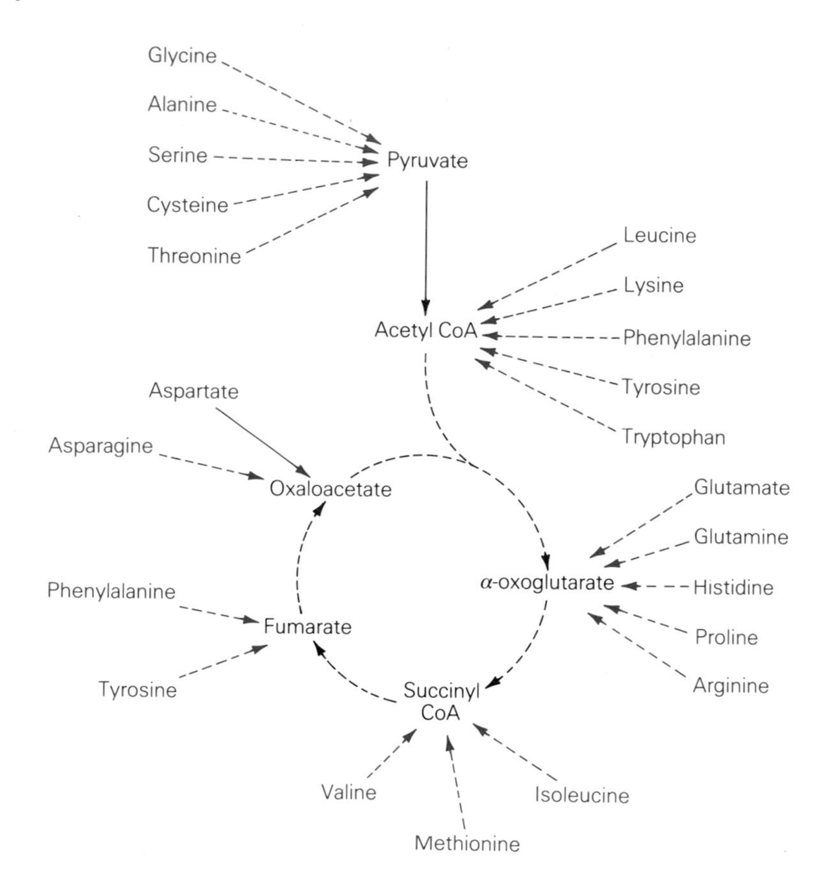

Fig. 4.23 Amino acid catabolism and the TCA cycle.

Dietary carbohydrate can give a net gain of body fat, but not vice versa

It is well-known that a high carbohydrate content in the diet can result in increased accumulation of body fat. Also, complete catabolism of fats requires a supply of carbohydrate. The reasons for these observations can now be understood.

Rapid carbohydrate catabolism results in the production of pyruvate and therefore of oxaloacetate and acetyl CoA, and consequently citrate and isocitrate are formed. However, because of the allosteric regulation of isocitrate dehydrogenase, the amounts of α-oxoglutarate and of subsequent TCA cycle intermediates do not increase in proportion. An alternative fate for

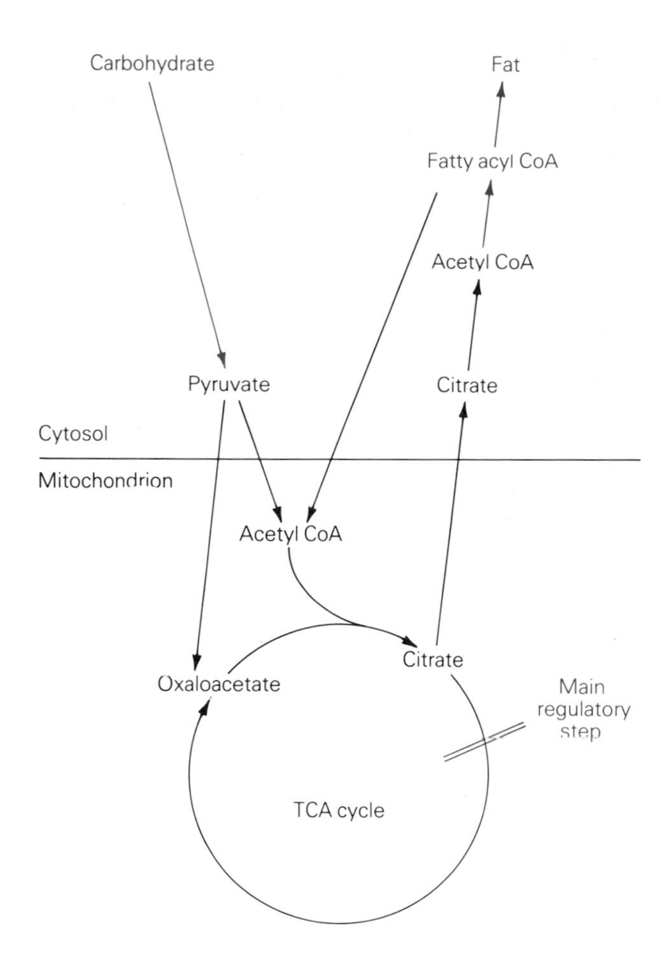

Fig. 4.24 The conversion of carbohydrate to fat.

the excess citrate is to be exported from the mitochondrion and used for the fatty acids synthesis. Consequently, it is observed that excess carbohydrate is converted to fat (Fig. 4.24).

Fat catabolism, on the other hand, results in the production of acetyl CoA but not of oxaloacetate, so that the rate of acetyl CoA production may exceed the rate at which it can be removed by the TCA cycle. Simultaneous catabolism of fatty acids and carbohydrates will overcome this problem, but when fatty acids are the major materials being catabolized, other mechanisms come into play. In principle, the excess acetyl CoA could simply accumulate until oxaloacetate production (largely from pyruvate and hence from carbohydrate) was sufficient to cope with it. However, in practice, very little coenzyme A is present in the cell and it needs to be regenerated all the time.

Box 4.5 *Ketosis*

When the rate of production of acetoacetate and β-hydroxybutyrate in a mammal exceeds the rate at which they are consumed by muscle tissue, they accumulate in the bloodstream and are excreted as anions through the kidneys. This leads to electrolyte and water loss and to a metabolic acidosis (a decrease of blood pH). This condition is known as ketosis, and can be clinically serious unless corrected. It is typical of diabetes mellitus, a condition in which carbohydrate stores are abnormally low and the body therefore relies more heavily on fat catabolism for energy.

Ketosis: *from ketone, and the suffix* -osis, *which can mean 'process' or 'condition'.*

$CH_3\,CO\,SCoA$ + $CH_3\,CO\,SCoA$

2 Acetylcoenzyme A

(reversible) $\rightarrow$ HSCoA

$CH_3COCH_2CHSCoA$

Acetoacetyl coenzyme A

$\downarrow$ $CH_3CO\,SCoA$

OH
|
O_2CCH_2—C—$CH_2CO\,SCoA$ + HSCoA
|
CH_3

β-Hydroxy-β-methylglutaryl CoA

$\downarrow$ H_2O → $CH_3CO\,SCoA$

$CH_3CO\,CH_2OO^-$

Acetoacetate

NADH + H^+ / NAD^+ (reversible) → $CH_3CH\,CH_2\,COO^-$ (with OH on CH) — β-Hydroxybutyrate

Spontaneous → CO_2 ; CH_3COCH_3 — Acetone

Fig. 4.25 Ketone body metabolism.

Exercise 8

Why can skeletal and cardiac muscle, as distinct from other tissues, use ketone bodies efficiently as energy sources? (To answer this question, it may help to look ahead to Chapter 6.)

This is achieved by the formation of the condensation and reduction products, acetoacetyl and β-hydroxybutyryl CoA, and their hydrolysis products, the free acids (Fig. 4.25). These so-called 'ketone bodies' are valuable energy sources. In mammals, for instance, they leave the liver cells in which they are synthesized and are transported in the blood to skeletal or cardiac muscle, where they are reconverted to acetyl CoA and completely oxidized via the TCA cycle, generating ATP.

Ketone bodies do not accumulate in plants and microorganisms that are equipped with the glyoxalate cycle enzymes. In these organisms, the excess acetyl CoA can be converted to malate and hence produce oxaloacetate.

Box 4.6
The stress response

In the overall process of conversion of protein to carbohydrate, the main rate-limiting step is transamination. In physiological or psychological stress, the body responds by increasing the blood glucose level and the carbohydrate stores at the expense of protein. This is achieved through the action of corticosteroids, which stimulate increased manufacture of transaminases and thereby accelerate protein catabolism and, in consequence, gluconeogenesis. Increased corticosteroid production also causes immunosuppression. This partly explains why corticosteroid treatment is used when it is medically necessary to suppress a patient's immune response (e.g. in transplant surgery). It may also explain why people under stress (e.g. those suffering from broken limbs or the threat of impending examinations) are more susceptible to infections than are others.

Reference Devlin, T.R. (1986) *Textbook of Biochemistry with Clinical Correlations*, 2nd edn, Wiley, New York, USA. A textbook for readers with medical interests.

Reference Benson, P.F. and Fensom, A.H. (1986) *Genetic Biochemical Diseases*, Oxford University Press, Oxford, UK. The biochemical bases of clinically significant inborn errors of metabolism are discussed. An understanding of the TCA cycle is relevant to an understanding of many disorders of this kind, although an absence of any of the TCA cycle enzymes themselves would result in a non-viable embryo.

Protein can be converted to carbohydrate

Carbohydrates and fats are often regarded as the main dietary sources of energy, with protein as an emergency back-up source. This may be approximately true of herbivores such as rabbits and omnivores such as humans, but it is not true of carnivores such as dogs and cats where protein is a main source of energy. Classical experiments performed by the pioneer of experimental physiology, Bernard, demonstrated this point. Bernard cannulated the hepatic portal vein and the hepatic vein of a well-fed dog. He observed that the blood entering the liver from the intestine contained a high concentration of amino acids but very little glucose. The blood leaving the liver, however, had a high glucose concentration and almost undetectable concentrations of amino acids. (Glucose is necessary for ATP production in many types of cells, notably brain cells.)

With a knowledge of the TCA cycle and of its interrelations with other metabolic pathways Bernard's observations are easily explained. Most of the amino acids entering the liver as a result of the digestion and absorption of the protein by the dog are catabolized, the nitrogen being removed and the carbon skeletons being largely converted to TCA cycle intermediates. These processes lead to a net production of malate and oxaloacetate, and thence, by gluconeogenesis, to a net production of carbohydrate (Fig. 4.26). Some of this carbohydrate is stored in the liver as glycogen; the rest is exported via the hepatic vein as glucose.

□ Most amino acids can cross the mitochondrial inner membrane by carrier-mediated processes, and many can be transaminated within the matrix. The mechanism for disposal of waste nitrogen in mammals, urea biosynthesis, is an intramitochondrial pathway. When you have studied Chapter 7 note the close relationship between the urea cycle intermediates and those of the TCA cycle.

□ Mitochondria reoxidize NADH but not NADPH, which tends to be reoxidized in anabolic processes. In many tissues, mitochondria can actually reduce cytosolic $NADP^+$ at the expense of NADH in the matrix; in some cases, this **transhydrogenase** reaction is a major source of NADPH for biosynthetic pathways.

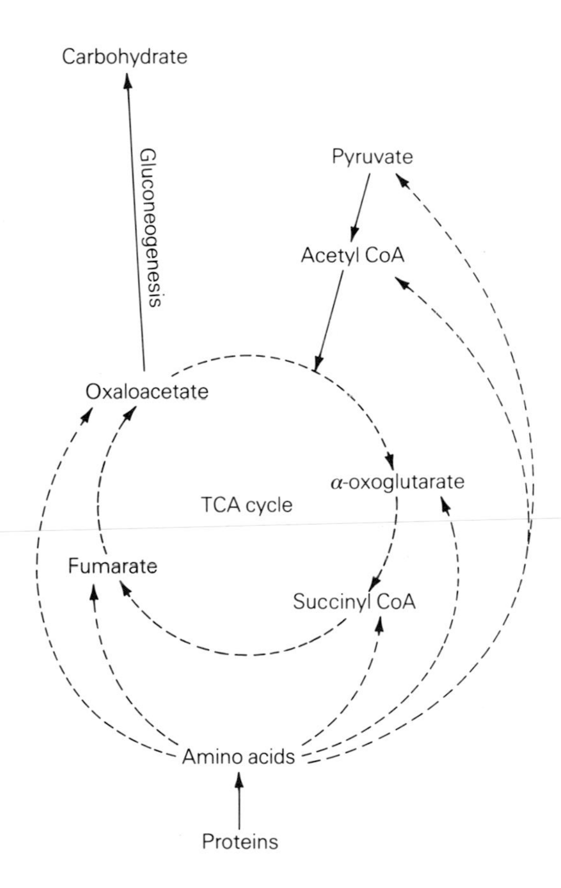

Fig. 4.26 Conversion of protein to carbohydrate.

4.5 *Overview*

The TCA cycle is the centrepiece of intermediary metabolism. The intermediates of the cycle are the starting-points for the biosynthesis of carbohydrates, fatty acids, many amino acids, and other compounds of biochemical importance, such as porphyrins. It is also the final common pathway for the aerobic oxidation of the products of carbohydrate, lipid and amino acid catabolism. During the process, hydrogen-carrying coenzymes are reduced, to be reoxidized via the electron transport chain with concomitant ATP production. The TCA cycle is responsible for most of the ATP production in aerobic cells. The overall rate of the pathway is tightly regulated largely by the intracellular ATP concentration.

The close association between the TCA cycle and electron transport is reflected in the fact that the TCA cycle enzymes are located on the mitochondrial inner membrane or in the mitochondrial matrix. The interrelationships between the cycle and other metabolic processes require that the cell can cope with the problems of transporting the relelvant metabolic intermediates across the mitochondrial membranes, particularly the inner membrane, which is the more significant permeability barrier. The interrelationships explain many physiological observations such as the fact that excess dietary carbohydrate is converted to body fat and that protein can be converted to carbohydrate.

Answers to exercises

1. Make extract of the 'used' muscle, and add it to a fresh muscle slice preparation to see if inhibition occurs.

2.

$$
\begin{array}{c}
COO^- \\
| \\
CH_2 \\
| \\
HO{-}C{-}CH_2{-}CO{-}COO^- \\
| \\
COO^-
\end{array}
$$

(a) nucleophilic attack by C-3 of pyruvate on C-2 oxaloacetate
(b) decarboxylation of what was C-1 of the pyruvate.

3.

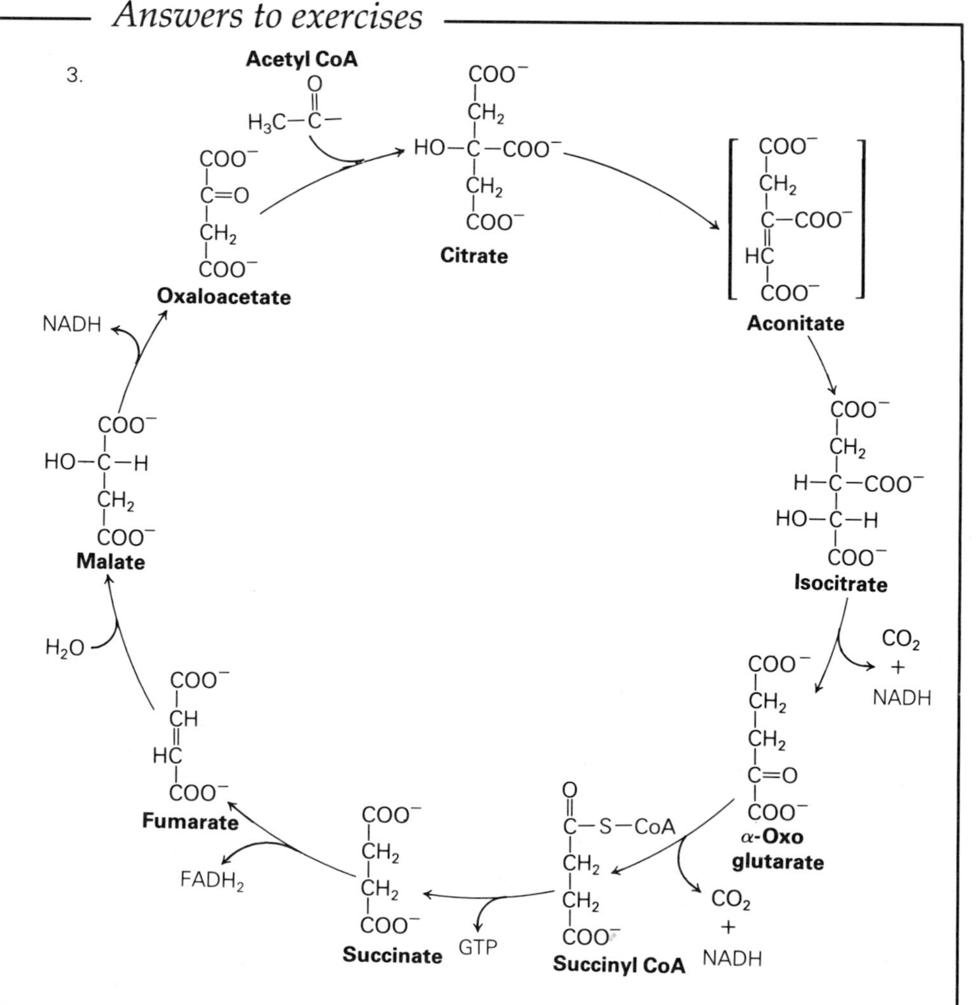

4. The conversions of fumarate to malate and oxaloacetate and acetyl CoA to citrate both require a water molecule. The succinate thiokinase reaction utilizes the third water.
5. C-2.
6. Small
7. Muscle, particularly cardiac, and brain both depend upon the aerobic catabolism of pyruvate.
8. Skeletal and cardiac muscle both store glycogen for their own use and therefore both have an 'in-house' supply of pyruvate.

QUESTIONS

FILL IN THE BLANKS

1. The TCA cycle is the major pathway for the _______ _______ of many types of compounds. Some of its intermediates also form starting points for the _______ of carbohydrates, lipids, many amino acids and other compounds, such as _______ .

The TCA cycle consists of _______ enzyme-catalysed reactions in which the _______ _______ of _______ _______ is oxidized to carbon dioxide. _______ _______ combines with _______ to form the C_6 _______ _______ , _______ . The reactions of the cycle operate to reform _______ , and produce one molecule each of _______ and _______ and _______ molecules of _______ . Thus the cycle produces the equivalent of _______ molecules of _______ for each _______ _______ oxidized.

The key regulating enzyme of the cycle is _______ _______ . Other enzymes which exert control are _______ _______ and _______ _______ . All three enzymes are subject to _______ _______ .

Choose from: three, eight, twelve, ATP, GTP, $FADH_2$, NADH, acetyl CoA (2 occurrences), acetyl residue (2 occurrences), aerobic catabolism, allosteric regulation, biosynthesis, citrate, isocitrate dehydrogenase, oxaloacetate (2 occurrences), porphyrins, pyruvate carboxylase, pyruvate dehydrogenase, tricarboxylic acid.

MULTIPLE-CHOICE QUESTIONS

2. State if the following are true or false:

A. The TCA cycle occurs in the cytosol.
B. The TCA cycle occurs in the cytoplasm.
C. The conversion of α-oxoglutarate to succinyl CoA.
D. CO_2 is released during the conversion of isocitrate to α-oxoglutarate and succinyl CoA to succinate.
E. GTP is produced during the reaction catalysed by succinyl thiokinase.
F. The conversion of citrate to isocitrate is catalysed by aconitase.
G. The conversion of malate to oxaloacetate is an anapleurotic reaction.

SHORT-ANSWER QUESTIONS

3. The following statements all have the form 'P *because* Q'. Mark them as follows:

A. if both P and Q are true and the causal connection is valid;
B. if both P and Q are true but the causal connection is *not* valid;
C. if P is true and Q is false;
D. if P is false and Q is true;
E. if both P and Q are false, but the causal connection would be valid if they were true;
F. if both P and Q are false, and the causal connection would be invalid even if they were true.

(a) TCA cycle in human liver mitochondrial is retarded by dietary ethanol *because* the intracellular ethanol concentrations make the active sites of the enzymes less accessible to their substrates.
(b) TCA cycle activity in human liver mitochondria is accelerated by dietary ethanol intake *because* the ethanol causes a marked decrease of the intracellular $NAD^+/NADH$ ratio.

(c) Beriberi results in inactive pyruvate carboxylase *because* pyruvate carboxylase is nicotinamide-dependent.
(d) Thyroxin accelerates the TCA cycle *because* it partially uncouples oxidative phosphorylation.
(e) Glycerol undergoes a net conversion to glycogen in liver *because* any compound that undergoes net conversion to a TCA cycle intermediate causes a net synthesis of glycogen.
(f) Thiazoline analogues inhibit isocitrate dehydrogenase *because* TPP is a cofactor for isocitrate dehydrogenase.

4. Why do so many of the intermediates of catabolism contain carboxylic acid groups?

5. How many ATP molecules are generated, and how many molecules of oxygen are consumed, during the complete oxidation of pyruvate to CO_2? (See Chapter 3 for the numbers of ATP molecules produced by oxidative phosphorylation when reduced cofactors are reoxidized by the electron transport system.)

6. If 3-chloropyruvate was added to a sample of actively respiring mitochondria, what product would you expect to accumulate? (Assume that the chlorine atom will not interfere with any reaction unless it substitutes for a hydrogen that is to be transferred to a cofactor.)

7. What would happen to the rate of oxidation of succinate or malate in a mitochondrial preparation if (a) cyanide, (b) dinitrophenol, (c) cyanide + dinitrophenol, (d) ethanol + alcohol dehydrogenase, (e) an ATP-generating system, or (f) oligomycin added? (Assume that oxidative phosphorylation remains coupled.)

ESSAY QUESTIONS

8. Using your knowledge of the TCA cycle and its relationships to other metabolic pathways, show how administration of $^{14}C_2$-pyruvate (i.e. pyruvate with carbon-2 labelled) to a mammal could lead to the production of (a) ^{14}C-labelled proteins, (b) ^{14}C-labelled carbohydrates, and (c) ^{14}C-labelled fats.

9. Suppose that in a distant part of the galaxy, on a planet with a reducing atmosphere consisting largely of ammonia, methane and nitrogen, there are animals that use carbohydrates as energy sources just as terrestrial animals do. Their 'glycolytic' pathway converts glucose to propionate. In their equivalent of mitochondria, there is an electron transport chain in which the final step is oxidation of ammonia to nitrogen, not reduction of oxygen to water. These organisms have a 'TCA cycle', by means of which they reduce propionate to methane. Invent a plausible 'TCA cycle' for them, keeping the analogy with the terrestrial TCA cycle as close as possible.

10. In the text of this chapter, it was stated that the 8-reaction pathway for oxidizing acetate to CO_2 ensured a more efficient generation of reduced cofactors and therefore of ATP than any possible direct oxidation pathway. Invent the most plausible pathway you can for direct oxidation of acetate to two molecules of CO_2, and compare the ATP yield (using your knowledge of electron transport and oxidative phosphorylation) with that of the TCA cycle.

5
Glycolysis

Objectives

After reading this chapter you should be able to:

- ☐ explain the significance of the elucidation of glycolysis in the emergence of biochemistry as a scientific discipline;
- ☐ describe the interrelations between the major pathways of carbohydrate metabolism;
- ☐ outline the relationship between the metabolism of carbohydrates and other metabolic processes;
- ☐ explain the biological significance of the major pathways for carbohydrate metabolism, and account for the clinical consequences of genetic defects which affect them.

5.1 Introduction

Carbohydrates are extremely versatile compounds that play a large number of roles in organisms. Carbohydrates are intimately linked with energy metabolism throughout almost the whole range of living things. For instance, they are the immediate products of photosynthesis, they are the most abundant components of the diet of many animals, and they act as vital energy storage compounds in a very wide range of organisms. Their polymeric forms (polysaccharides) are utilized as structural materials. Examples include chitin in fungi and arthropods, cellulose in most plant taxa and chondroitin sulphate in mammals.

This chapter describes the main aspects of the catabolism and interconversions of carbohydrates; details of the metabolism of highly specialized compounds are not given. The greatest emphasis is placed on glycolysis, the main pathway of glucose catabolism, partly because this pathway plays a major role in energy metabolism throughout the biosphere, and partly because the elucidation of this pathway is more or less coextensive with the maturation of biochemistry as a scientific discipline.

This chapter emphasizes the historical approach in order to show how things were discovered. The elucidation of glycolysis was of great importance to the development of biochemistry and occupied a great part of its early history.

5.2 Early history of the study of glucose metabolism

A 'metabolic pathway', such as glycolysis, is a piece of information, and any piece of information is useful only in so far as it provides an answer to a

Reference Fruton, J.S. (1972) *Molecules and Life: Historical Essays on the Interplay of Chemistry and Biology*, Wiley, New York. This book describes the early history of glycolysis in great detail and contains information relevant to the development of enzymology.

Reference Kalckar, H.M. (ed.) (1969) *Biological Phosphorylations: Development of Concepts*, Prentice-Hall, New Jersey. This book describes the later history and contains translations of several of the classical papers. It delineates the evolution of the concept that transphosphorylation reactions play a key part in energy metabolism.

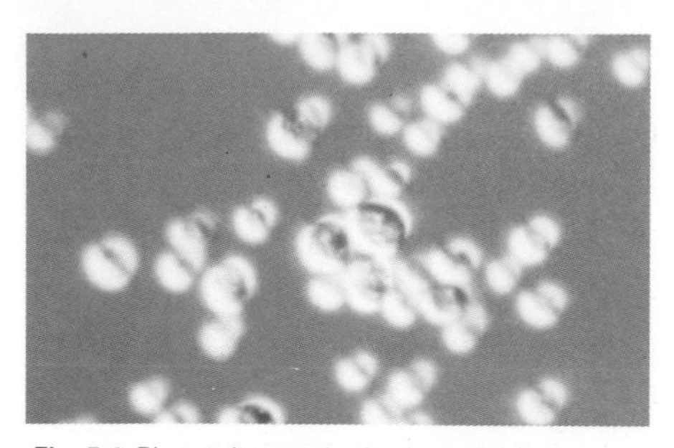

Fig. 5.1 Photomicrograph of yeast cells (*Saccharomyces cerevisiae*) (× 720). Courtesy of M.J. Hoult, Department of Biological Sciences, Manchester Polytechnic.

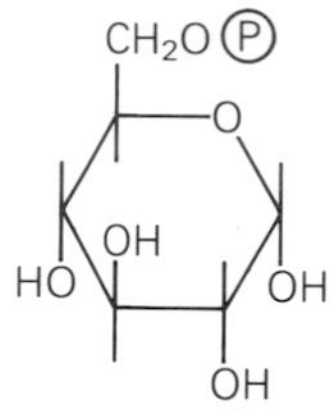

question. Glycolysis was elucidated in order to answer not one, but two apparently very different questions:

- How do yeasts produce ethanol from sugars under ***anaerobic*** conditions?
- How do vertebrate skeletal muscles generate the energy for contraction under ***anaerobic*** conditions?

The first question (the mechanism of ***fermentation***) was a problem for organic chemists; the second (the energy ***transduction*** mechanism in muscle contraction) was a problem for physiologists. Both questions occupied the attention of many leading researchers in the early part of the twentieth century.

Pasteur first established fame (1850s) by demonstrating, contrary to Liebig's belief, that fermentation required living yeast cells (Fig. 5.1). In 1897, however, the Buchner brothers discovered that cell-free yeast extracts could ferment glucose to ethanol. This discovery, which established fermentation as a 'chemical' rather than a 'vital' process, marks the beginning of biochemistry as a distinct science. In 1905, studies on this cell-free fermentation phenomenon by Harden and Young led to the discovery that, although a heat-labile factor was involved (***zymase***, now known to be a collection of enzymes), there were also heat-stable factors. One of these factors was phosphate, and further experiments demonstrated that phosphate esters of sugars were produced during the fermentation reactions (Fig. 5.2). Other factors, which came to be known as 'coenzymes', were more complex.

Glucose 6-phosphate

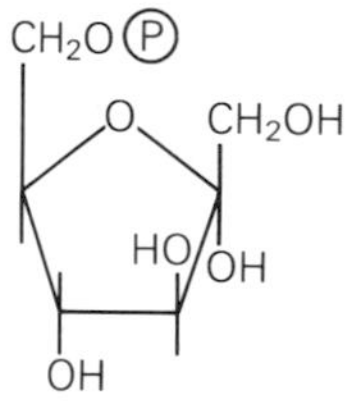

Fructose 6-phosphate

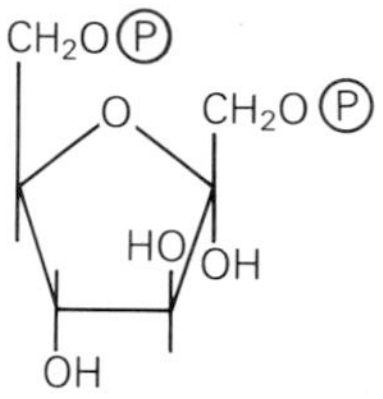

Fructose 1,6-bisphosphate

Fig. 5.2 The hexose phosphates found in yeast extracts supplemented with glucose and phosphate. Note that fructose derivatives also occur as pyranose rings though usually the furanose form predominates in aqueous solution.

EXPERIMENTAL PHYSIOLOGY. The pioneers of experimental physiology included French (Bernard), German (Müller and Helmholtz) and British (Foster) scientists, all clinicians by training. By 1900, the efforts of these pioneers had initiated several lines of inquiry, of which two proved particularly important for the study of carbohydrate metabolism: (a) how glucose is formed, utilized and stored in the body; (b) how muscles obtain the energy for contraction.

Two British physiologists, Fletcher and Hopkins, performed a series of experiments on contractions of frog muscles during the first decade of the twentieth century. They discovered, to their surprise, that an oxygen supply was not necessary for contraction, only for *recovery* from contraction. Immediately before the First World War, Hill pursued these studies further and established reproducible quantitative relationships between the force of contraction, the muscle tension and the heat generated in the process. After the war, Hill's further studies benefited from the collaboration of a German physiological chemist, Meyerhof. Hill and Meyerhof demonstrated that when

Box 5.1 Chemical and vital processes

'Vitalism' is the belief that organisms are fundamentally distinct from the non-living world; they are characterized by a 'vital force' that is not reducible to physicochemical forces. 'Mechanism' is the denial of this view. These were the different views held in nineteenth century biology. Lamarck's model of evolution is inherently vitalistic. Darwin's is mechanistic. Müller's pioneering work in physiology betrayed a vitalistic perspective, but his pupil Helmholtz was a mechanist. It is sometimes said that Wöhler's synthesis of urea from ammonium cyanate put paid to vitalism; this is manifestly false, since it predated much of the controversy. Nevertheless, it is fair to say that Wöhler's achievement marked the beginning of modern organic chemistry, in much the same sense that Buchner's discovery marked the beginning of biochemistry.

Aerobic: *a contracted form of aerobiotic; from the Greek* aer, *air, and* bios, *life.* ***Anaerobic:*** *aerobic plus the Greek prefix* a(n), *not/none.*
Fermentation: *from the Latin* ferveo, *boil; a reference to the tendency of fermenting liquids to froth and bubble.*

Transduction: *the conversion and transmittance of energy in one form to another.*
Zymase: *collection of enzymes extractable from yeast. From the Greek* zyme, *leaven.*

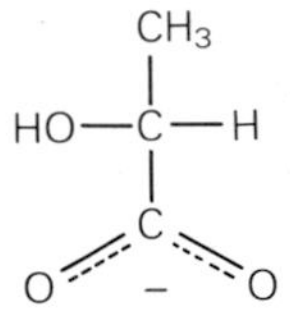

Fig. 5.3 Lactate.

a muscle contracted in the absence of oxygen, its ***glycogen*** stores were depleted and lactate (Fig. 5.3) accumulated. Moreover, the extent of lactate accumulation correlated with the heat and work output of the muscle.

Subsequently, Embden was primarily responsible for discovering the fundamental relationships between ethanolic fermentation in yeast, the chemistry of muscle contraction, and the mechanism of glucose utilization in mammals. Embden developed methods for studying chemical processes in mammalian cells *in vitro*, for example, by incubating thin slices of liver in buffers containing physiological ion concentrations. Using such techniques, he extended the findings of Hill and Meyerhof. He found that during anaerobic muscle contraction, sugar phosphates were produced that were identical to those found by Harden and Young in cell-free yeast extracts. The same sugar phosphates were also found in liver during glucose utilization. Gradually, Embden accumulated enough evidence to be able to conclude that the series of chemical reactions involved in yeast fermentation, anaerobic muscle contraction and glucose utilization by the liver were all essentially the same (Fig. 5.4).

This discovery marked the birth of the concept of the 'metabolic pathway', which may be defined as a series of enzyme-catalysed reactions that has physiologically significant beginning and end points, and need not be restricted to a single type of cell. It also, incidentally, helped to vindicate Bernard's conviction that the liver played a central role in mammalian metabolism.

Embden was to add yet another important discovery to an already impressive list. He identified another cofactor involved in anaerobic glucose catabolism, a phosphate ester of adenosine. This eventually led to the isolation of ATP (Fig. 5.5) from muscle and its chemical characterization by a Danish physiological chemist, Lundsgaard, in 1929. In 1925, Meyerhof succeeded in demonstrating the conversion of glycogen to lactate *in vitro* using cell-free muscle extracts. This method was the precursor of subcellular fractionation techniques, and was another major methodological breakthrough without which, perhaps, metabolic pathways could not have been elucidated.

Exercise 1

In the 1920s, a popular explanation for muscle contraction was the 'colloid aggregation hypothesis'. According to this hypothesis, the lactate produced from glycogen precipitated and aggregated the muscle proteins (colloids), thus decreasing the intracellular volume and causing the muscle to contract. This process was reversed when the lactate acid was removed. How could you refute this hypothesis experimentally?

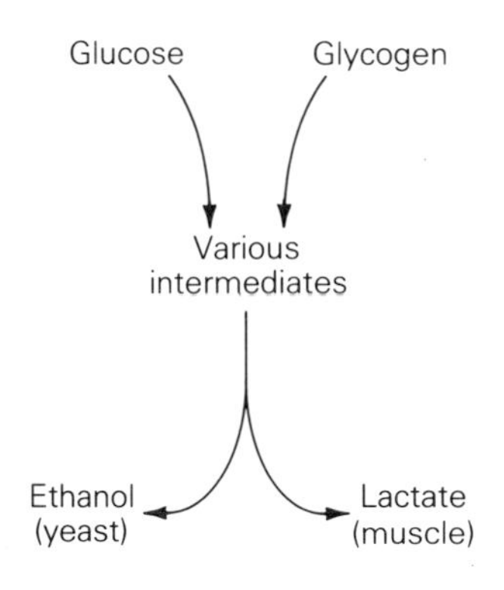

Fig. 5.4 Commonality between yeast and mammalian muscle with respect to anaerobic glucose catabolism.

☐ Embden became director of the Medical Clinic at Frankfurt in 1904; by 1907, he had added a set of physiological chemistry laboratories to the Clinic; and in 1914, when Frankfurt University was founded, he obtained a professorship. Therefore, at the time when Hill and Meyerhof were performing their Nobel Prizewinning work, he was an established, senior figure in the field.

☐ Embden's use of thin tissue slices immersed in physiological buffers was at the time a revolutionary methodological advance. It was the first technique by which *in vitro* functional studies could be carried out on insoluble biological material, as opposed to soluble components such as cell-free yeast extracts.

☐ It is worth recalling that at this same period of history, the first crystallization of an enzyme and its identification as a protein were performed (Sumner, 1926; the enzyme was urease).

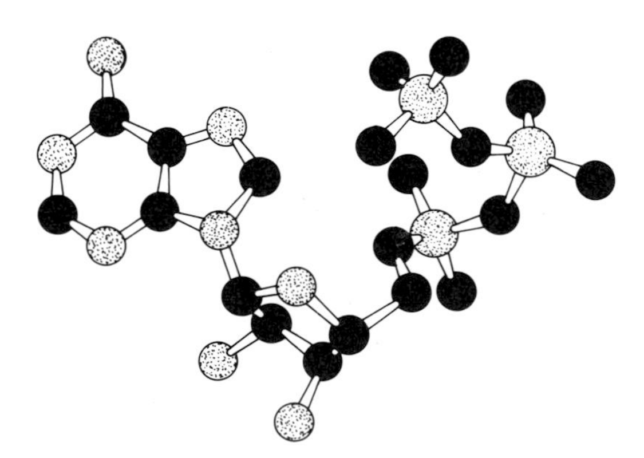

Fig. 5.5 Model of adenosine triphosphate (ATP).

Glycogen: *from the Greek* glykr, *sweet, plus* gen *from the verb* gennaein, *to produce. This alludes to the production of sweet-tasting glucose upon hydrolysis of glycogen.* ***Glycogenolysis:*** *glycogen, plus the Greek* lysis, *dissolution.*

Reference Baldwin, E. (various editions) *Dynamic Aspects of Biochemistry*, Cambridge University Press. A simple historical outline. The first edition (1947) and the second (1949) are particularly valuable, because they were written while knowledge of central metabolic pathways was still evolving.

5.3 The Embden–Meyerhof pathway

By 1929, the two initial questions (see above) that had concerned the chemical mechanism of fermentation and the underlying chemistry of anaerobic muscle contraction, could be reformulated as a single question; by what series of enzyme-catalysed reactions is glucose converted to lactate or ethanol under anaerobic conditions? The steps were painstakingly worked out in the early 1930s and the complete details of the pathway of **glycolysis**, alternatively known as the **Embden–Meyerhof** pathway after the two pioneers of its elucidation, were published in 1937.

□ Hill and Meyerhof shared the 1922 Nobel Prize for Medicine in recognition of their work on muscle metabolism. The formation of glycerol from glucose by yeasts in the presence of iodoacetate or sulphite was one of Neuberg's main contributions to the research. Carl and Gerty Cori were awarded the 1947 Nobel Prize for Medicine for their work identifying glucose 1-phosphate and the enzymes of glycogen synthesis and breakdown. Fritz Lipmann first formalized the concept of 'high-energy compounds' and delineated their significance in biochemistry; he was awarded the 1953 Nobel Prize for Medicine for discovering, isolating and identifying the structures of coenzyme A and its acetyl derivative. David Needham made several important contributions to understanding the physiology of muscle contraction. Embden, incidentally, did not live to see the maturation of the work he had initiated: he died in 1933.

Glycolysis

The overall strategy employed in the conversion of glucose (C_6) to two molecules of pyruvate (C_3) is summarized in Fig. 5.6. Glycolysis begins with the metabolic activation of the glucose molecule by its phosphorylation. In fact, two phosphorylations occur early in glycolysis leading to the formation of a hexose sugar with two phosphate groups, one on the C-1 and one on the C-6 atoms. Subsequent cleavage between C-3 and C-4 generates two different but interconvertible C_3-P molecules.

Fig. 5.6 Strategy of glycolysis.

THE REACTIONS OF GLYCOLYSIS are summarized in Figures 5.7 and 5.8 and Table 5.1. The first and third steps, the ATP-dependent phosphorylations of glucose and of fructose 6-phosphate, respectively, have equilibria that greatly favour the products and are effectively irreversible. The second step is the interconversion between the phosphate esters of the aldohexose, glucose, and the corresponding ketohexose, fructose. Step 4 (aldolase catalysed) splits the hexose diester into two triose phosphates, which are interconverted in step 5. Step 6, the glyceraldehyde 3-phosphate dehydrogenase reaction, is complex: it has an absolute requirement for NAD^+ and inorganic phosphate. The product, 1,3-bisphosphoglycerate, contains one phosphate ester linkage (on the 3 position) and one phosphate carboxylate anhydride (on the 1 position). The latter has much the greater standard free energy of hydrolysis. This fact is exploited in step 7, when the anhydride phosphate is transferred to ATP. Step 8 isomerizes one phosphate ester of phosphoglycerate to another, which by dehydration (step 9) is converted to the phosphate ester of the enol form of pyruvate. Because this too has a high standard free energy of hydrolysis, the phosphate can also be transferred to ADP. This reaction (the pyruvate kinase reaction, step 10) is, like steps 1 and 3, effectively irreversible. All other steps in glycolysis are readily reversible.

□ Pyruvate undergoes keto–enol tautomerism in aqueous solution, the keto form being favoured:

$$CH_3{-}C({=}O){-}COO^- \rightleftharpoons CH_2{=}C(OH){-}COO^-$$

When the enol form is esterified by phosphate, the pyruvate is 'fixed' in this normally unfavoured form. Removal of the phosphate allows the normal predominance of the keto form to be re-established. This largely accounts for the high standard free energy of hydrolysis (40–45 kJ/mol) of phosphoenolpyruvate.

Table 5.1 *The glycolytic enzymes*

Step	Enzyme	$M_r \times 10^{-3}$	Subunits	Cofactors
1	Hexokinase	97	2	Mg^{2+}, ATP
2	Phosphohexose isomerase	62	1	
3	Phosphofructokinase	340	4	Mg^{2+}, ATP
4	Aldolase	150	3	
5	Triosephosphate isomerase	62	1	
6	Glyceraldehyde 3-phosphate dehydrogenase	144	4	P_i, Mg^{2+}, NAD^+
7	Phosphoglycerate kinase	45	1	ADP, Mg^{2+}
8	Phosphoglyceromutase	57	2	
9	Enolase	85	2	
10	Pyruvate kinase	220	4	Mg^{2+}, K^+, ADP

Reference *The Enzymes of Glycolysis: Structure, Activity and Evolution* (a compilation of papers presented at a Royal Society Symposium) *Philosophical Transactions of the Royal Society B*, **293** (1981). This contains esoteric enzymological and protein-chemical detail. A glance through it is useful because it illustrates how research in this area has advanced; it also gives some very interesting insights into the evolution of enzymes and of metabolic pathways.

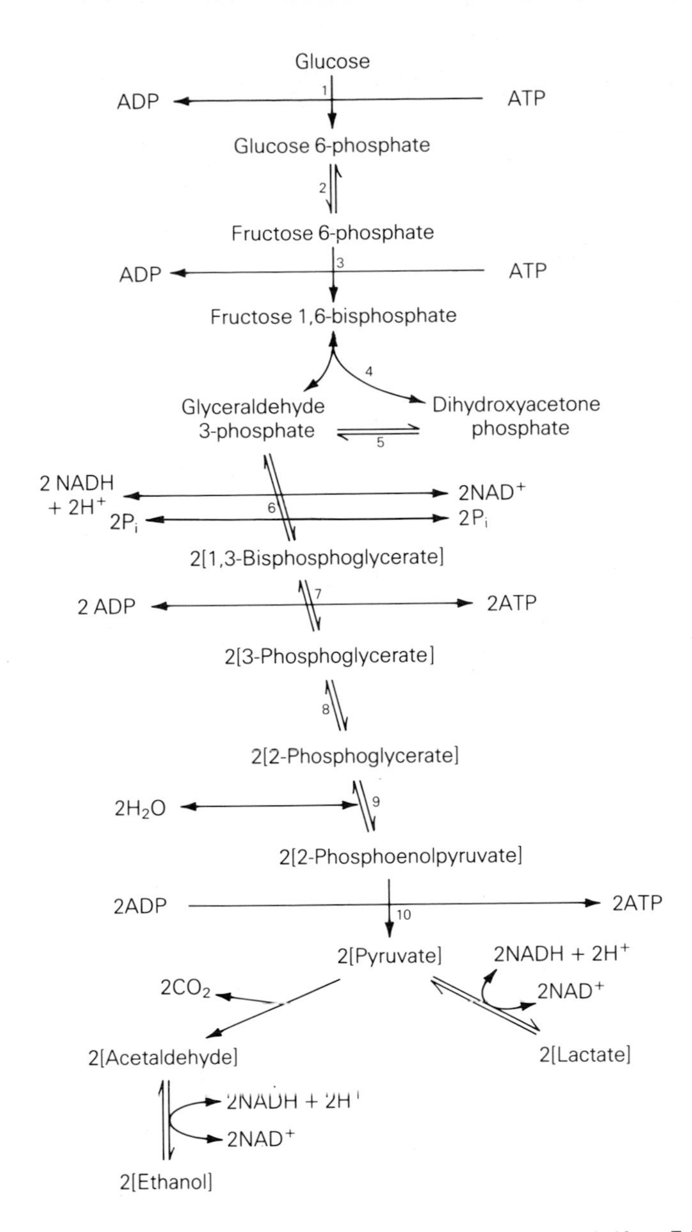

Fig. 5.7 The Embden–Meyerhof pathway (glycolysis) (for names of enzymes at steps 1–10 see Table 5.1).

THE METHODOLOGICAL APPROACHES available at the time explain why it took so long to establish the details of the pathway. It is important to remember that in the 1920s and 1930s, biochemical techniques that are taken for granted nowadays were not available. Preparative ultracentrifugation had not yet been developed, and therefore subcellular fractionation was not possible. Gel electrophoretic methods and chromatographic procedures for separating polypeptides; radioactive labelling; and mass spectrometric, X-ray crystallographic or magnetic resonance methods for rapidly elucidating the structures of organic molecules were not available. Meyerhof and his colleagues had to rely on traditional organic chemical methods for the purification, analysis and synthesis of individual compounds. Any intermediates that accumulated in the pathway were purified and analysed. Sometimes, intermediates could be made to accumulate by adding enzyme inhibitors. For example, iodoacetate or sulphite (Fig. 5.9) inhibited one enzyme in the pathway, glyceraldehyde-3-phosphate dehydrogenase.

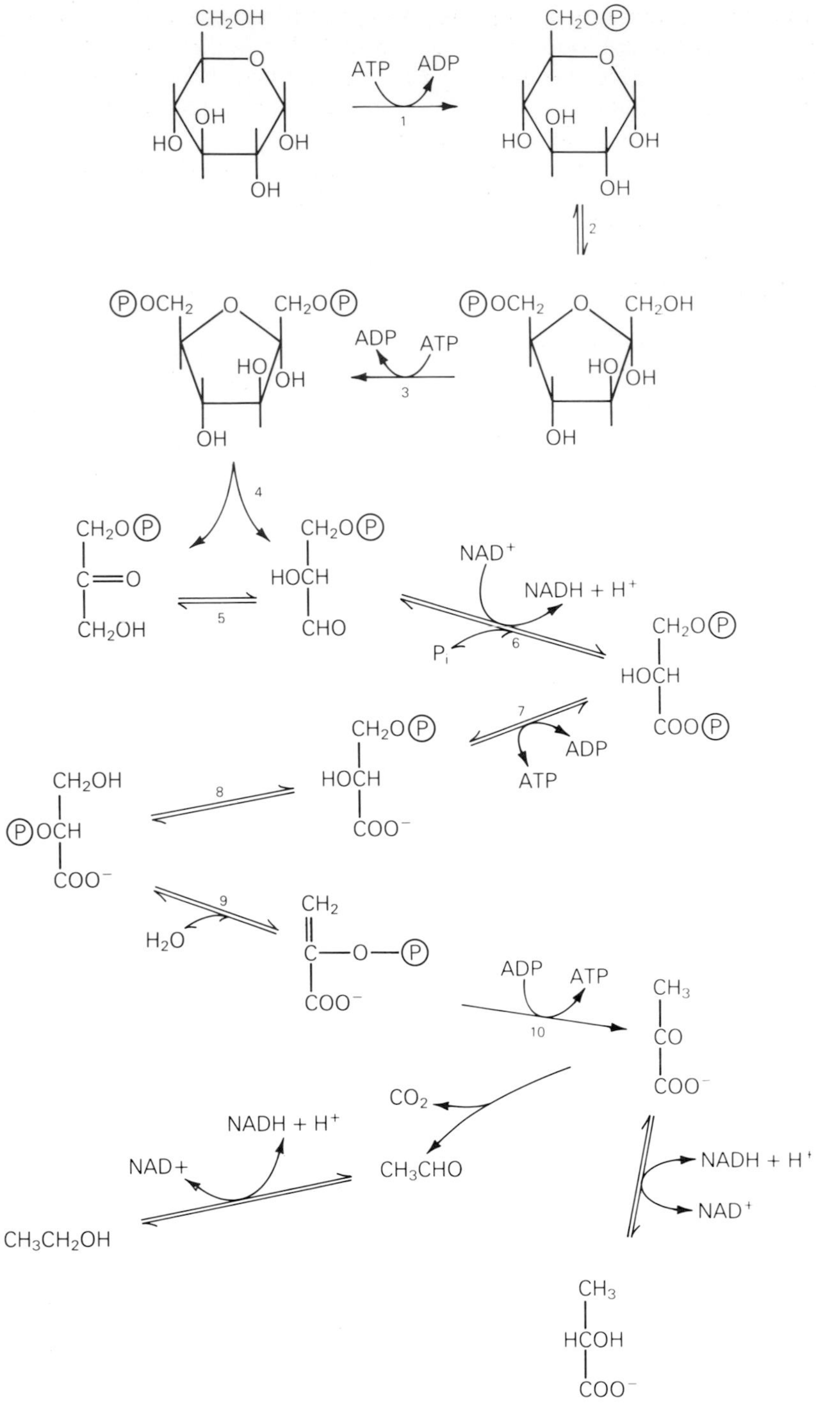

Fig. 5.8 The glycolytic pathway represented by the chemical structures involved.

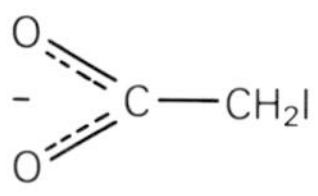

Fig. 5.9 Inhibitors of glycolysis: (a) iodoacetate, (b) sulphite, (c) fluoride.

Therefore when iodoacetate was added to skeletal muscle or yeast extract, fructose 1,6-bisphosphate, dihydroxyacetone phosphate, glycerol and a trace of glyceraldehyde 3-phosphate accumulated. Iodoacetate also inhibited alcohol dehydrogenase in yeast and in mammalian liver. Fluoride (Fig. 5.9) inhibited enolase and caused the accumulation of 2-phosphoglycerate.

When muscle extracts were dialysed, the loss of factors of low molecular weight (such as ATP, NAD^+ and inorganic phosphate) prevented several reactions from occurring, leading to the accumulation of intermediates. To make sure that the compound that accumulated really was an intermediate and not a side-product, the investigators synthesized it chemically, added it to a fresh tissue preparation, and demonstrated that the end-product of the pathway (lactate), accumulated at the normal rate.

Using these approaches, the investigators identified several intermediates; but there were gaps between them. Around 1934 a 'gap' was known to exist between glyceraldehyde 3-phosphate and 2-phosphoglycerate. The former was clearly converted to the latter, but not directly. What intervening intermediate(s) could there be? The investigators used their knowledge of organic chemistry to make reasonable guesses. In this case, the most obvious guess was 3-phosphoglycerate. Then they synthesized the postulated 'gap-filling' intermediate chemically, added it to a tissue preparation, and looked for production of the normal end-product of the pathway at the normal rate. (The formation of glycerate 1,3-bisphosphate took much longer to establish because of the complexity of the glyceraldehyde 3-phosphate dehydrogenase reaction.) Alternative possibilities were investigated similarly; and as controls, obviously incorrect guesses such as the 'wrong' stereoisomers, were synthesized and tested in parallel experiments.

5.4 *Biological significance of the glycolytic pathway*

This work was immediately significant in two ways. Firstly, it consolidated the view that the disciplines within which the initial questions had been formulated, physiology and organic chemistry, could be unified ***epistemologically*** and ***ontologically***. Secondly, it showed important commonalities at the metabolic level between organisms in widely different taxa: mammals and fungi. It became apparent over the subsequent years that the glycolytic pathway is ubiquitous in nature: almost all cells in plants, animals and most groups of prokaryotes have the enzymes of glycolysis.

Anaerobic oxidation

Under aerobic conditions, pyruvate obtained from carbohydrates such as glucose is oxidized completely to CO_2 and water, and the energy released in the process is used to form ATP from ADP and P_i via the TCA cycle and electron transport/oxidative phosphorylation. The key to this process is the repeated reductions of hydrogen-carrying cofactors such as NAD^+, with concomitant oxidation of the carbohydrate carbons to CO_2, followed by reoxidation of the reduced cofactors at the expense of oxygen (electron transport). Under anaerobic conditions, this reoxidation of reduced cofactors is not possible because oxygen is not available. Instead, the pyruvate obtained from glycolysis acts as an electron acceptor itself, either directly (producing lactate) or indirectly (producing ethanol after decarboxylation of the pyruvate to acetaldehyde). Figure 5.10 illustrates the processes involved.

See Chapters 3 and 4

□ In the overall (aerobic) reaction:

$$\text{glucose} + 6O_2 \rightarrow 6CO_2 + 6H_2O$$

the energy yield is around 2800 kJ mol^{-1}. In principle, the net ATP yield is 36 moles per mole of glucose oxidized in the anaerobic reaction:

$$\text{glucose} \rightarrow 2 \text{ lactate}$$

the energy yield is around 180 kJ mol^{-1}. The net ATP yield is 2 moles per mole of glucose oxidized. The evolution of aerobic catabolism entailed a substantial increase in the efficiency with which usable energy was obtained from glucose.

The yield of ATP

During the conversion of glucose to lactate (or ethanol), there is a net yield of two molecules of ATP. One molecule of ATP per glucose catabolized is used in each of the kinase reactions (steps 1 and 3 in Figures 5.7 and 5.8). After steps 4 and 5, two molecules of glyceraldehyde 3-phosphate are formed per molecule of glucose. The equilibrium of the reaction catalysed by triose phosphate isomerase greatly favours dihydroxyacetone phosphate formation, but the glyceraldehyde 3-phosphate is removed efficiently by

Epistemology: *the field of philosophy that investigates the nature and limits of knowledge. From the Greek* episteme, *knowledge.*
Ontology: *the science of reality. From the Greek* ontos, *being; i.e. the theoretical and the practical.*

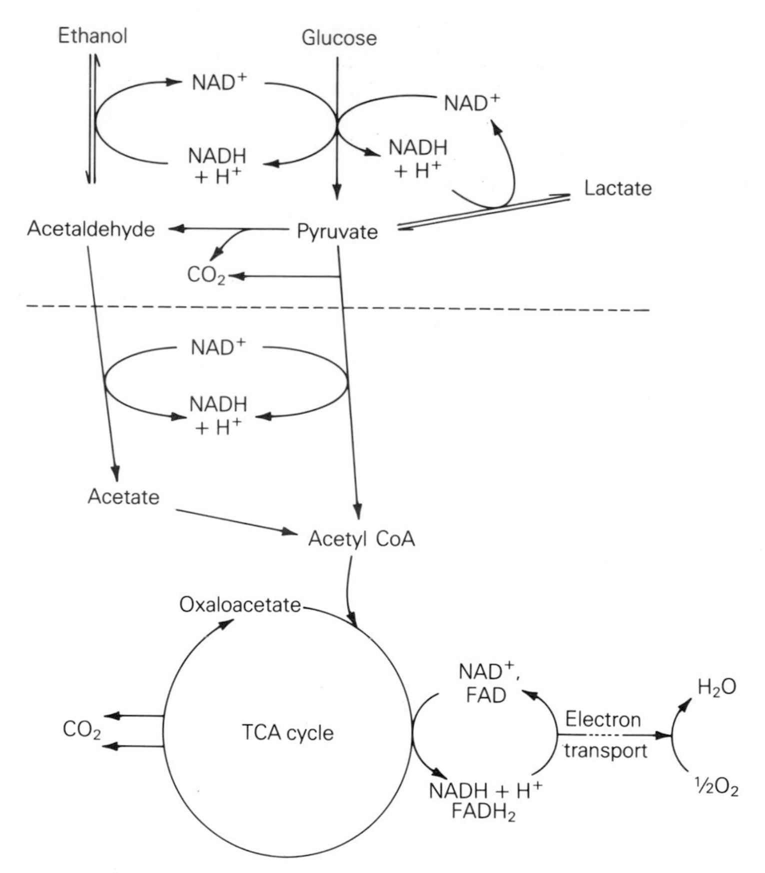

Fig. 5.10 The relationship between anaerobic (above the line) and aerobic (below the line) pathways of glucose catabolism. Under anaerobic conditions, the NAD^+/NADH cycle couples the glyceraldehyde 3-phosphate dehydrogenase reaction to the lactate dehydrogenase or alcohol dehydrogenase reactions in mammalian tissues and yeasts respectively.

Exercise 2

Demonstrate from your knowledge of glycolysis, the TCA cycle and electron transport/oxidative phosphorylation that a net 36 molecules of ATP are synthesized per glucose molecule fully aerobically oxidized.

subsequent glycolytic reactions. During the conversion of each molecule of glyceraldehyde 3-phosphate to lactate, a molecule of ATP is generated at step 7 and another at step 10. Therefore, for every two molecules of glyceraldehyde 3-phosphate converted to lactate, i.e. for each molecule of glucose catabolized, four molecules of ATP are synthesized. Hence the *net* yield is $4 - 2 = 2$ molecules of ATP.

The Pasteur effect

Pasteur observed that glucose is catabolized more rapidly by yeasts in anaerobic conditions than in aerobic conditions. Warburg and Meyerhof confirmed this observation, but were unable to offer a convincing explanation. Since about 1960, ideas about the control of metabolism at the molecular level have developed, including the concept of **allosterism**, which allows the Pasteur effect to be fully explained. Phosphofructokinase (step 3 in Figures 5.6 and 5.7) is an allosteric enzyme for which ATP is a weak allosteric inhibitor, and AMP an allosteric activator. The AMP/ATP ratio increases when cells become ***hypoxic*** or ***anoxic*** because 36 molecules of ATP can be generated from a molecule of glucose under aerobic conditions but only 2 under anaerobic conditions (see above). Therefore, when the oxygen supply is decreased, phosphofructokinase is activated allosterically. This enzyme is rate-limiting in glycolysis, so the effect is to increase the overall rate of glycolysis (see Fig. 5.11).

Hypoxia: *low O_2 intake.* ***Anoxia:*** *a deficiency of O_2.*

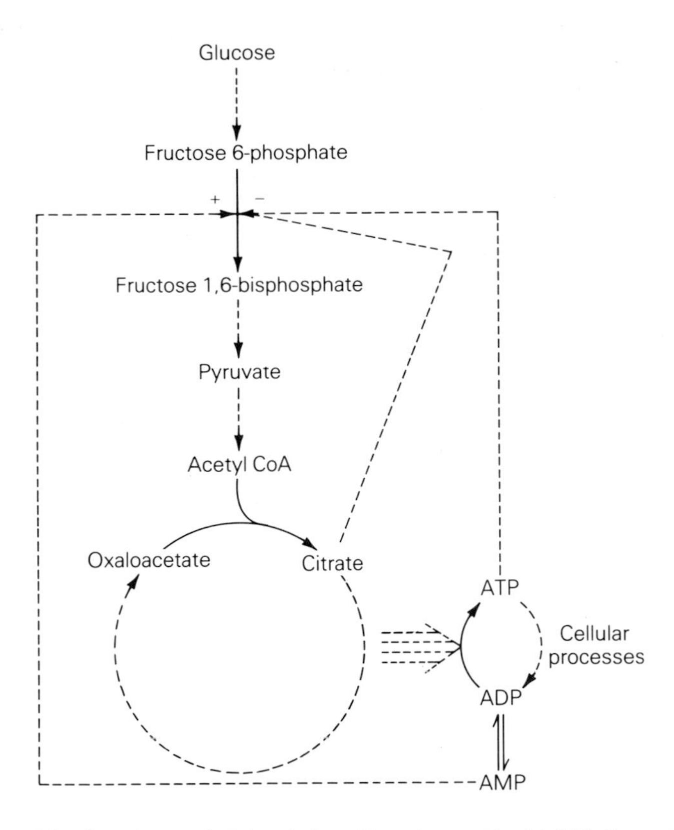

Fig. 5.11 Significance of the allosteric control of phosphofructokinase (see text for details) in the regulation of the glycolytic pathway.

There are other allosteric regulators of phosphofructokinase. Citrate enhances the inhibition by ATP. When citrate builds up in the cytoplasmic compartment, the TCA cycle will be well-fuelled, and the consequent down-regulation of glycolysis is a good example of negative feedback control. Another allosteric activator, fructose 2,6-bisphosphate, was discovered by Hers and van Schaftingen in 1980. The hormone glucagon apparently down-regulates glycolysis by inhibiting the synthesis, and promoting the breakdown, of this regulator.

> **Exercise 3**
>
> What is the structure of fructose 2,6-bisphosphate? How is it metabolized? (An attempt to work out the answers before consulting other text sources would be a good test of understanding.)

The versatility of glycolytic intermediates

The generation of pyruvate and its various products is not the only possible fate of the compounds participating in glycolysis. Glucose 6-phosphate, for example, can have a variety of fates, as illustrated in Figure 5.12. UDP-glucose, which may be produced from glucose 6-phosphate, also has a number of fates, some of which are discussed below. Fructose 6-phosphate can accept an amino group from the amide of glutamine to form glucosamine 6-phosphate. Dihydroxyacetone phosphate can be converted to glycerol 1-phosphate, forging a link between carbohydrate and lipid metabolism (Fig. 5.13). Pyruvate, itself, can be converted to amino acids, most directly to alanine, forming a direct link between carbohydrate and amino acid metabolism. There are many such interconnections in metabolism. These examples suffice to emphasize that no individual metabolic pathway can be regarded as isolated from all the other processes taking place in the same cell or in other cells of the same organism.

> **Exercise 4**
>
> When $^{14}C_3$-labelled alanine, i.e. alanine radiolabelled on the methyl carbon, is infused into the hepatic portal vein of a mammal, some radiolabel is recovered in the muscle glycogen. How does this happen? (Reference to Chapters 4 and 7 may be helpful.)

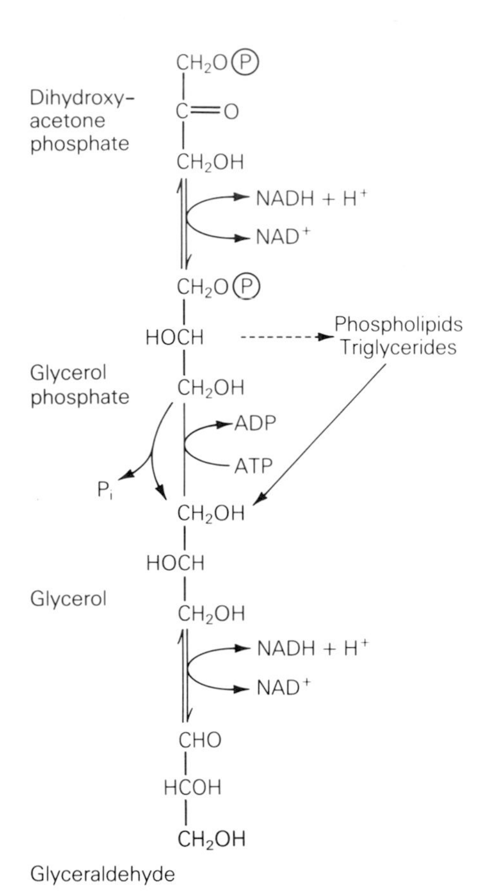

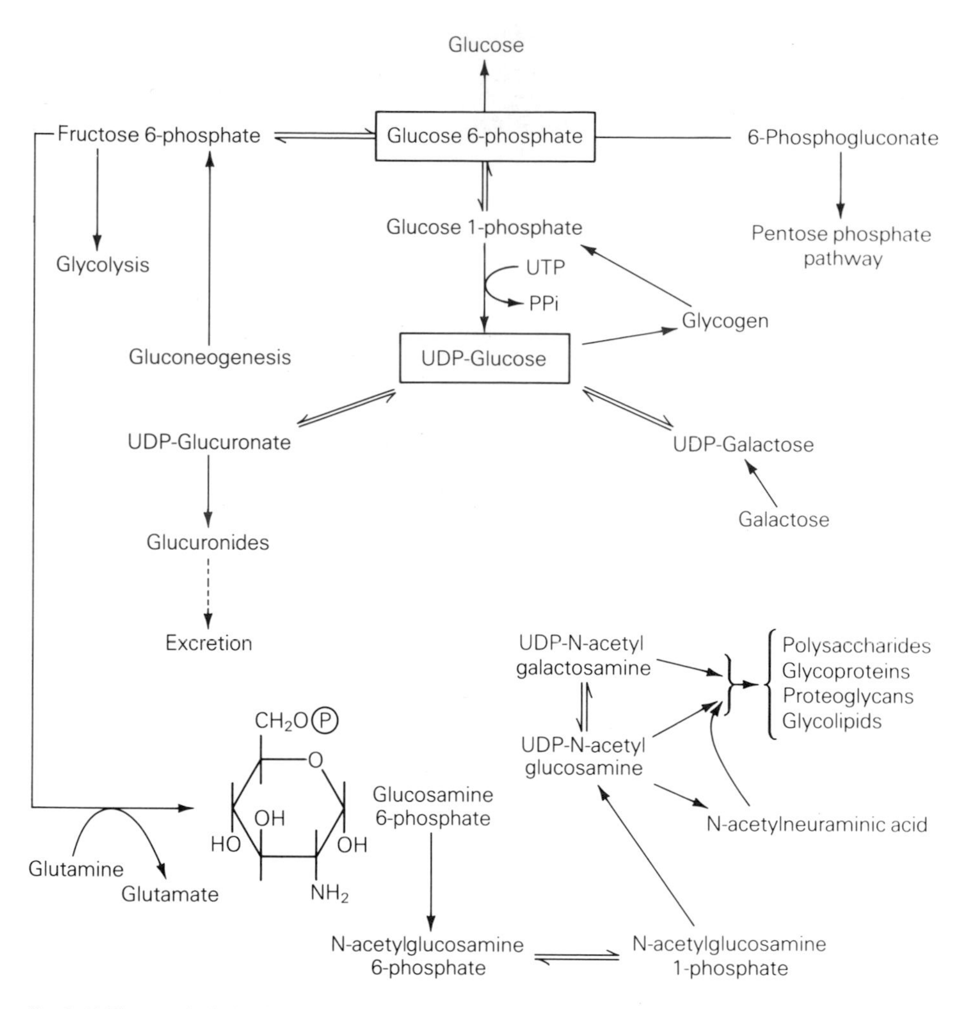

Fig. 5.12 The metabolic fates of glucose 6-phosphate and UDP-glucose. Galactose and glycogen metabolism, the pentose phosphate pathway, gluconeogenesis and the assembly of complex polysaccharides are discussed later in this chapter. Glucuronide formation is a frequently exploited method for increasing the water-solubility of waste products and xenobiotics, thus facilitating their excretion in the urine.

5.5 Glycogen and other polysaccharides

The concentrations of monosaccharides such as glucose can be as high as 10 mmol dm^{-3} in extracellular fluids such as the blood plasma in mammals, but they are much lower inside cells. Cells store carbohydrates as polysaccharides such as starch in higher plants and glycogen in animals.

Understanding of glycogen synthesis and breakdown (Fig. 5.14) began with the work of Cori and Cori. Glycogen synthesis will be discussed in *Biosynthesis*, Chapter 3. Glycogen breakdown involves a debranching enzyme and glycogen phosphorylase. Different forms of glycogen phosphorylase exist in the major mammalian glycogen storing tissues, liver and skeletal muscle. Adrenalin (muscle and liver) and glucagon (mainly liver) are among the hormones that indirectly activate this enzyme. The process (Fig. 5.15) involves several steps and is a 'cascade'; it amplifies a small signal to produce a large response. In the case of skeletal muscle, the biological value of a cascade is easy to see. For example, a grazing antelope attacked by a predator needs rapidly to increase the rate of ATP production in its leg muscles if it is to escape and survive. The stimulus received by the muscles to elicit this change is a tiny amount of adrenalin.

Exercise 5

Why do cells store polysaccharides rather than monosaccharides?

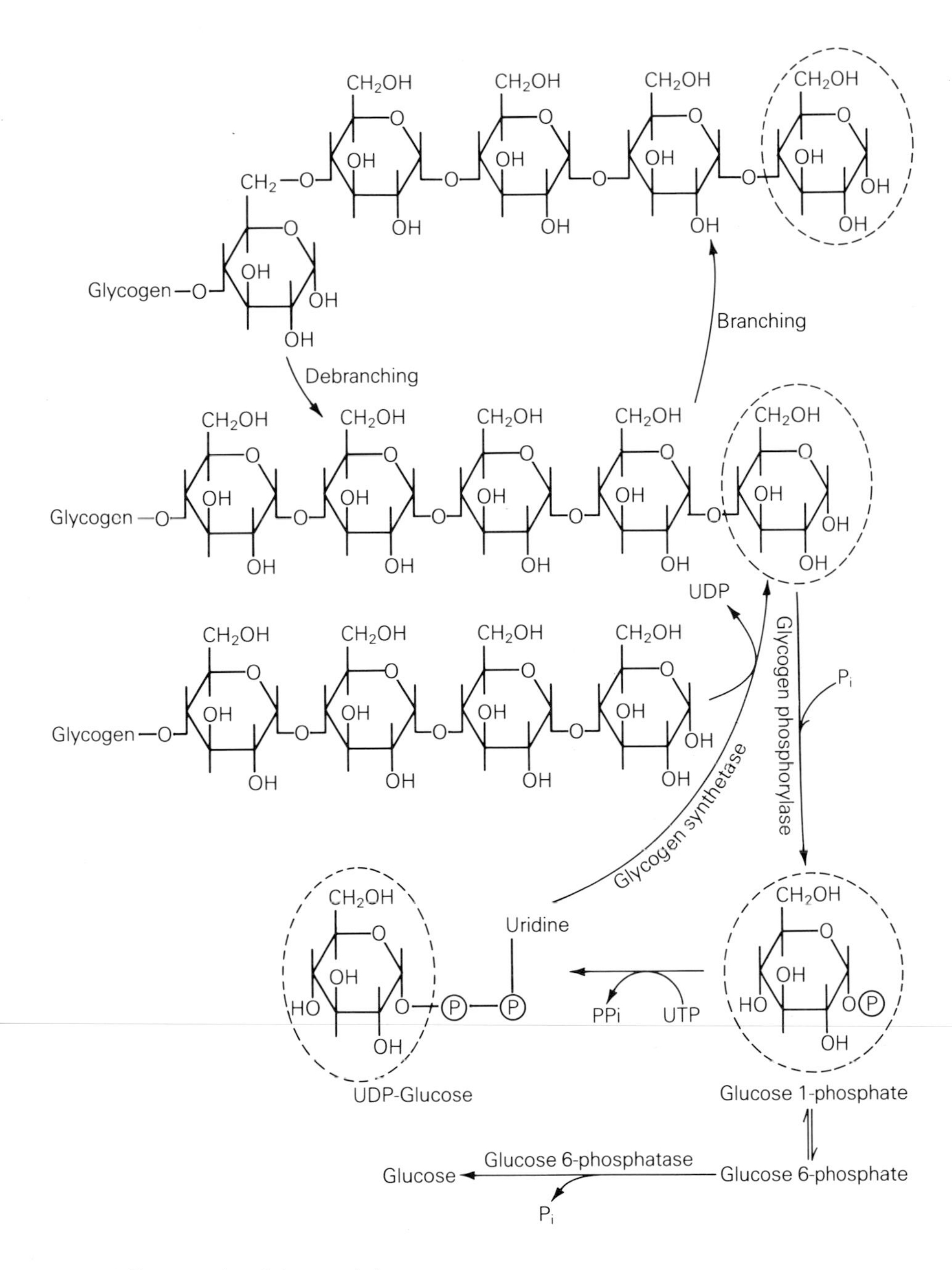

Fig. 5.14 Glycogenesis and glycogenolysis.

Box 5.2 Glycogen storage diseases

Any gene defect that results in the absence or impaired function of an enzyme in central metabolism (glycolysis or the TCA cycle, for example) is almost certain to result in a non-viable cell, at least in homozygotes. Therefore, patients with such deficiencies are almost never seen: they could not have survived the zygote stage of development. Gene defects in less central parts of metabolism, however, although they can have very damaging effects, can permit survival beyond the fetal stage. Among these are the glycogen storage diseases, in which one or other of the enzymes involved in glycogenesis or glycogenolysis is absent. Details of these diseases are given in the table. Some, such as Pompe's disease, lead to severe incapacity and death in early childhood. Others, such as McArdle's disease, cause serious muscle weakness and discomfort but no other major symptoms, and do not necessarily lead to an unduly short life expectancy.

Name	Defective enzyme	Effects
von Gierke's disease	Glucose-6-phosphatase	Fasting hypoglycaemia; lactate acidaemia; hyperlipidaemia; uricaemia and gout
Pompe's disease	Lysosomal glycogen hydrolase	Massive glycogen accumulation but no serious hypoglycaemia; neuron damage, muscle hypotonia, cardiomegaly, cardiac failure. Early death
Cori's disease	Debranching enzyme	Hepatomegaly. Similar to von Gierke's but milder
Andersen's disease (amylopectinosis)	Branching enzyme	Abnormal (sparsely branched) glycogen in liver and other tissues; liver cirrhosis; death in early infancy
McArdle's disease	Muscle glycogen phosphorylase	Painful muscle cramps and weakness. Blood contains myoglobin and muscle enzymes
Hers's disease	Liver glycogen phosphorylase	Hepatomegaly; slight hypoglycaemia; fairly benign
	Glycogen synthetase	Almost no liver glycogen; severe fasting hypoglycaemia

Schwartz, V. (1984) *A Clinical Companion to Biochemical Studies*, 2nd edn, Freeman, New York. Approaches metabolic diseases (including two of relevance to this chapter), from a medical case-history perspective.

Hexokinase and glucokinase

Hexokinase (Fig. 5.16) has a low Michaelis constant for glucose, but is inhibited by its own product, glucose 6-phosphate. When glycolysis is down-regulated by allosteric inhibition of phosphofructokinase, fructose 6-phosphate and glucose 6-phosphate accumulate in the cell causing hexokinase to be inhibited. This situation arises when glucose is plentiful in the diet or in the blood; it is physiologically desirable to replenish glycogen stores rather than to catabolize the abundance of nutrient. However, glucose can only be converted to glycogen if glucose 6-phosphate is formed. This problem is solved by cells having another enzyme, glucokinase, which has a high Michaelis constant for glucose and therefore becomes active only at higher glucose concentrations. It catalyses exactly the same reaction as hexokinase, but is less susceptible to inhibition by the product.

The Cori cycle

Gluconeogenesis is the formation of glucose and glycogen from non-carbohydrate precursors, including TCA cycle intermediates (Fig. 5.17). It is mentioned here because it is necessary for understanding the inter-conversions between muscle and liver glycogen.

Reference Stanbury, J.B., Wyngaarden, J.B., Frederickson, D.S., Goldstein, J.L. and Brown, M.S. (1983) *The Metabolic Basis of Inherited Diseases* 5th edn, McGraw-Hill, New York. A massive compilation, for reference only. Chapters 5, 7 and 73 are relevant to inborn errors of carbohydrate metabolism.

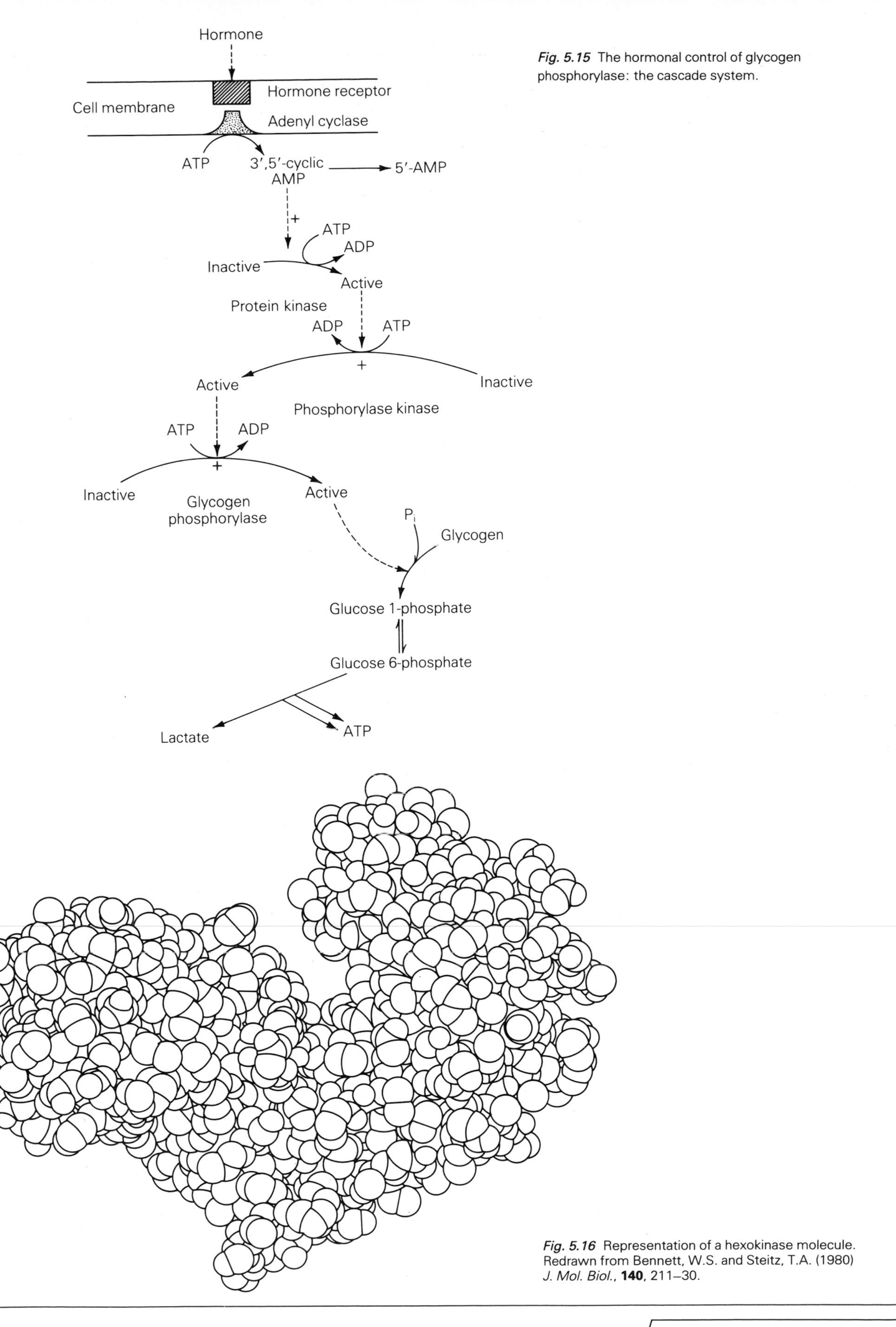

Fig. 5.15 The hormonal control of glycogen phosphorylase: the cascade system.

Fig. 5.16 Representation of a hexokinase molecule. Redrawn from Bennett, W.S. and Steitz, T.A. (1980) *J. Mol. Biol.*, **140**, 211–30.

☐ The equilibrium of the phosphoglucomutase-catalysed reaction (see Fig. 5.12) lies in favour of glucose 6-phosphate. This is because glucose 6-phosphate is a genuine phosphate ester (the C-6 of glucose bears a primary alcohol group), but glucose 1-phosphate is not (the C-1 is an aldehyde group, predominantly in the hemiacetal form) and has a substantially greater free energy of hydrolysis. The intracellular concentration of glucose 1-phosphate is very low.

☐ Most amino acids are catabolized ultimately to TCA cycle intermediates and therefore protein and amino acid catabolism lead to a net synthesis of oxaloacetate. This results in a net synthesis of glucose/glycogen through the pathway of gluconeogenesis. Because of this, hormones that affect protein catabolism (usually by accelerating transamination, the rate-limiting step in catabolism of most amino acids), affect blood glucose and liver glycogen levels. The primary physiological effect of glucocorticoids, which elevate the blood glucose level at the expense of protein, can be explained in this way.

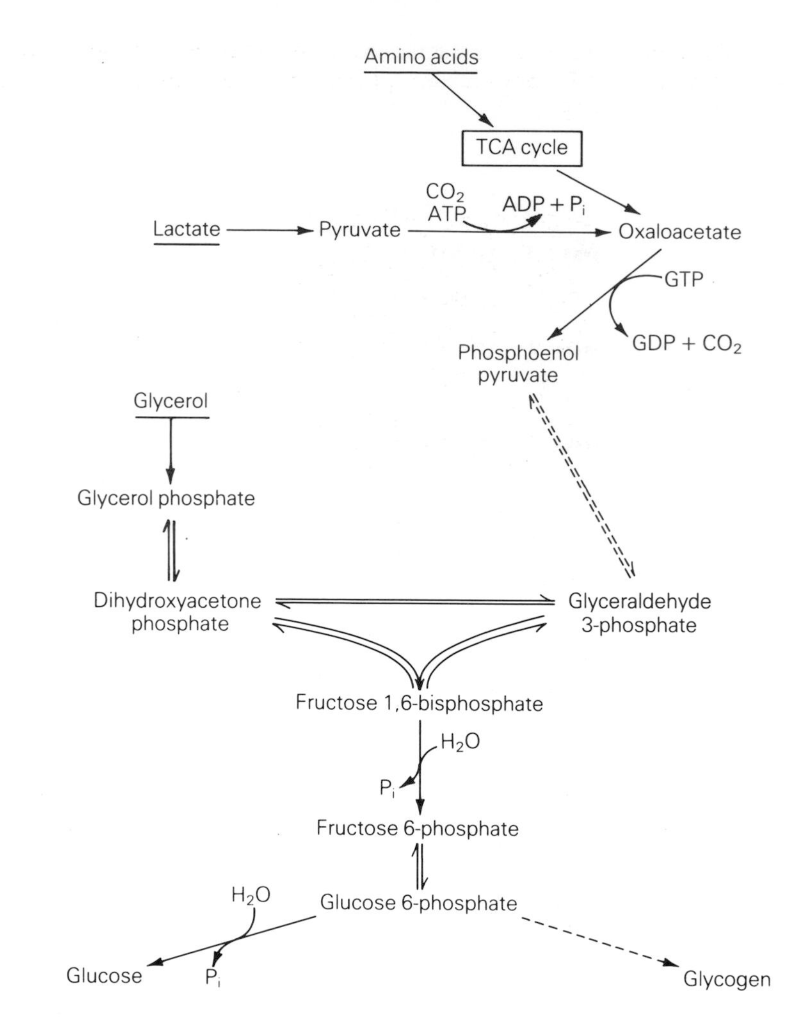

Fig. 5.17 Gluconeogenesis.

☐ Hill and Meyerhof demonstrated that in frog sciatic and gastrocnemius muscles, about 80% of the lactate formed during anaerobic contraction is reconverted to glycogen during the recovery phase and the remaining 20% is oxidized to CO_2. The Coris showed that in mammalian muscle most of the lactate diffuses out into the blood stream. The Cori cycle therefore applies much more to mammals (and birds) than to ectothermic vertebrates. This point emphasizes the dangers of extrapolating uncritically from a knowledge of the biochemistry and physiology of one species to the situation in another, notwithstanding the importance of apparent 'biological universals' such as glycolysis.

Glucose 6-phosphatase is present in mammalian liver and kidney, but is absent from skeletal muscle. Consequently, although liver glycogen can act as a store of glucose that is available to the whole body, muscle glycogen is for consumption within the muscle alone, since it cannot be converted to free glucose for blood transport. During anaerobic muscle contraction, lactate is generated. Some of this is reconverted to pyruvate when the ***oxygen debt*** is paid off, and of this pyruvate, some is completely oxidized to CO_2, but the majority is reconverted to glycogen (not glucose). The rest of the lactate enters the bloodstream and, of this, most is taken up by the liver and is reconverted there to glycogen. Muscle glycogen stores can also be replenished from blood glucose at the expense of mobilized liver glycogen stores. This set of reactions constitutes the Cori cycle (Fig. 5.18).

Creatine phosphate and muscle contraction

Rapidly contracting skeletal muscle requires an immediate and large supply of ATP. The cascade control of glycogen phosphorylase is important here, but the further catabolism of glucose 6-phosphate involves utilization rather than generation of ATP in order to produce fructose 1,6-bisphosphate. In order to cope with this there is another mechanism for immediate rapid ATP

Oxygen debt: *the physiological condition produced during temporary anoxia. Intermediary metabolism is switched to an anaerobic mode, producing compounds that can be stored until sufficient O_2 becomes available to complete oxidative processes.*

production. Mammalian skeletal muscles contain **creatine kinase**, which catalyses the reversible ATP-dependent phosphorylation of the guanido compound **creatine** (Fig. 5.19).

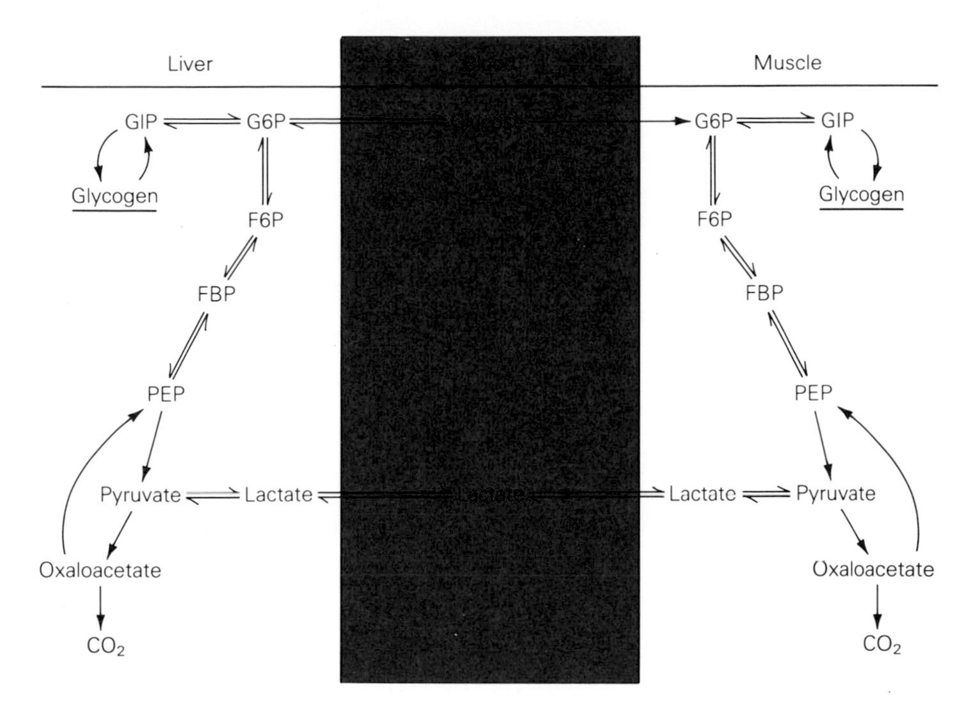

Fig. 5.18 The Cori cycle. Key: G6P, glucose 6-phosphate; G1P, glucose 1-phosphate; F6P, fructose 6-phosphate; FBP, fructose 1, 6-bisphosphate; PEP, phosphoenolpyruvate.

Exercise 6

The accompanying figure shows two glucose tolerance curves, one obtained from a normal human subject, one obtained from a patient with diabetes mellitus. These curves are obtained by giving the overnight-fasted patient, who is resting, 50 g of glucose orally, then taking blood samples every 30 min to measure the glucose concentration in the plasma. Can you explain: (a) why the normal curve has the shape it has, and (b) why the diabetic curve is different?

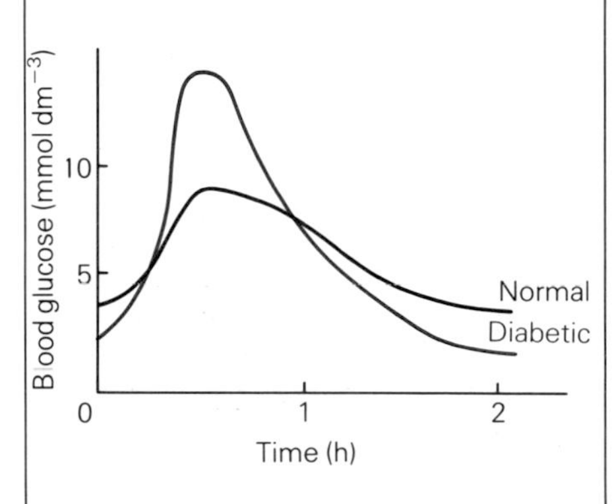

Glucose tolerance curves. The fasting subjects/patients received 50 g of glucose orally at time zero.

ATP concentrations in skeletal muscle are low, partly because of its wide-ranging metabolic fates and allosteric effects, and partly because it is a powerful chelator of the Mg^{2+} and Ca^{2+} necessary for the structural and functional integrity of the cell. Also, intracellular total concentrations of ATP and ADP do not usually exceed 1–2 mmol dm^{-3}. However, these considerations do not apply to creatine phosphate, which may be stored at high concentration. Creatine kinase reconverts ADP to ATP almost as soon as the ADP becomes available using the phosphate from creatine phosphate. This reaction was discovered by Lohmann in the early 1930s, and almost immediately explained a previously mysterious finding, namely why glycogenolysis and glycolysis continues rapidly for some 30–60 seconds after muscle contraction has terminated. The ATP that is being generated during this postcontraction period is being used to replenish the depleted stock of creatine phosphate.

☐ Creatine phosphate is the 'phosphagen', i.e. the means for the rapid rephosphorylation of ADP, in the muscles of most vertebrates. Throughout the animal kingdom, muscle tissues that are capable of rapid contraction have phosphagens, and in all cases they seem to be guanidophosphates, with the general structure:

R
N–Ⓟ
HN=C
NR′R″

Many invertebrates use arginine phosphate for this purpose; annelids utilize an alternative compound, lombricine phosphate, and there are several other examples.

5.6 *Other monosaccharides*

Monosaccharides other than glucose are also widespread. For example, mannose and sialic acids are found in glycoproteins and glycolipids. Mannose is also found in appreciable quantities in components of the cell walls of yeasts. In mammals, the main dietary hexoses are glucose, fructose and galactose. Figure 5.20 shows the structures of the major hexose sugars and chief pathways for metabolizing these sugars are shown in Figures 5.21 and 5.22.

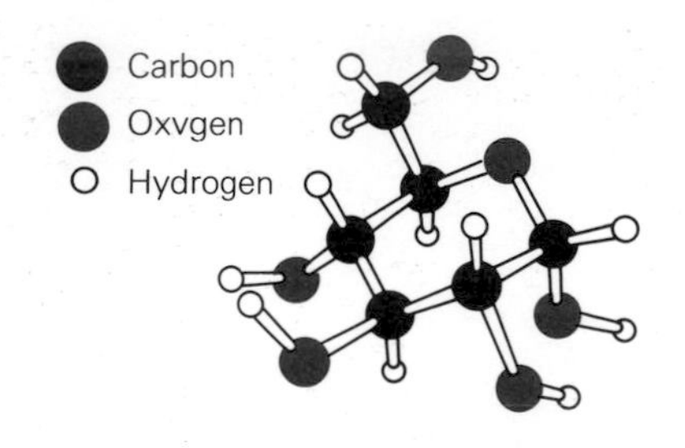

(a)

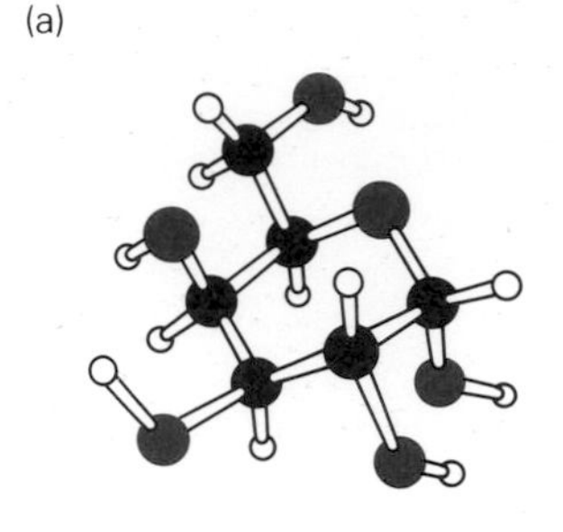

(b)

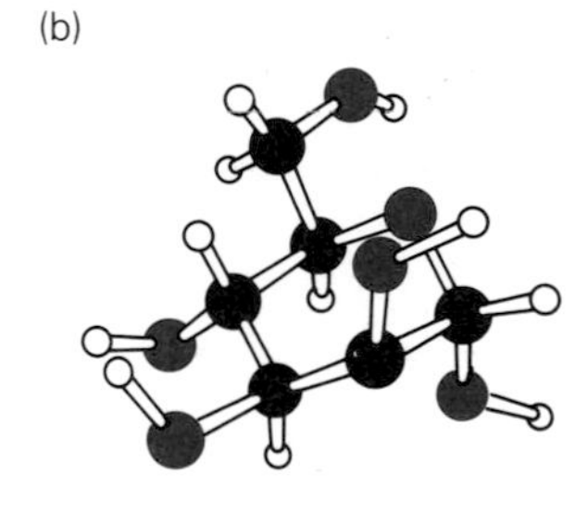

(c)

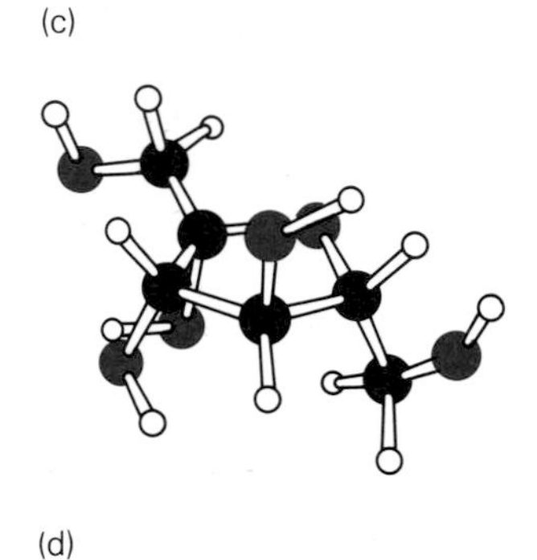

(d)

Fig. 5.20 Structures of common hexose sugars: (a) α-D-glucose; (b) α-D-galactose; (c) α-D-mannose; (d) β-D-fructose.

$$HN{=}C(NH_2){-}N(CH_3){-}CH_2COO^- + ATP \rightleftharpoons HN{=}C(NH\text{Ⓟ}){-}N(CH_3){-}CH_2COO^- + ADP$$

Fig. 5.19 Phosphorylation of creatine.

GALACTOSE is converted to its 1-phosphate derivative by the activity of galactokinase, and the product is converted to UDP-galactose. An enzyme, which contains an NAD^+-prosthetic group, catalyses the transformation to UDP-glucose which is incorporated into glycogen (Fig. 5.21).

FRUCTOSE is mainly converted to its 1-phosphate by fructokinase, which is converted by fructose 1-phosphate aldolase, to dihydroxyacetone phosphate, a glycolytic intermediate (Figs 5.7 and 5.8) and glyceraldehyde. This can be reduced to glycerol, which can then be phosphorylated to glycerol 1-phosphate a starting point for the synthesis of glycerol-containing lipids. Fructose can also be converted to the 6-phosphate derivative by hexokinase, but at a much slower rate than glucose (Fig. 5.22). However, in some tissues such as adipose tissue (Fig. 5.23), where the glucose supply is low, the phosphorylation of fructose by hexokinase is significant.

RIBOSE (Fig. 5.24) is the most abundant pentose in the diet. It can be phosphorylated in the 5-position, and its subequent metabolism is described later.

5.7 *The pentose phosphate pathway*

Warburg and Christian identified and chemically characterized 'coenzyme II' which is now known as $NADP^+$. Although it is very similar in structure to NADH, this coenzyme is not directly oxidized by the mitochondrial electron transport system. Subsequent research, extending into the 1960s, revealed that NADPH supplies the reducing equivalents needed for a range of anabolic processes, especially those involved in generating fatty acids and steroids. When the chemistry of nucleic acids became known, the question arose of what is the origin of the pentoses that form the backbone of nucleic acid structure, since the dietary supply of pentose sugars is insufficient? The work

Galactose —(ATP → ADP; Galactokinase)→ Galactose 1-phosphate —(Uridyl transferase; UTP → PP_i)→ UDP Galactose ⇌(4′-Epimerase) UDP Glucose

(CH_2OH, HO, O, OH, OⓅ; O—Ⓟ—Ⓟ—Uridine)

Fig. 5.21 Galactose metabolism.

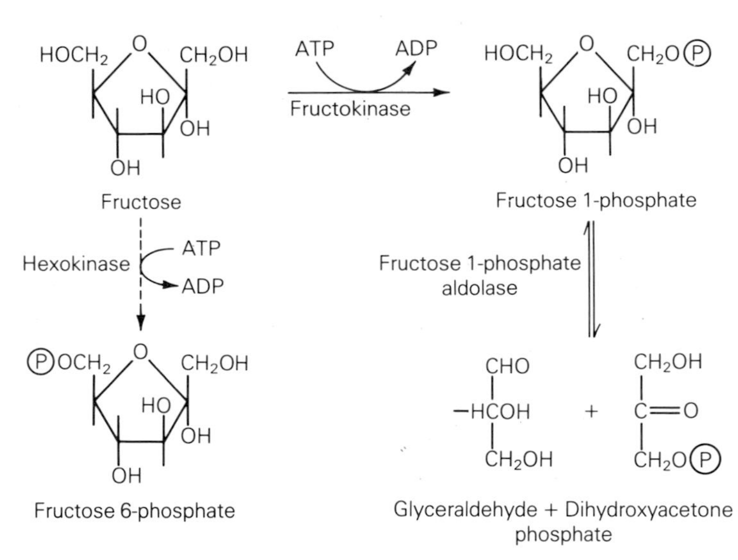

Fig. 5.22 Fructose metabolism.

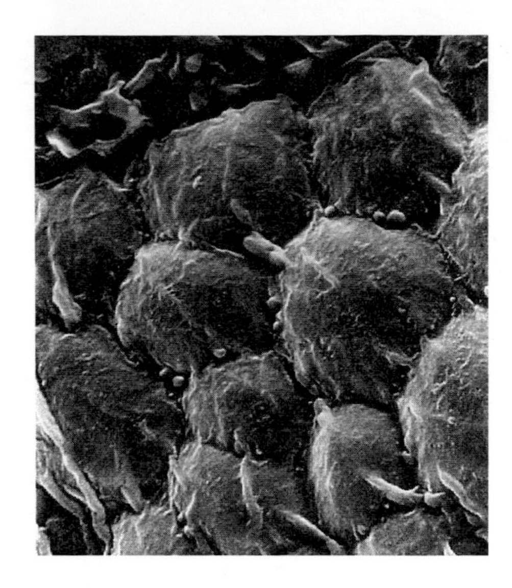

Fig. 5.23 Electron micrograph of cells of adipose tissue from rat epididymis. Courtesy of Dr N.O. Nilsson, Department of Physiological Chemistry, University of Lund.

of pioneers, such as Dickens, revealed a novel set of enzyme-catalysed reactions usually grouped into a 'pathway', variously known as the **pentose phosphate pathway**, the **hexose monophosphate shunt** or the **phosphogluconate pathway**, and which could generate pentose phosphates.

The pentose phosphate pathway is the major pathway for generating the reducing power, in the form of NADPH, necessary for biosynthetic reactions and also provides the major route for the formation and utilization of the pentose sugar, ribose. Several of the reactions of the pentose phosphate pathway are common to those of the dark reactions of photosynthesis where hexoses are generated from CO_2.

See *Biosynthesis*

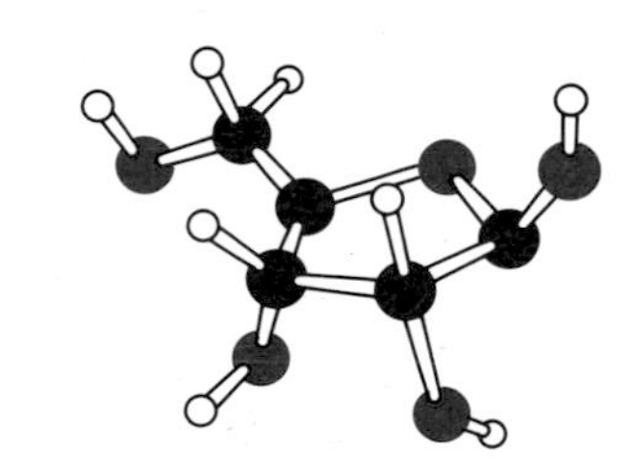

Fig. 5.24 β D Ribose.

The reactions of the pentose phosphate pathway are complicated and may be summarized as follows:

1. Glucose 6-phosphate is oxidized by glucose 6-phosphate dehydrogenase to 6-phosphogluconolactone. A lactonase hydrolyses this to 6-phosphogluconate (Fig. 5.12), which in turn is oxidized to ribulose 5-phosphate

Box 5.3 Defects in hexose metabolism

Although no fetus with a metabolic defect in glycolysis could survive, this is not true of individuals with defects in galactose or fructose metabolism. Absence of galactokinase results in **galactosuria**, and absence of the uridyl transferase that generates UDP-galactose results in **galactosaemia**. These names are rather arbitrary: both result in detectable levels of galactose in both blood and urine. Both conditions are managed by galactose-free diets. Galactosaemia involves the formation of a toxic hexitol, galactitol, from the unmetabolized galactose 1-phosphate in cells. The main effects are kidney, liver and brain damage. The symptoms in infants with the condition include vomiting, diarrhoea, failure to gain weight, jaundice, cataract and severe mental retardation. Galactosuria is rather less serious, but cataracts are still among the symptoms. There are corresponding defects in fructose metabolism: **fructosaemia** or **fructose intolerance** (absence of the fructose 1-phosphate aldolase) and **fructosuria** (absence of fructokinase). The latter is asymptomatic; the only sign is that fructose is detectable in the urine (and fructose will not alleviate hangovers, see Box 5.4). The former is associated with ***hepatomegaly***, ***hypophosphataemia***, ***hypoglycaemia***, and vomiting. Removal of fructose (or fructose-containing nutrients such as sucrose) from the diet abolishes the symptoms.

Hepatomegaly: *enlarged liver. From the Greek* hepar, *liver, and* megas, *large.*
Hypophosphataemia: *low serum concentration of phosphate. From the Greek* hypo, *under.*

Hypoglycaemia: *low serum concentration of glucose. From the Greek* hypo, *under.*

Box 5.4
Hangovers and honey

Since the enzyme alcohol dehydrogenase is plentiful in mammalian liver, large amounts of dietary alcohol greatly elevate the $NADH/NAD^+$ ratio in the liver, impairing glycolysis, which requires free NAD^+. This is among the metabolic disturbances characteristic of the pathological condition commonly known as hangover. Because a major product of fructose metabolism is glyceraldehyde, which is reduced by NADH to glycerol (Fig. 5.21), dietary fructose alleviates at least this aspect of the condition. Therefore, a biochemically sound way to treat a hangover is to eat copious quantities of honey, which contains about 50% fructose.

and CO_2 by 6-phosphogluconate dehydrogenase. Both of these reactions entail the reduction of $NADP^+$ (Fig. 5.25) and indeed these reactions are the main metabolic sources of NADPH.

Ribulose 5-phosphate is converted to xylulose 5-phosphate, or is converted to ribose 5-phosphate. This latter is the main metabolic source of the ribose used in nucleotide and nucleic acid biosynthesis. 2-Deoxyribose 5-phosphate for DNA biosynthesis is made from ribose 5-phosphate. These reactions are summarized in Figure 5.26.

The rest of the 'pathway' converts the xylulose 5-phosphate into intermediates of glycolysis. No further major energy changes and no oxidation–reduction reactions are involved. The intermediates in the conversion process include the C_7 ketose seduheptulose 7-phosphate and the C_4 aldose, erythrose 4-phosphate.

2. The fragment of xylulose 5-phosphate comprising carbons 1 and 2 is transferred to C-1 of ribose 5-phosphate by transketolase with production of glyceraldehyde 3-phosphate and seduheptulose 7-phosphate (Fig. 5.27). In a rather similar reaction, transaldolase transfers carbons 1–3 of seduheptulose 7-phosphate to C-1 of glyceraldehyde 3-phosphate, generating fructose 6-phosphate and erythrose 4-phosphate (Fig. 5.28).
3. In a second transketolase reaction, the carbons 1 and 2 of another molecule of xylulose 5-phosphate are transferred to the C-1 of the erythrose

Box 5.5
Glucose-6-phosphate dehydrogenase deficiency

In humans and many other mammals, the gene for glucose 6-phosphate dehydrogenase is carried on the X chromosome. Genetic deficiency in this enzyme is therefore a sex-linked recessive condition. It is usually asymptomatic, because newly formed erythrocytes at least have sufficient of the enzyme to support the reductive machinery despite the defect. However, men with this condition develop massive haemolytic anaemias when the reductive machinery of the erythrocyte is damaged further, such as by antimalarial drugs. The malarial parasite requires the reductants in the host cell to survive, and the most commonly used antimalarial drugs act by decreasing the ratio of reduced to oxidized glutathione. In normal individuals this does not seriously affect the viabilities of erythrocytes, but in men with the enzyme defect all but the youngest erythrocytes become damaged and many of them lyse.

Stanbury, J.B., Wyngaarden, J.B., Frederickson, D.S., Goldstein, J.L. and Brown, M.S. (1983) *The Metabolic Basis of Inherited Diseases*, 5th edn, McGraw-Hill, New York. A massive compilation, for reference only. Chapters 5, 7 and 73 are relevant to inborn errors of carbohydrate metabolism.

Benson, P.F. and Fensom, A.H. (1986) *Genetic Biochemical Disorders*, Oxford University Press, Oxford. A highly informative compilation and includes references to the recently elucidated mucopolysaccharidoses. The discussion of the biochemical relationship between aetiology and symptoms is excellent.

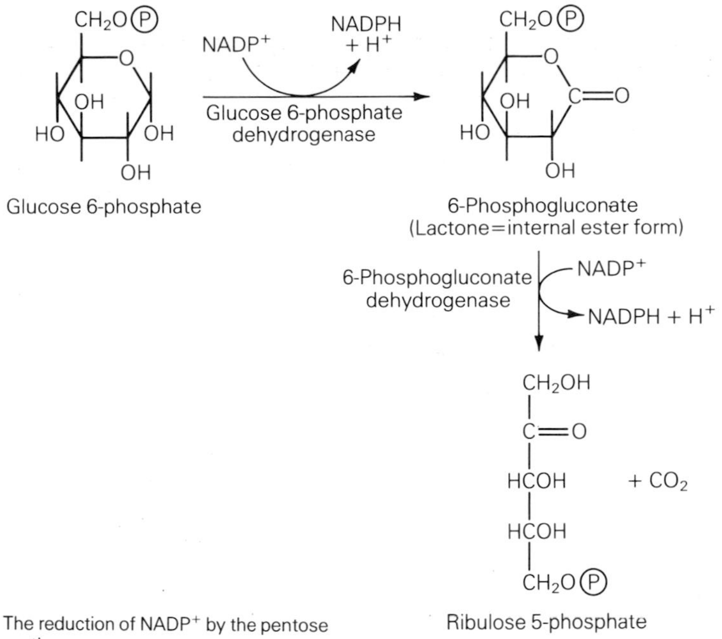

Fig. 5.25 The reduction of $NADP^+$ by the pentose phosphate pathway.

4-phosphate, generating another molecule of fructose 6-phosphate and glyceraldehyde 3-phosphate (Fig. 5.29).

When these reactions are combined with those catalysed by phosphohexose isomerase, triose phosphate isomerase and aldolase (Fig. 5.8 and Table 5.1), and with the fructose 1,6-bisphosphatase involved in the pathway of gluconeogenesis (Fig. 5.10), the overall process can be presented as a cycle of reactions (Fig. 5.30) of overall equation.

$$6 \text{ glucose 6-phosphate} + 12\ NADP^+ \rightarrow 5 \text{ fructose 6-phosphate} + 6\ CO_2 + 12\ NADPH + 12\ H^+$$

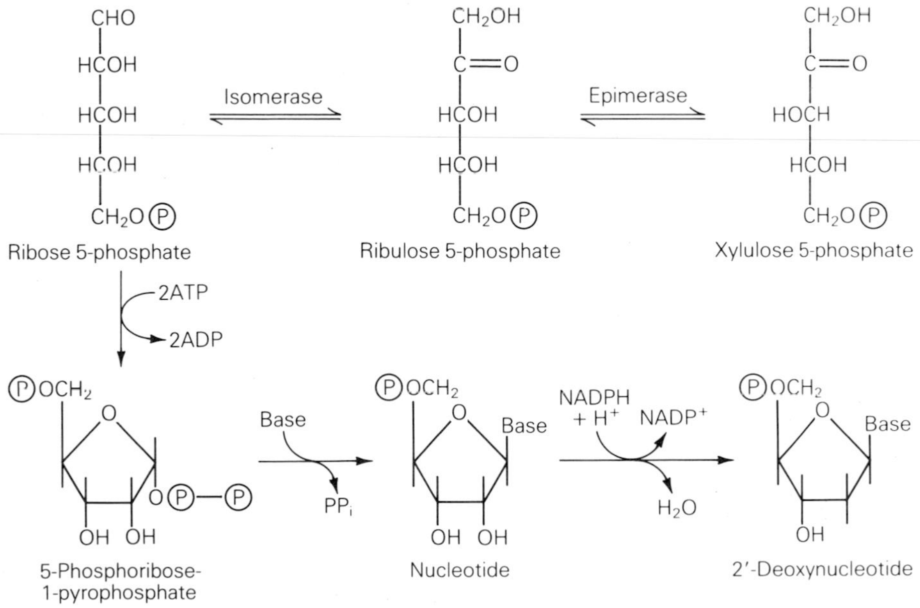

Fig. 5.26 Interconversions of pentose phosphates; formation of nucleotides and 2′-deoxynucleotides.

Exercise 7

In different tissues the enzymes of the pentose phosphate pathway have different activities relative to the enzymes of glycolysis. The table illustrates this point. What is the biological significance of these differences?

Percentage glucose metabolized via glycolysis and the pentose phosphate pathway (PPP)

Tissue	Glycolysis	PPP
Skeletal muscle	100	0
Liver	50	50
Adipose	30	70
Thyroid, gonads, adrenal cortex	20	80

Reference Devlin, T.R. (1986) *Textbook of Biochemistry with Clinical Correlations*, 2nd edn, Wiley, New York. One of the best textbook accounts of carbohydrate metabolism (Ch. 7, pp. 261–328). Of all the standard textbooks, this is perhaps the one that is most useful to medical students.

Reference Stryer, L. (1988) *Biochemistry*, 3rd edn, Freeman, New York. Probably the best general biochemistry textbook available. Carbohydrate metabolism is treated from a predominantly enzymological standpoint, with the aid of beautiful molecular models Ch. 15, pp. 349–72.

The pentose phosphate pathway shows a striking similarity with the dark reaction pathway of photosynthesis. This similarity implies an evolutionary connection between the two processes.

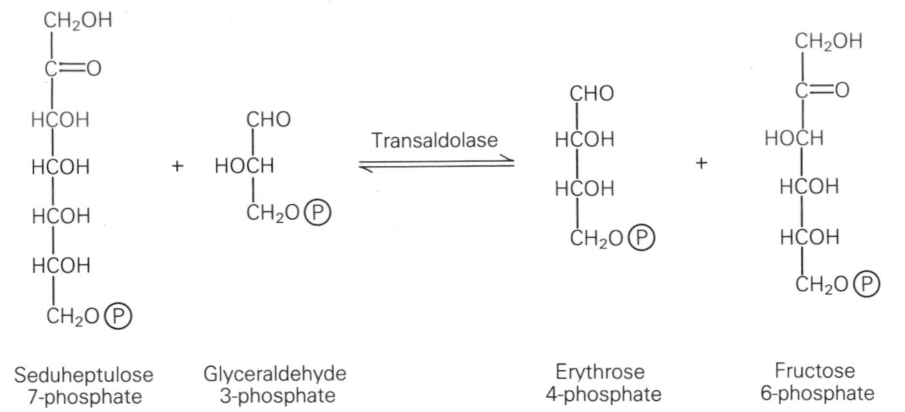

Seduheptulose 7-phosphate + Glyceraldehyde 3-phosphate ⇌ (Transaldolase) Erythrose 4-phosphate + Fructose 6-phosphate

Fig. 5.27 The transketolase reaction of the pentose phosphate pathway.

Xylulose 5-phosphate + Erythrose 4-phosphate ⇌ (Transketolase) Glyceraldehyde 3-phosphate + Fructose 6-phosphate

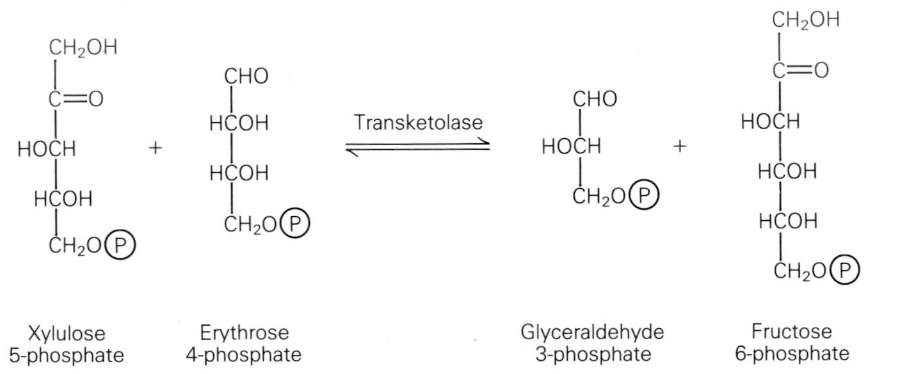

Fig. 5.28 The transaldolase reaction of the pentose phosphate pathway.

Xylulose 5-phosphate + Ribose 5-phosphate ⇌ (Transketolase) Seduheptulose 7-phosphate + Glyceraldehyde 3-phosphate

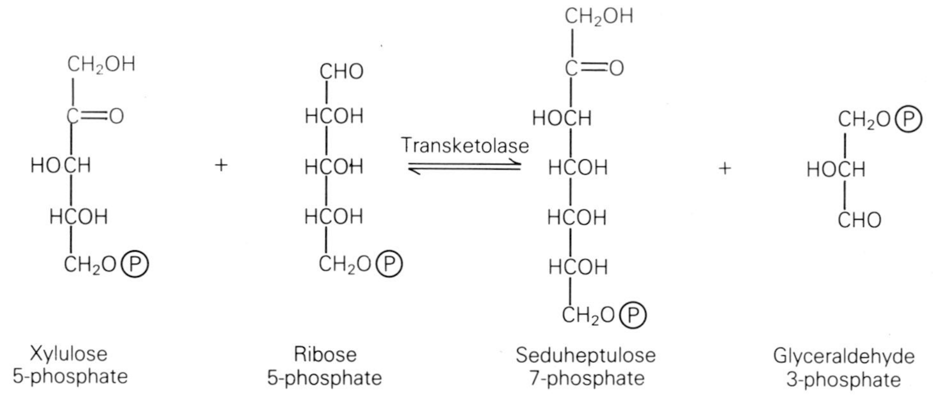

Fig. 5.29 The second transketolase reaction.

Box 5.6 The problems of being an erythrocyte

Erythrocytes, so most textbooks claim, survive for 120 ± 20 days (mean ± SD) in the circulation. Actually, this value is valid only for normal adult human males. It is not necessarily valid for other species, or for human females (where at least the variance is greater), or for human males with haematological disorders. Nevertheless, it is generally the case that circulating erythrocytes do normally continue to function for several months. This is remarkable because erythrocytes spend their time not only receiving a continuous mechanical battering but also carry large loads of oxygen; and invariably, when oxygen and water are present together with transition metal ions (such as the ferrous ion in haemoglobin), peroxides and their highly reactive and toxic derivatives, such as the superoxide radical O_2^-, are generated. These substances cause widespread damage to biological materials, converting unsaturated fatty acids to highly reactive aldehydes, modifying and cross-linking protein side-chains, oxidizing carbohydrates to physiologically useless derivatives and converting haemoglobin to its functionless oxidation product, the Fe^{3+}-containing methaemoglobin. Therefore, because mammalian erythrocytes have no protein synthetic machinery by which this wide-ranging damage can be repaired, their long-term viability seems very surprising.

The solution to the puzzle lies in the fact that erythrocytes are equipped with a battery of protective reductive mechanisms. They contain glycolytic enzymes, which supply them not only with sufficient ATP for their needs (such as ion pumping) but also with NADH, and with enzymes that use the NADH to reduce spontaneously oxidized haemoglobin and membrane components. It is perhaps significant that the NADH-generating enzyme of glycolysis, glyceraldehyde-3-phosphate dehydrogenase, seems to be largely associated with the erythrocyte membrane. They also contain superoxide dismutase, an enzyme ubiquitous in aerobic organisms that converts superoxide radicals to molecular oxygen and peroxide. Finally, they contain pentose phosphate pathway enzymes, which provide NADPH. This provides the reducing equivalents for the enzyme glutathione reductase. Together with catalase activity glutathione provides the erythrocyte's principal mechanism for reducing hydrogen peroxide to water, a reaction catalysed by the enzyme glutathione peroxidase.

Interference with any part of this reductive machinery (summarized in the figure) leads to loss of erythrocyte function, and usually to oxidative damage so extensive as to precipitate haemolysis. For example, many arsenic compounds cause haemolysis by reacting with the functionally essential sulphydryl group of reduced glutathione.

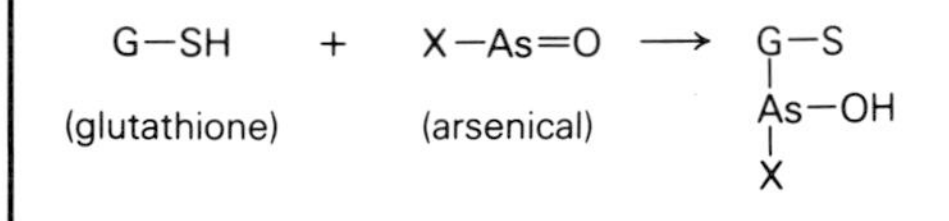

In most, if not all, organisms, the conversion of 3-phosphoglycerate to 2-phosphoglycerate requires a cofactor, 2,3-bisphosphoglycerate, which is produced in trace amounts from 1,3-bisphosphoglycerate and is itself subsequently converted to 3-phosphoglycerate (figure). Human erythrocytes contain far larger concentrations of this cofactor and of the enzymes catalysing its formation and removal than other cells. It binds to haemoglobin, enhancing the Bohr effect, i.e. the decrease of oxygen binding to haemoglobin by acidification. This ensures that the haemoglobin delivers its bound oxygen to respiring tissues when the erythrocyte enters a capillary, because the end-products of metabolism, such as lactate and CO_2 acidify the content of the red cell.

The reductive machinery of mammalian erythrocytes. Key: G6P, glucose 6-phosphate; FBP, fructose 1,6-bisphosphate; GSH, reduced glutathione; GSSG, oxidized glutathione; Hb Fe^{2+}, haemoglobin; Hb–Fe^{3+}, methaemoglobin; A, glycolysis; B, pentose phosphate pathway; c, methaemoglobin reductase; d, glutathione reductase; e, glutathione peroxidase; s, spontaneous reaction.

5.8 Overview

Glycogenolysis and glycolysis are the main processes by which the chemical energy of carbohydrates is made available to the cell. The metabolism of 4-, 5-, and 7-carbon monosaccharides and of hexoses other than glucose can best be understood by relating them to these central pathways. Figure 5.30 sums up these relationships in a simplified form. Gene defects that impair enzymes in the central pathways are highly likely to produce non-viable cells; defects that impair enzymes in the more peripheral pathways do not, but are usually associated with more or less serious pathological conditions. Different cell types exploit carbohydrate metabolism in ways that relate to the specific functions of those cells: for instance, endocrine cells often have high activities of pentose phosphate pathway enzymes in order to produce the NADPH needed for steroid hormone synthesis, and erythrocytes couple the production of reduced coenzymes to mechanisms for defence against oxidative damage.

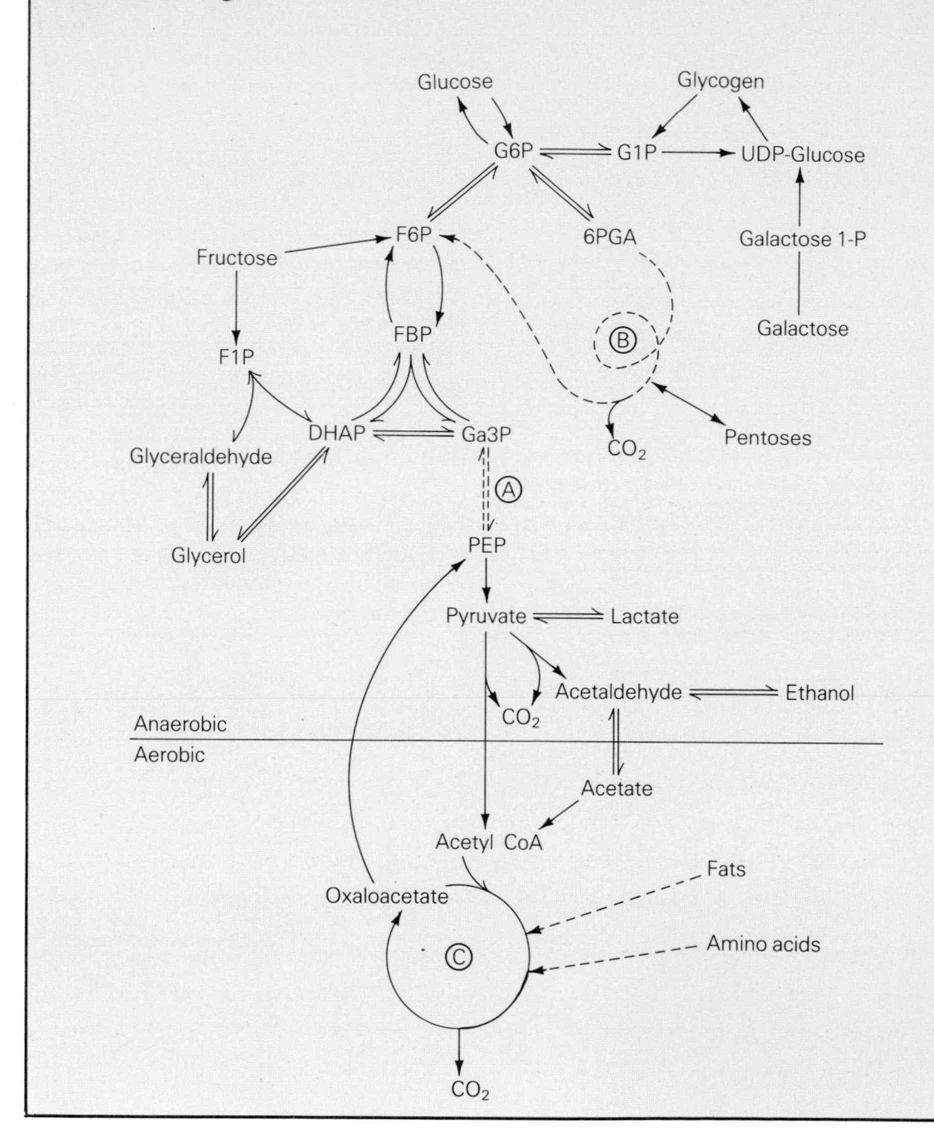

Fig. 5.30 Summary of main pathways of carbohydrate metabolism. Key: G6P, glucose 6-phosphate; G1P, glucose 1-phosphate; F6P, fructose 6-phosphate; FBP, fructose 1,6-bisphosphate; PEP, phosphoenolpyruvate; Ga3P, glyceraldehyde 3-phosphate; DHAP, dihydroxyacetone phosphate; A, glycolysis; B, pentose phosphate pathway; C, TCA cycle.

Answers to exercises

1. For example, inject a variety of protein precipitants into muscles of experimental animals and see if muscle contracts. Alternatively, add inhibitors of glycolysis and then see if injection of lactate still causes contraction.

2.

Reaction	*Anaerobic*	*Aerobic*
glucose → glucose 6-phosphate	−1	−1
fructose 6-phosphate → fructose 1,6-bisphosphate	−1	−1
glyceraldehyde 3-phosphate → 1,3-bisphosphoglycerate	0	+4 (as 2* NADH)
1,3-bisphosphoglycerate → 3-phosphoglycerate	+2	+2
phosphoenolpyruvate → pyruvate	+2	+2
pyruvate → acetyl CoA	0	+6 (as 2 NADH)
tricarboxylic acid cycle and electron transport	0	24
Total	2	36

* Since formed in the cytosol.

3. ⓅOCH_2 ... CH_2OH ... OⓅ (fructose 2,6-bisphosphate ring structure; H, H HO, HO H)

Fructose 2,6-bis Ⓟ $\xrightarrow{H_2O}$ P_i + Fructose 1-Ⓟ $\xrightarrow{H_2O}$ P_i + Glycolysis

4. $^{+}NH_3-CH(^{14}CH_3)-COO^-$ ⟶ $COO^--C(=O)-{}^{14}CH_3$ ⟶

Alanine ⟶ Pyruvate

$[^{14}C]$-glucose ⟶ UDP-$[^{14}C]$-glucose ⟶ $[^{14}C]$-glycogen

5. Avoids osmotic problems.

6. (a) Glucose absorbed, therefore concentration rises. Secretion of insulin promotes a rapid cellular uptake, therefore concentration falls.
(b) Diabetic does not secrete, or secretes reduced amounts, of insulin therefore glucose concentration rises sharply, and falls due to renal excretion.

7. Pentose phosphate pathway activity is highest in those tissues that synthesize lipids and therefore have a great need for NADPH.

QUESTIONS

FILL IN THE BLANKS

1. _______ and _______ are the main fuel molecules in organisms. Glucose is degraded by a series of reactions called _______. The first enzyme active in the pathway is _______, which requires the cofactors _______ and _______. Feedback control of the regulatory enzyme _______ decreases glycolysis if the concentration of _______ is high. _______ is inhibited by glucose 6-phosphate. However, the enzyme _______ is not inhibited so strongly by this metabolite. The latter enzyme also differs from _______ in having a _______ Michaelis constant for _______.

_______ molecules of ATP per glucose molecule are required to initiate glycolysis. In the absence of _______ a net gain of only _______ molecules of ATP for each glucose molecule is possible. In an absence of oxygen the end product of glycolysis is _______, in yeast _______ and _______ _______.

Choose from: ATP (2 occurrences), carbohydrates, carbon dioxide, ethanol, glucokinase, glucose, glycolysis, hexokinase (3 occurrences), high, lactate, lipids, Mg^{2+}, oxygen, phosphofructokinase, two (2 occurrences).

MULTIPLE-CHOICE QUESTIONS

2. Which is the odd one out?

A. hexokinase
B. pyruvate kinase
C. adenylate kinase
D. phosphofructokinase

3. Doubling of the intracellular glucose concentration has little effect on the rate of glycolysis because:

A. hexokinase is more or less saturated even at lower glucose concentrations.
B. of the special properties of phosphofructokinase.
C. the rate of removal of the end-products of glycolysis increases in proportion to the substrate supply.
D. the supply of ATP and of NAD^+ becomes limiting.

SHORT-ANSWER QUESTIONS

4. Levels of particular enzymes in blood plasma are elevated when cells that contain large quantities of these enzymes are damaged and become leaky. In what pathological conditions would you expect to find elevated levels in human plasma of: (a) creatine kinase, (b) lactate dehydrogenase?

5. If glucose labelled with ^{14}C at (a) the carbon-1 position only, (b) the carbon-2 position only were supplied to growing yeast cells under anaerobic conditions, would you expect the radioisotope to appear in the CO_2 produced? Which carbon(s) in the ethanol would be labelled, if any?

6. Intravenous infusion of glucose raises the blood lactate concentration only slightly, but intravenous infusion of fructose raises the blood lactate level about four-fold. (a) Why? (b) Why is intravenous glucose infusion preferable to fructose infusion?

7. A one-year-old child has hepatomegaly, severe fasting hypoglycaemia, and mental retardation. Liver biopsy reveals an abnormally high accumulation of glycogen with normal structure. Clinical chemical investigations reveal an elevated blood lactate, acidosis, and uricaemia. (a) What is the diagnosis? (b) Would the symptoms be alleviated, aggravated, or neither by (i) glucagon or (ii) cortisol treatment?

8. People who wish to lose weight frequently adopt low-carbohydrate diets. How can this practice be justified biochemically?

ESSAY-TYPE QUESTIONS

9. How would you approach the task of elucidating the glycolysis pathway in liver tissue using techniques that are available now (as opposed to the techniques that were available in Meyerhof's day)?

10. In Chapter 4, a question was given that concerned an organism on an imaginary planet with a reducing atmosphere. It was suggested that this organism had a form of glycolysis that converted glucose to propionate. Can you outline, with reasoned explanations, such a form of glycolysis? What do you think would be its 'anaerobic' end-product (the analogue of lactate in terrestrial organisms)? Would this pathway give a net synthesis of ATP when glucose was reduced to its 'anaerobic' end-product?

11. Write an essay with the title 'the role of carbohydrate metabolism in mammalian erythrocyte function'.

6

Lipids: breakdown of fatty acids; brown fat; ruminant metabolism

Objectives

After reading this chapter you should be able to:

- ☐ describe the pathway of β-oxidation of fatty acids;
- ☐ compare the ways in which fatty acids are oxidized to provide ATP and heat by a variety of different types of tissues and organisms;
- ☐ outline how cellulose is used as a source of energy by ruminants.

6.1 *Introduction*

Triacylglycerols are significant energy stores in many organisms. Their hydrolysis releases free fatty acids whose subsequent oxidation yields about twice as much energy as the oxidation of similar weights of carbohydrate or protein.

The major pathway by which fatty acids are oxidized was investigated by Knoop in 1904 by feeding dogs fatty acids in which the terminal methyl group was substituted with a phenyl group (Fig. 6.1). When the fatty acid administered had an even number of carbon atoms its breakdown product was phenylacetate. However, if it had an odd number of carbon atoms benzoate was formed (Fig. 6.1). Knoop therefore suggested that fatty acids are degraded in a stepwise fashion, with oxidation occurring at the β-carbon atom and each step removing a two-carbon fragment from the fatty acid. The process was repeated until phenylacetate or benzoate were released (Fig. 6.1).

During the 1940s and 1950s, studies on the metabolic pathways of fatty acids established the details of their oxidation in organisms.

The oxidation of fatty acids does not necessarily lead to the production of metabolic energy in the form of ATP. In brown adipose tissue the process is deliberately inefficient and energy is released as heat.

Ruminants differ from many other mammals in that they obtain most of their energy from cellulose. This is digested by gut bacteria which eventually release short-chain fatty acids which are the immediate sources of energy for the ruminant.

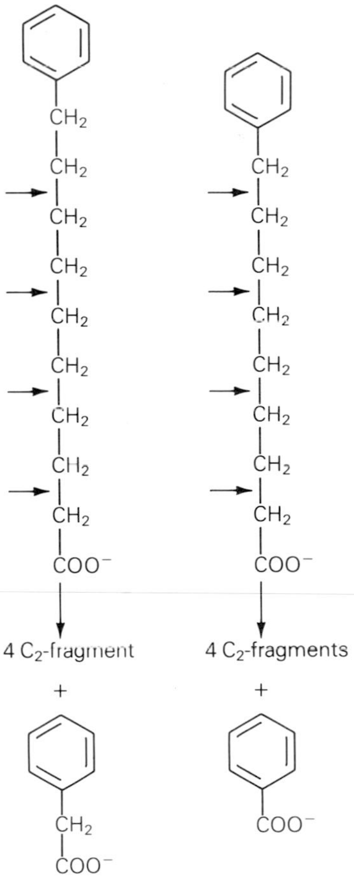

Fig. 6.1 Schematic view of the oxidation of phenyl derivatives of fatty acids. The sites of cleavage of the fatty acids are indicated by arrows.

6.2 *Triacylglycerols as energy reserves*

Triacylglycerols are stored in a virtually anhydrous form as globules in adipocytes (Fig. 6.2), forming a concentrated energy store. The first stage in preparing the stored fats for oxidation is their hydrolysis by lipase (Fig. 6.3). The action of lipase is under hormonal control. **Adrenalin** and **glucagon** stimulate fatty acid release (Fig. 6.3), the activation occurring by a **cyclic AMP** cascade mechanism.

See *Cell Biology*, Chapter 8

Reference Gurr, M.I. and James, A.T. (1980) *Lipid Biochemistry*, 3rd edn, Chapman and Hall, London, UK. A very good, quite in-depth coverage of lipid biochemistry.

Reference Gunstone, F.D., Harwood, J.L. and Padley,F.B. (eds) (1986) *The Lipid Handbook*, Chapman and Hall, London, UK. Splendid reference book covering most aspects of the chemistry, biochemistry and industrial and medical uses of lipids.

□ The use of cascade activation systems to amplify a signal occurs in the activation of the enzyme adenylate cyclase by certain hormones. This allows a weak signal, that is a low concentration of hormone in the serum, to produce gross effects within the cell.

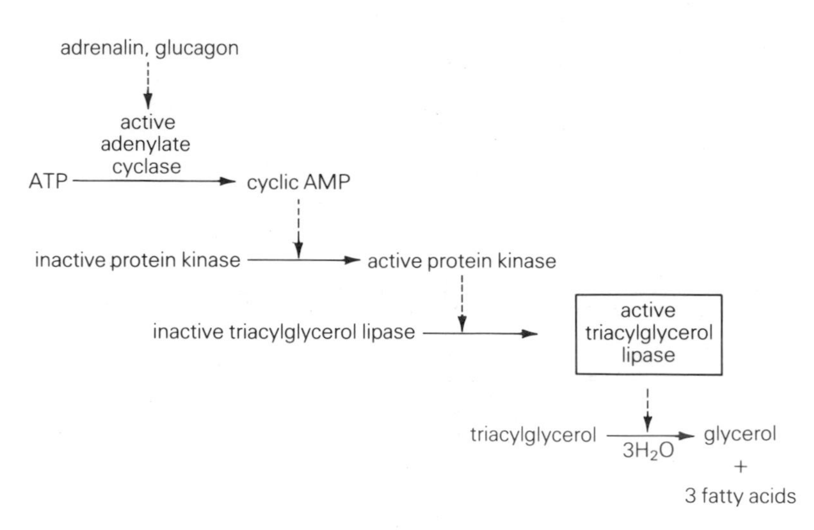

Fig. 6.3 Cyclic AMP-mediated hormone activation of triacylglycerol lipase.

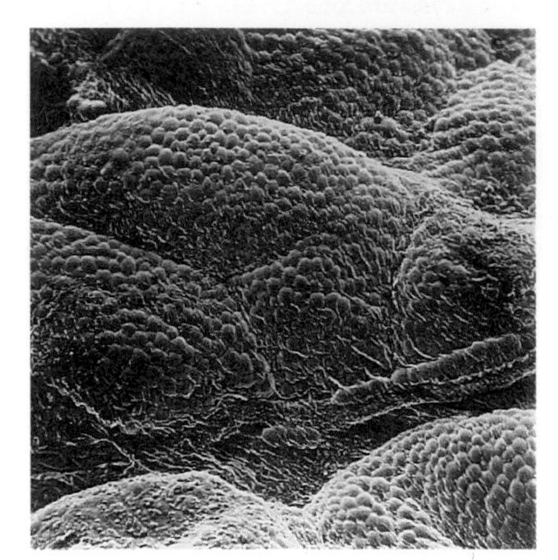

Fig. 6.2 Scanning electron micrograph of white adipose tissue from rat epididymis (magnification × 50). Courtesy of Dr N.O. Nilsson, Department of Physiological Chemistry, University of Lund.

Conversely, **insulin** increases the deposition of fats. Glycerol, released as a consequence of triacylglycerol breakdown, can be fed into the **glycolytic pathway** after being converted via glycerol phosphate to **dihydroxyacetone phosphate** (Fig. 6.4). However, there is no glycerol kinase in white adipose tissue. Consequently, the glycerol is released into the bloodstream and passes to the liver for phosphorylation where it is gluconeogenic. The fatty acids are transported bound to **serum albumin** to the tissues, chiefly skeletal muscle and liver, for oxidation.

6.3 *Oxidation of fatty acids*

The breakdown of fatty acids to **acetyl CoA** with the production of **reduced coenzymes** occurs in the **mitochondrial matrix**. Catabolism occurs by a series of reactions sometimes referred to as the **β-oxidation spiral**. This series of reactions shortens the fatty acid chain by two carbons for every turn of the spiral. The two-carbon fragments are released as acetyl CoA units, which may be oxidized by the TCA cycle. Before catabolism can commence the fatty acids must be converted to their coenzyme A derivatives, forming a ***thioester*** acyl CoA (Fig. 6.5). This activation process involves the participation of **ATP**, in which **pyrophosphate**, PP_i, is produced. The subsequent hydrolysis of PP_i to two molecules of inorganic phosphate effectively makes the formation of the acyl CoA irreversible. The activation of fatty acids occurs at the outer mitochondrial membrane, which is permeable, and so transfer to the intermembrane space is easily effected. However, the inner membrane is impermeable and the carrier molecule, **carnitine** (Fig. 6.6), is employed to transfer the acyl CoA to the matrix. β-Oxidation of the acyl CoA occurs within the matrix.

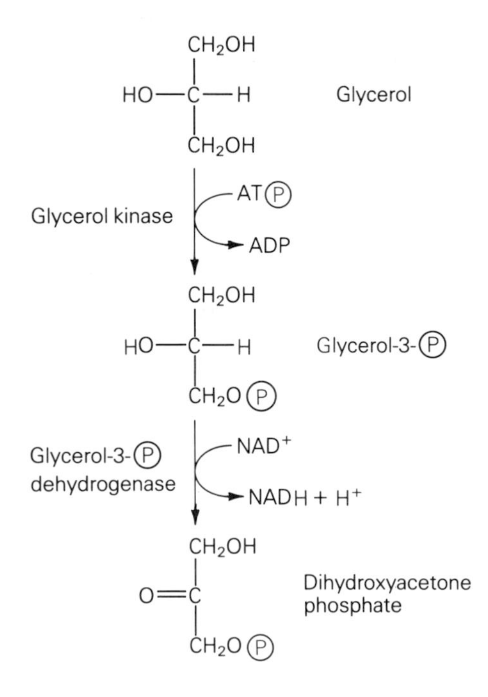

Fig. 6.4 Conversion of glycerol to dihydroxyacetone phosphate.

β-OXIDATION consists of the cyclic repetition of four reactions (Fig. 6.7). The first is a **dehydrogenation** involving **FAD**, yielding an unsaturated **acyl CoA**, containing a *trans* double bond, and one molecule of $\mathbf{FADH_2}$. The unsaturated acyl CoA is then stereospecifically hydrated to the **L-hydroxyacyl derivative** in a reaction catalysed by **enoyl CoA hydratase**.

The third reaction is another dehydrogenation, this one using $\mathbf{NAD^+}$ and catalysed by **L-3-hydroxyacyl CoA dehydrogenase** which is non-specific for the length of the acyl chain, but is specific for the L-isomer. The **oxoacid** produced is the substrate for **thiolase**, the enzyme catalysing the fourth

Thioesters: *are esters containing an S rather than an O atom (—C(=O)—S—). From the Greek,* theions, *sulphur.*

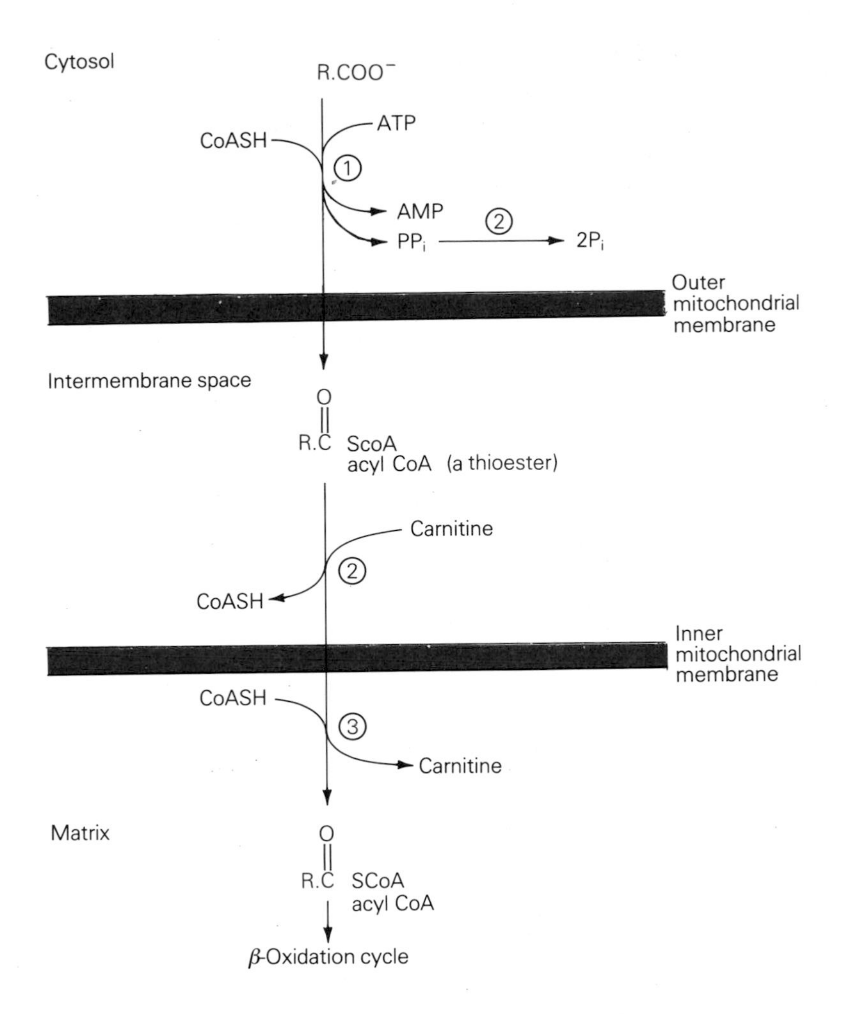

Fig. 6.5 Overview of activation of fatty acids, and their transport into the mitochondrial matrix.

□ The production, and subsequent hydrolysis of pyrophosphate, is a common strategy for driving biosynthetic reactions to completion. The standard free energy change for the hydrolysis of pyrophosphate to orthophosphate is similar to that for the hydrolysis of ATP to ADP and P_i. Thus, the reaction proceeds readily, removing PP_i from the reaction in which it was formed and bringing it to completion. Other examples of pyrophosphate hydrolysis are found in protein and nucleic acid biosyntheses (see *Molecular Biology and Biotechnology* and *Biosynthesis*).

reaction of the spiral. In the last step acetyl CoA is cleaved, leaving an acyl CoA shorter by two carbons. The series of reactions is then repeated until the final thiolysis of a four-carbon acyl CoA is cleaved to give two molecules of acetyl CoA.

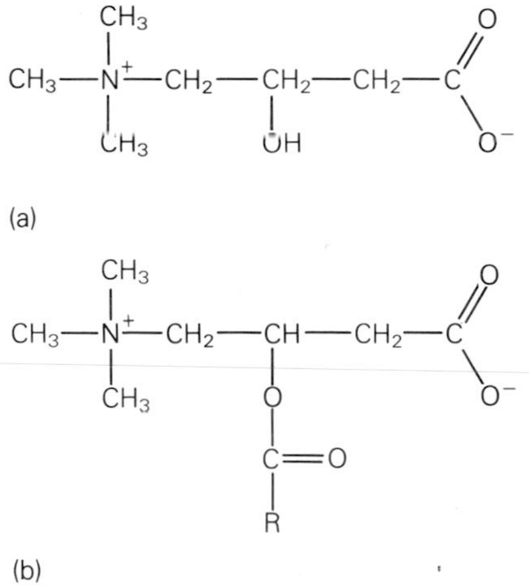

Energy balance of β-oxidation

The β-oxidation of palmitate (C_{16}) requires the oxidative steps to operate *seven* times. The reactions may be summarized:

$$\text{palmitoyl CoA} + 7\,NAD^+ + 7\,FAD + 7\,CoASH + 7\,H_2O \rightarrow 8\,\text{acetyl CoA} + 7\,NADH + 7\,H^+ + 7\,FADH_2$$

Acetyl CoA, NADH and $FADH_2$ are oxidized by the TCA cycle and the electron transport chain. The oxidation of each acetyl CoA yields 12 ATP; $NADH + H^+$ produces 3 ATP and $FADH_2$ gives 2 ATP. The total yield of ATP from palmitate is therefore:

8 acetyl CoA	=	96 ATP
$7\,NADH + H^+$	=	21 ATP
$7\,FADH_2$	=	14 ATP
Total	=	131 ATP

Fig. 6.6 (a) Carnitine and (b) fatty acyl carnitine derivatives.

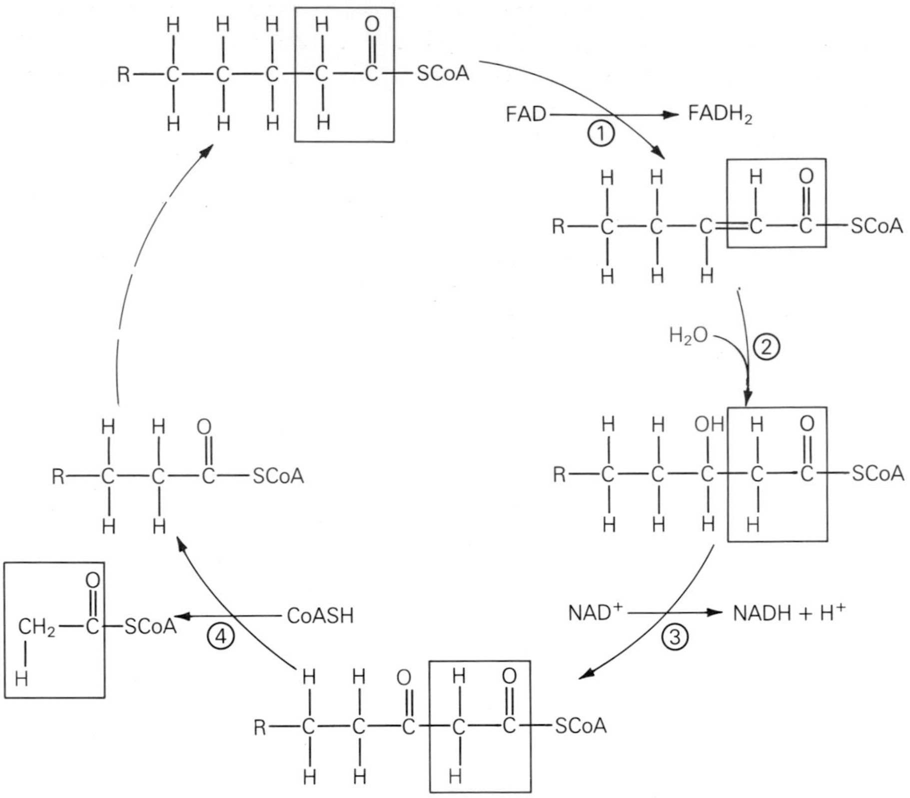

Fig. 6.7 Outline of the four reactions involved in the β-oxidation of fatty acids. Step 1 is catalysed by acyl CoA dehydrogenase, of which there are four enzymes. Step 2 is catalysed by enolyl CoA hydratase (single enzyme). Step 3 is catalysed by L-3-hydroxyacyl CoA dehydrogenase, and finally Step 4 is catalysed by thiolase. There are several thiolases, each specific for different acyl lengths.

Exercise 1

What would be the difference in the number of $FADH_2$ molecules produced by the β-oxidation of oleate ($C_{18:1}$) compared with stearate ($C_{18:0}$)?

The initial activation of the fatty acid, however, requires the expenditure of the equivalent of 2 ATP (ATP $\rightarrow$ AMP + PP_i; $PP_i \rightarrow 2\ P_i$) and so the complete yield of ATP from palmitate is 129 molecules.

The complete catabolism of one glucose molecule generates 36 ATP molecules or 6.0 ATP per carbon, while β-oxidation of palmitate produces the equivalent of 8.1 ATP per carbon. This is one of the advantages of triacylglycerols over carbohydrates as energy stores. Another is that the triacylglycerols are stored as virtually anhydrous globules in the cell. In contrast, storage of carbohydrate such as glycogen entails the simultaneous storage of a considerable amount of water too.

Fat metabolism and ketone bodies

The catabolism of fatty acids produces more molecules of acetyl CoA than does the complete catabolism of glucose. For example, the catabolism of one mole of stearic acid ($C_{18:0}$) produces 9 moles of acetyl CoA, while the catabolism of 3 moles of glucose (C_6) produces only 6 moles of acetyl CoA. Acetyl CoA must react with **oxaloacetate** to enter the TCA cycle. However, oxaloacetate is present in lower concentrations than other intermediates of the TCA cycle and may quickly be depleted if large quantities of acetyl CoA become available. When this happens, the excess acetyl CoA accumulates. Typical cases leading to this situation are starvation when fat stores of the body are mobilized, and in diabetes mellitus when glucose cannot enter extrahepatic tissue cells and the body must use stored fat. In liver cells the excess acetyl CoA is converted to the **ketone bodies: acetoacetate, β-hydroxybutyrate** and **acetone** (Fig. 6.8). Small amounts of ketone bodies are continually produced by the liver in healthy individuals, but if excess accumulates they are excreted in the urine. The concentration of ketone bodies in urine may rise from the normal value of about 125 mg dm^{-3} to as much as 5 g dm^{-3}. Both acetoacetate and β-hydroxybutyrate are moderately strong acids, and are therefore lost in the urine as their sodium salts. This leads to a depletion of Na^+ in the tissues. Acetone is sufficiently volatile to be lost through the lungs, giving the 'sweet breath' characteristic of untreated diabetes.

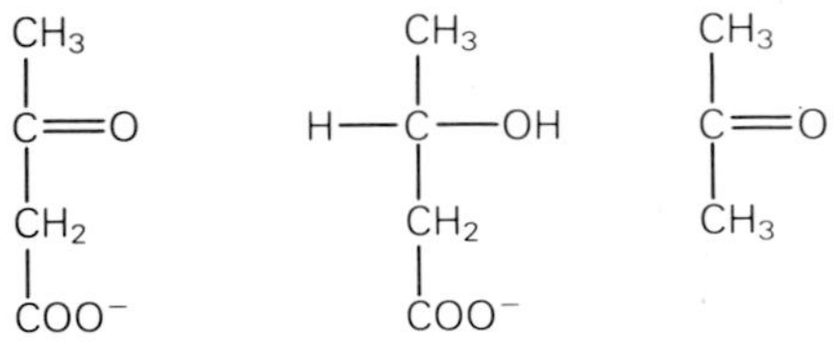

(a) CH_3–C(=O)–CH_2–COO^-

(b) CH_3–CH(OH)–CH_2–COO^-

(c) CH_3–C(=O)–CH_3

Fig. 6.8 The 'ketone bodies': (a) acetoacetate, (b) β-hydroxybutyrate and (c) acetone.

☐ Ketone bodies is a rather silly name, since they are chemical compounds not 'bodies' and since β-hydroxybutyrate is not even a ketone! Both acetoacetate and β-hydroxybutyrate are produced by the liver and transported in the plasma to the muscles. Here they serve as a source of energy and may be thought of as part of the extramitochondrial acetyl CoA pool.

Other types of oxidation

Unsaturated fatty acids are degraded by the β-oxidation spiral, but two additional enzymes are required. Naturally occurring unsaturated fatty acids generally have a *cis* configuration, rather than the *trans* structure produced by β-oxidation (Fig. 6.7). The hydratase will hydrate *cis*-double bonds but generates a **D-2-hydroxyacyl CoA**, which is not a substrate for the subsequent enzyme (i.e. the second dehydrogenation). Also the double bond occurs between C-3 and C-4, rather than C-2 and C-3 as required for β-oxidation. These problems are eliminated by an **enoyl CoA racemase** which can interconvert D- and L-hydroxyacyl CoA isomers, and an **enoyl CoA isomerase** able to shift the position of double bonds. The operation of these two enzymes in the oxidation of linoleoyl CoA is illustrated in Fig. 6.9.

Fatty acids with odd numbers of carbon atoms are degraded in the usual way by β-oxidation but the final thiolysis generates a molecule of **propionyl CoA** as well as of acetyl CoA. The propionyl CoA may be converted to succinyl CoA, one of the intermediates of the TCA cycle (see ruminant metabolism, section 6.5). The oxidation of odd-numbered fatty acids is of greatest importance in bacteria.

α-OXIDATION is a pathway of fatty acid oxidation that occurs typically in germinating seeds. The carboxyl carbon is lost as carbon dioxide. The second carbon atom is oxidized firstly to an aldehyde, and then to a carboxylic acid group giving a fatty acid with an odd number of carbon atoms (see Box 6.1).

ω-OXIDATION is a minor pathway for fatty acid oxidation. The pathway produces a **dicarboxylic** acid, with carboxylic acid groups at the first and last carbon atoms. ω-Oxidation occurs in the endoplasmic reticulum of many tissues, although its function is uncertain.

6.4 *Thermogenesis*

The production of heat, ***thermogenesis***, is a physiological response to cold. Heat may be generated by shivering but **non-shivering thermogenesis** (NST) occurs in **brown adipose tissue** (BAT), particularly in young animals. Hibernating animals on arousal and cold-adapted adult mammals also carry out non-shivering thermogenesis.

In newborn mammals at least 60% of NST is due to the metabolic activities of BAT, easing the transition from intrauterine life at constant temperature to one of external cold stress. Adult animals, adapted to a cold environment have a greatly increased capacity for NST. When cold-adapted rats are infused

Thermogenesis: *from the Greek* thermos *and* genesis *for heat and origin, i.e. origin of heat.*

Reference McGarry, J.D. and Foster, D.W. (1980) Regulation of hepatic fatty acid oxidation and ketone body production. *Annual Review of Biochemistry* **49**, 395-420. Somewhat old, but still useful reading on its self-explanatory title.

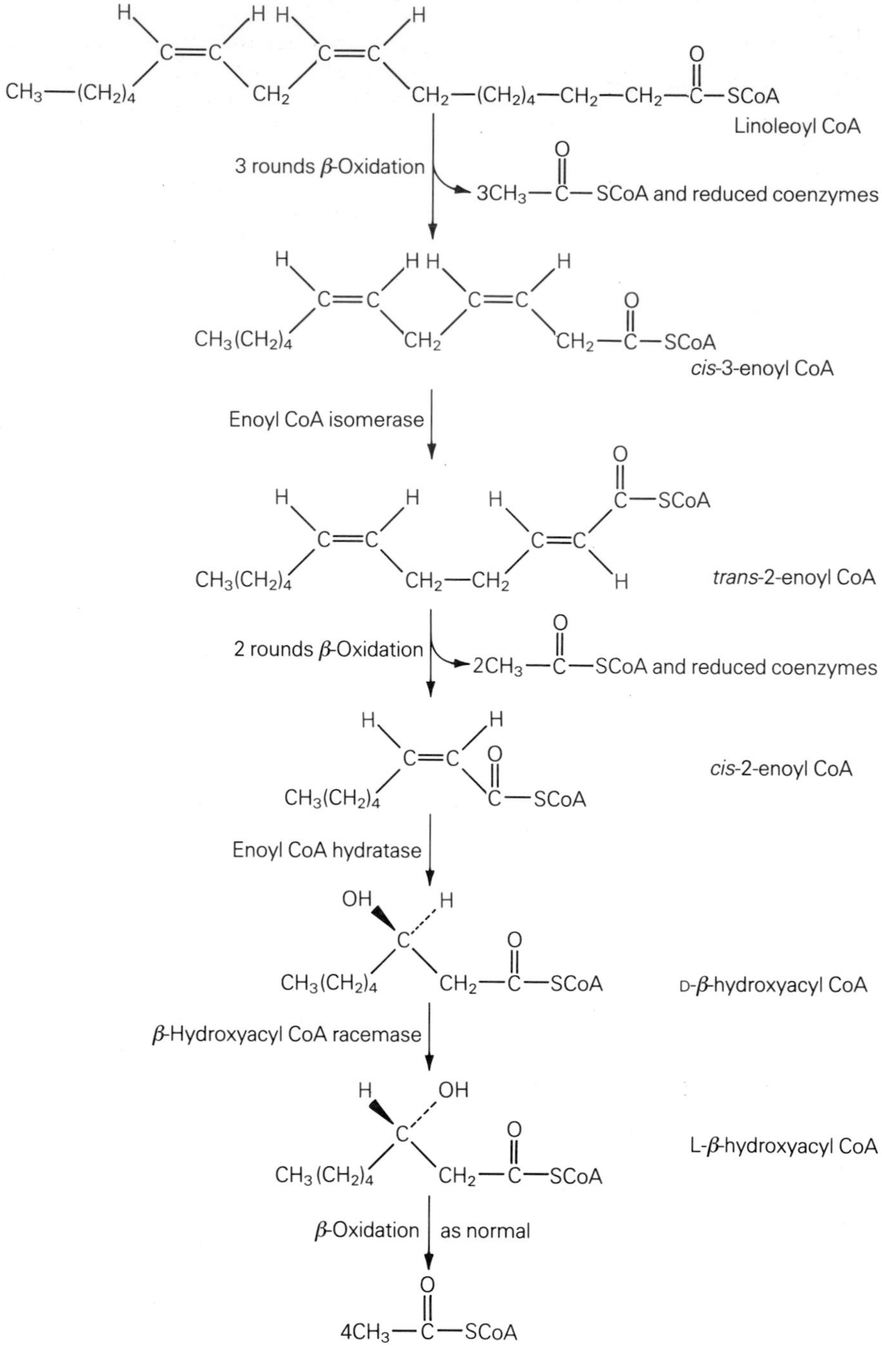

$CH_3-(CH_2)_4-CH=CH-CH_2-CH=CH-CH_2-(CH_2)_4-CH_2-CH_2-C(=O)-SCoA$

Linoleoyl CoA

3 rounds β-Oxidation → $3CH_3-C(=O)-SCoA$ and reduced coenzymes

$CH_3(CH_2)_4-CH=CH-CH_2-CH=CH-CH_2-C(=O)-SCoA$

cis-3-enoyl CoA

Enoyl CoA isomerase

$CH_3(CH_2)_4-CH=CH-CH_2-CH_2-CH=CH-C(=O)-SCoA$

trans-2-enoyl CoA

2 rounds β-Oxidation → $2CH_3-C(=O)-SCoA$ and reduced coenzymes

$CH_3(CH_2)_4-CH=CH-C(=O)-SCoA$

cis-2-enoyl CoA

Enoyl CoA hydratase

$CH_3(CH_2)_4-CH(OH)-CH_2-C(=O)-SCoA$

D-β-hydroxyacyl CoA

β-Hydroxyacyl CoA racemase

$CH_3(CH_2)_4-CH(OH)-CH_2-C(=O)-SCoA$

L-β-hydroxyacyl CoA

β-Oxidation as normal

$4CH_3-C(=O)-SCoA$

Fig. 6.9 Oxidation of linoleoyl CoA. Redrawn from Devlin, T.M. (1986) *Textbook of Biochemistry*, 2nd edn, Wiley & Sons, UK, (Fig. 9.21)

Exercise 2

Calculate the number of grams of lipids oxidized by 100 g of brown adipose tissue operating at maximum rate, and assuming a 100% efficiency in energy conservation.
Assume 1 kW = 3.6×10^6 J, and oxidation of 1 g fat yield about 38 kJ of energy.

with **noradrenalin** (the hormone that stimulates NST), there is a 300% increase in their rate of respiration. This enormous increase allows the tissue to generate 300–400 W kg^{-1} body weight, compared with an average of about 1 W kg^{-1} for resting tissues.

The heat-producing properties of BAT were first clearly demonstrated in marmots arousing from hibernation. During hibernation, the body temperature is close to zero. Hibernation is broken when heat production in BAT is initiated, raising the general body temperature several degrees. Further heat is then generated by violent shivering.

Box 6.1 Refsum's disease

Refsum's disease or **phytanic acid storage disease** is a rare inborn disorder of lipid metabolism, which was first described by Refsum in 1946. Major symptoms of the condition include poor night vision, motor and sensory neuron pathology, elevated concentrations of protein in cerebrospinal fluid, deafness and skeletal abnormalities.

The disease is associated with elevated levels of the C_{20} branched fatty acid, phytanate in tissues and plasma. The normal catabolism of phytanate occurs by an α-oxidation, followed by a series of β-oxidations (see figure). The α-oxidation occurs in two steps. Firstly, the conversion of phytanatic acid to give α-hydroxyphytanate, a reaction catalysed by the enzyme phytanate α-hydroxylase. The second step is the decarboxylation of the α-hydroxyphytanate, catalysed by phytanate α-oxidase, to yield pristanate (see figure).

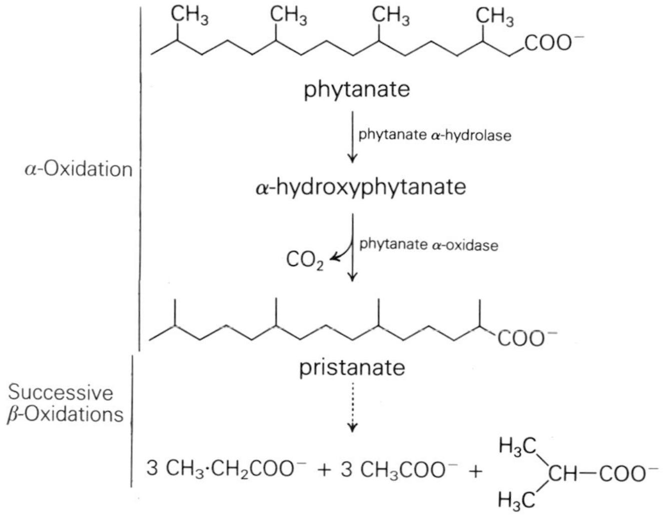

Sufferers of Refsum's disease oxidize phytanate at less than 5% of the normal rate. Studies using cultured fibroblasts have shown that the condition arises from an enzymic defect leading to decreased phytanate α-hydroxylase activity.

The endogenous biosynthesis of phytanate in mammals has not been demonstrated. Phytanate is exogenous, that is of dietary origin, being largely derived from the phytol chain of chlorophyll. Treatment of sufferers of Refsum's disease is by administering a dietary regime low in fats.

Box 6.2 Peroxisomes

Peroxisomes are membrane-bound organelles, which oxidize fatty acids as well as some amino acids. Peroxisomes have a β-oxidation cycle, consisting of the same four reactions found in mitochondria but the reactions are catalysed by a different set of enzymes. Rather than completely catabolizing the fatty acid, as in mitochondria, peroxisome activity fragments the acyl chain: for example $C_{16:0} \rightarrow 5 \times C_2 + 1 \times C_6$. The reduced coenzymes generated by this activity are also reoxidized by a different mechanism to the mitochondrial system.
In peroxisomes, oxygen is used as a hydrogen acceptor to produce hydrogen peroxide, H_2O_2. Alternatively, the reduced NAD may be oxidized by an unknown intermediate, X:

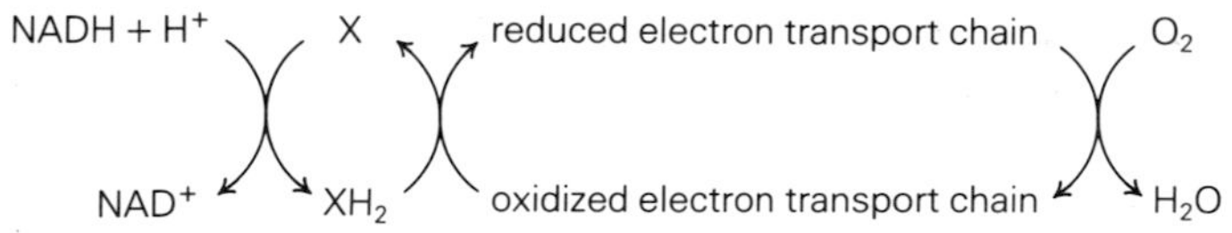

XH_2 passes out to the cytosol, eventually to be oxidized in mitochondria.

Reference Stanbury, Wyndgaardenm, J.B., Fredrickson, D.S., Goldstein, J.L. and Brown, M.S. (eds) (1983) *The Metabolic Basis of Inherited Disease*, 5th edn, McGraw-Hill, New York. Covers many aspects of individual inherited disorders. Many chapter are devoted to genetic diseases involving lipids. Very good reference text.

Brown adipose tissue

Brown adipose tissue consists of adipocytes 25–40 μm in diameter. These are smaller than white adipose tissue cells (Fig. 6.10) and have a different structure (Fig. 6.11). BAT cells contain many small lipid droplets and appreciable numbers of mitochondria (Fig. 6.12). In contrast, cells of white adipose tissue usually contain a single large lipid droplet and relatively few mitochondria.

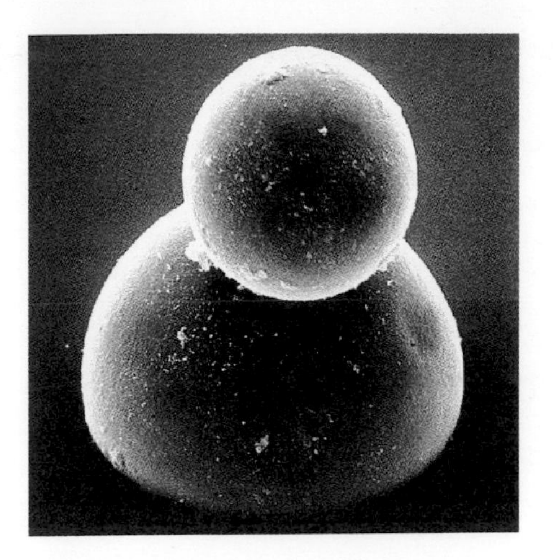

Fig. 6.10 Scanning electron micrograph of two isolated adipocytes from white adipose tissue (magnification ×670). Courtesy of Dr N.O. Nilsson, Department of Physiological Chemistry, University of Lund.

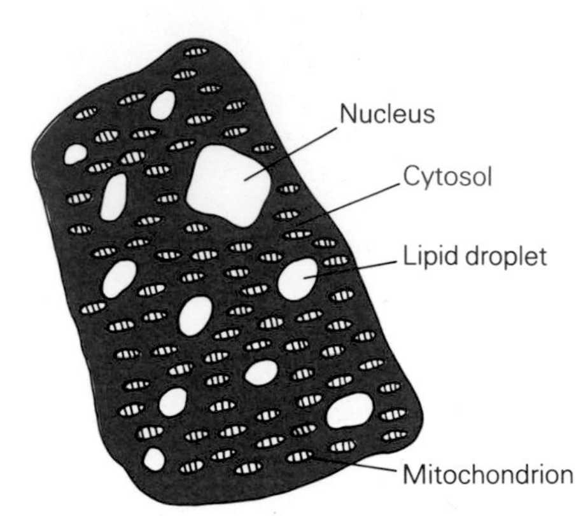

Fig. 6.11 Representation of an adipocyte from brown adipose tissue. Note the many relatively small lipid droplets, and the vast number of mitochondria.

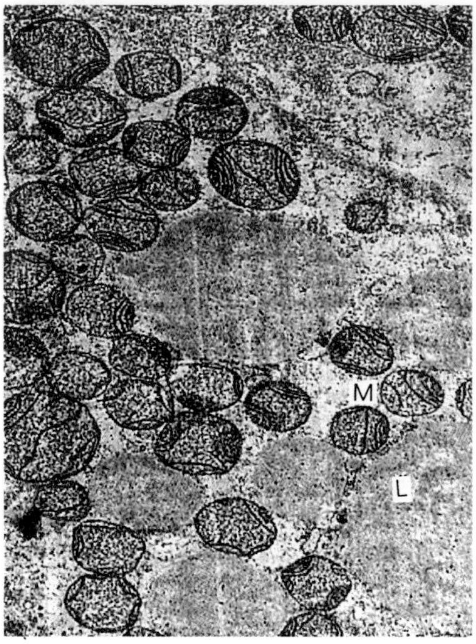

Fig. 6.12 Electron micrograph of brown adipose tissue: L, lipid droplet; M, mitochondria (magnification ×8000).

☐ Catecholamines are hormones based on catechol.

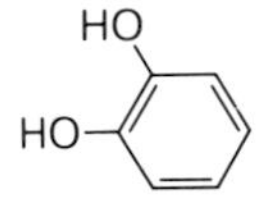

HO–C$_6$H$_4$–OH (catechol)

They are derived from the amino acid tyrosine, and include the hormones adrenalin

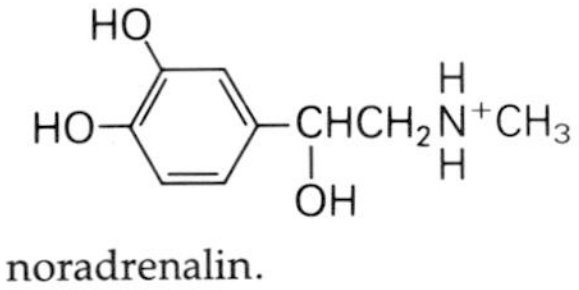

(HO)$_2$C$_6$H$_3$–$CH(OH)CH_2N^+H_2CH_3$

and noradrenalin.

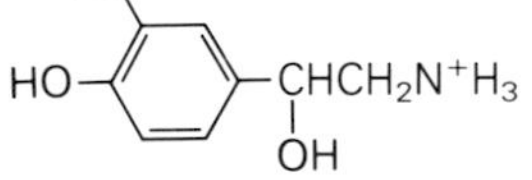

(HO)$_2$C$_6$H$_3$–$CH(OH)CH_2N^+H_3$

See *Cell Biology*, Chapter 8

Brown adipose tissue is usually found dorsally between the scapulae, and as two symmetrical lobes in the cervical region. Deposits are also found in contact with blood vessels in the thorax, neck and abdomen. Brown adipose tissue is extensively vascularized, a feature that is largely responsible for its colour. If BAT is perfused with a haemoglobin-free balanced salt solution, the tissue colour fades to a light tan, as the blood is washed away. The faint yellow-brown colour of the perfused tissues is mainly due to the high flavin and cytochrome content of the cells, a consequence of their large number of mitochondria. However, it is only in the case of young animals and hibernators that BAT is macroscopically distinguishable from white adipose and connective tissue.

In the majority of animals, BAT comes to resemble white adipose tissue as the animal ages. It is possible that BAT in adults may be reactivated but the whole area is controversial. Certain pathological conditions that lead to elevated levels of **catecholamines** in the circulation are usually associated with increased activity of brown adipose tissue.

Substrates for energy production

Triacylglycerol reserves in BAT are mobilized by a lipase, which is activated by noradrenalin, operating a cyclic AMP cascade, similar to that of white adipose tissue (Fig. 6.3). The fatty acids are activated within the brown adipose tissue and degraded by β-oxidation within the mitochondria of the BAT cells. The large mitochondria of brown adipocytes are morphologically

Reference Girandier, L. and Stock, M.J., (eds) (1983) *Mammalian Thermogenesis*, Chapman and Hall, London, UK. Covers many aspects of heat production. Emphasis more physiological than biochemical.

distinct from those of liver (Fig. 6.12). BAT mitochondria have straight cristae, which are tightly packed and traverse the full width of the mitochondrion. β-Oxidation within the mitochondria supplies large amounts of acetyl CoA and NADH and $FADH_2$ to support thermogenesis.

Mechanism of heat production

Electron transport is used to increase the concentration of protons (H^+) in the mitochondrial intermembrane space compared with the matrix and this **proton gradient** across the inner membrane constitutes a store of energy. In most tissues the energy is tapped by allowing the protons to flow back to the matrix through specific enzymes, the flow of protons being coupled to the synthesis of ATP. **Uncoupling agents** (such as dinitrophenol) are weak lipid-soluble acids, which transport H^+ across the inner mitochondrial membrane, by-passing the ATP-generating machinery. Mitochondria of brown fat are poor at synthesizing ATP, although they do generate proton gradients. For example, mitochondria from liver cells have about 20 times more ATP-generating ability for an equivalent electron transport capacity than those from BAT. Also, the inner mitochondrial membranes of BAT cells contain a natural uncoupler, a protein of M_r 32 000, called **thermogenin**. This acts as a **proton transporter**, allowing protons to move from the intermembrane space to the matrix without the concomitant synthesis of ATP. Consequently, the energy stored in the proton gradient appears as heat.

See Chapter 3

□ 2,4-Dinitrophenol

is the best known uncoupling agent, compounds which uncouple oxidative phosphorylation from electron transport. This results in elevated rates of respiration and increased temperature, weight loss and eventually death.

6.5 *Ruminant metabolism*

Ruminants depend for most of their metabolic energy on cellulose but they cannot directly digest the cellulose in their diet. Specialized gut microflora digest the cellulose and other carbohydrates anaerobically to produce short-chain fatty acids and it is these that are used by the ruminant as primary sources of energy.

The stomach of the ruminant is divided into four compartments (Fig. 6.13). The first two, the **rumen** and **reticulum**, are often considered together as the reticulorumen; the **omasum** and **absomasum** are the third and fourth components.

The ruminant stomach contains about 10^{10} bacteria per cm^3 and 10^6 protozoa per cm^3 (Fig. 6.14). These organisms digest ingested **cellulose**, **hemicellulose** and **starch** to produce single sugars, which are immediately

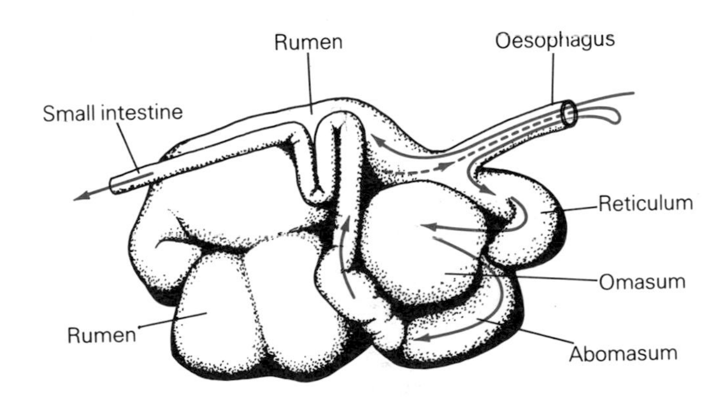

Fig. 6.13 The divisions of the ruminant stomach. The pathway taken by ingested food is indicated. Redrawn from Rogers, T.A. (1958) *Scientific American*, **198**, 34.

Reference Cannon, B. and Nedergaard, J. (1985) The biochemistry of an inefficient tissue: brown adipose tissue. In *Essays in Biochemistry*, Vol. 20 (eds Campbell, P.N. and Marshall, R.D.), Academic Press, London, UK. A quite lengthy account of the biochemistry of brown adipose tissue. Complements previous reference on the topic.

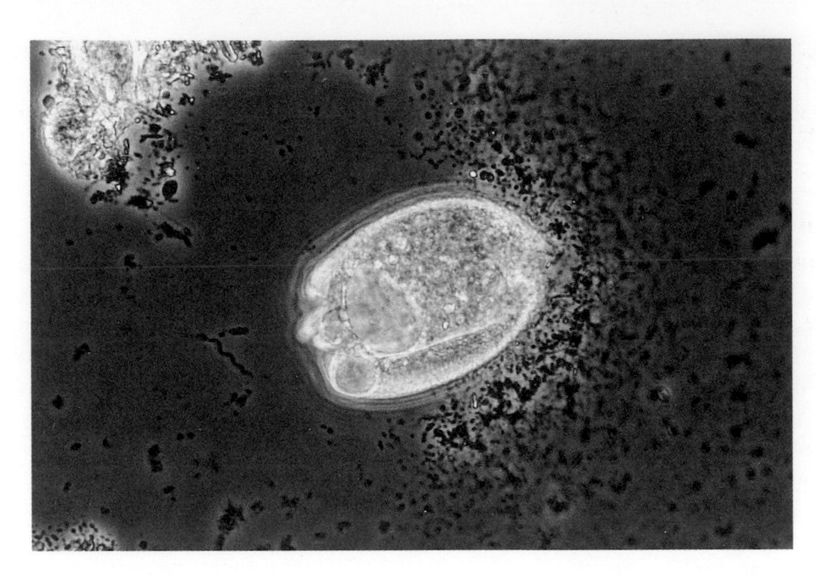

Fig. 6.14 Microorganisms from a ruminant stomach. Courtesy of Dr K.T. Holland, Department of Microbiology, University of Leeds, UK.

taken up by the bacteria inhabiting the rumen. In the anaerobic environment of the gut, the bacteria metabolize the sugars by a variety of fermentation pathways, producing a range of short-chain fatty acids including **acetate**, **propionate** and **butyrate** (Fig. 6.15) as well as **methane** and **carbon dioxide**. Other fatty acids, such as **isobutyrate**, **2-methylbutyrate** and **3-methylbutyrate**, may also be formed in the rumen by the deamination of amino acids. The total concentration of volatile fatty acids in the rumen liquor is 2–15 g dm^{-3}. These products of digestion are absorbed across the rumen wall for use by the ruminant.

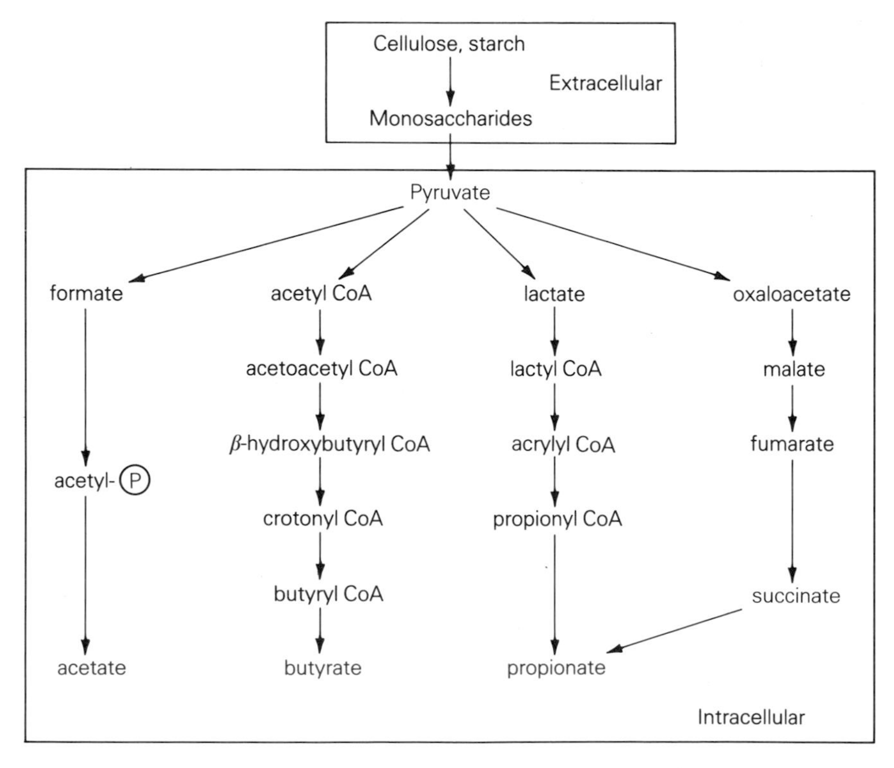

See Chapter 7

Fig. 6.15 Overview of the conversion of carbohydrates to volatile fatty acids during ruminant digestion.

Reference Lassiter, J.W. and Edwards Jr, H.M., (1982) *Animal Nutrition*, Reston Publishing Co., Reston, Virginia. Broad, but readable, coverage of both ruminant and non-ruminant digestion. Useful chapters on micro- as well as macronutrients.

Propionate as an energy source

The propionate produced in the rumen is carried to the liver via the hepatic portal vein. Here the propionate is converted to **succinyl CoA** (Fig. 6.16), a reaction which requires a cobamide coenzyme. The succinyl CoA can enter the citric acid cycle or be converted to glucose as shown in Fig. 6.17. NADH (equivalent to 3 ATP) is produced as a by-product of the conversion of propionate to glucose.

CH_3–CH_2–COO^- (Propionate) + CoASH, ATP → AMP (Thiokinase) → CH_3–CH_2–CO CoA (Propionyl CoA) + CO_2, ATP → AMP → CH_3–CH COO–CO CoA (Methylmalonyl CoA)

Rearrangement

COO^-–$(CH_2)_2$–CO CoA (Succinyl CoA) —to→ Citric acid cycle

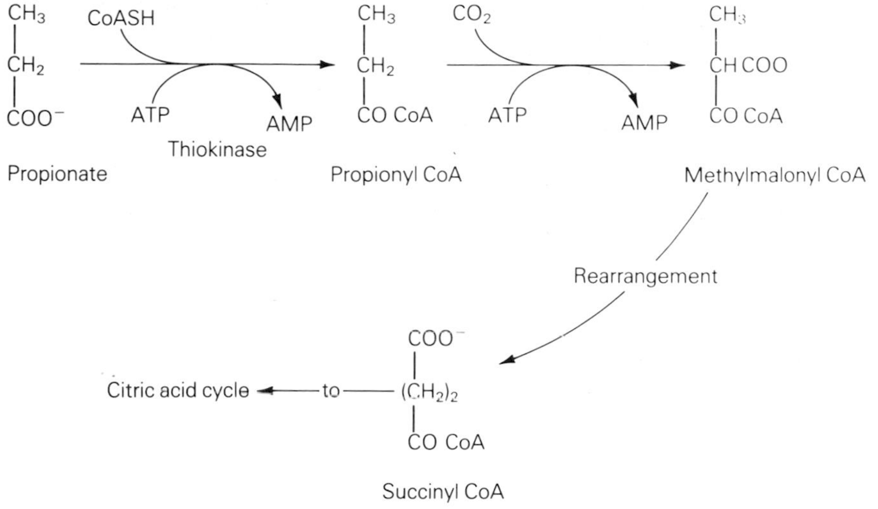

Fig. 6.16 Conversion of propionate to succinyl CoA.

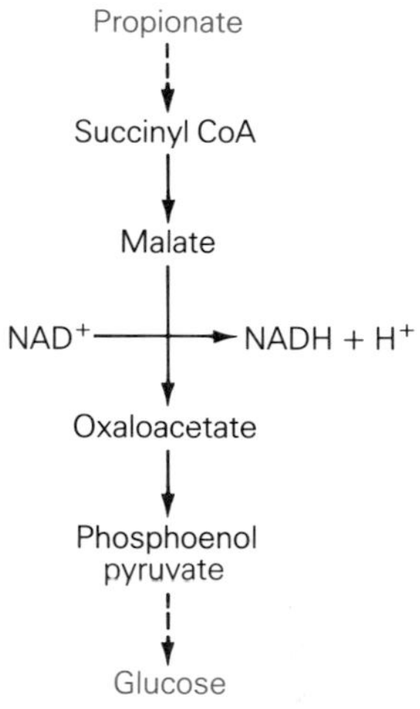

Propionate
↓
Succinyl CoA
↓
Malate
↓ (NAD^+ → NADH + H^+)
Oxaloacetate
↓
Phosphoenol pyruvate
↓
Glucose

Fig. 6.17 Conversion of propionate to glucose.

Box 6.3 Cobalamin (vitamin B_{12})

Cobalamin (vitamin B_{12}) contains **cobalt** in the position corresponding to that occupied by iron in haem. Bonded to the cobalt is a molecule of 5′-deoxyadenosine. Other cobamide coenzymes are known in which the 5,6-dimethylbenzimidazole is replaced by other bases. Cobalamins are generally isolated complexed to cyanide hence were called cyanocobalamins, but the cyanide ion may be replaced by OH^-, NO_3^-, Cl^- or SO_4^{2-}. All derivatives are equally effective biologically.

Neither animals nor plants can synthesize vitamin B_{12}, although both require it, and microorganisms are the major source. In human beings the requirement for vitamin B_{12} is about 1 μg per day and this is so low that deficiency of the vitamin is rare (it occurs occasionally in vegans who do not eat any animal produce.)

See Box 7.9

One reaction involving vitamin B_{12} occurs in the metabolism of propionyl CoA. Propionyl CoA arises from breakdown of odd-numbered or branched-chain fatty acids and it is metabolized to L-methylmalonyl CoA:

$$\text{propionyl CoA} + \text{ATP} + CO_2 \xrightarrow[Mg^{2+}]{\text{propionyl CoA carboxylase}} \text{D-methylmalonyl CoA} + \text{ADP} + P_i$$

$$\text{D-methylmalonyl CoA} \xrightarrow[\text{racemase}]{\text{methylmalonyl CoA}} \text{L-methylmalonyl CoA}$$

$$\text{L-methylmalonyl CoA} \xrightarrow[\text{mutase}]{\text{methylmalonyl CoA}} \text{succinyl CoA}$$

The reaction is of particular importance in ruminant metabolism.

Exercise 3

Calculate the energy generated, as ATP molecules, by the catabolism of 1 mole of propionate.

Butyrate as an energy source

Butyrate is converted to β-hydroxybutyrate during its passage across the rumen wall and muscles can use this directly as an energy source. Alternatively, it can be metabolized in the liver to acetyl CoA (Fig. 6.18), which can be oxidized by the TCA cycle.

Acetate as an energy source

Acetate is the major product of carbohydrate digestion in the rumen. It can be used by all tissues as an energy source after conversion into acetyl CoA, by a process previously described.

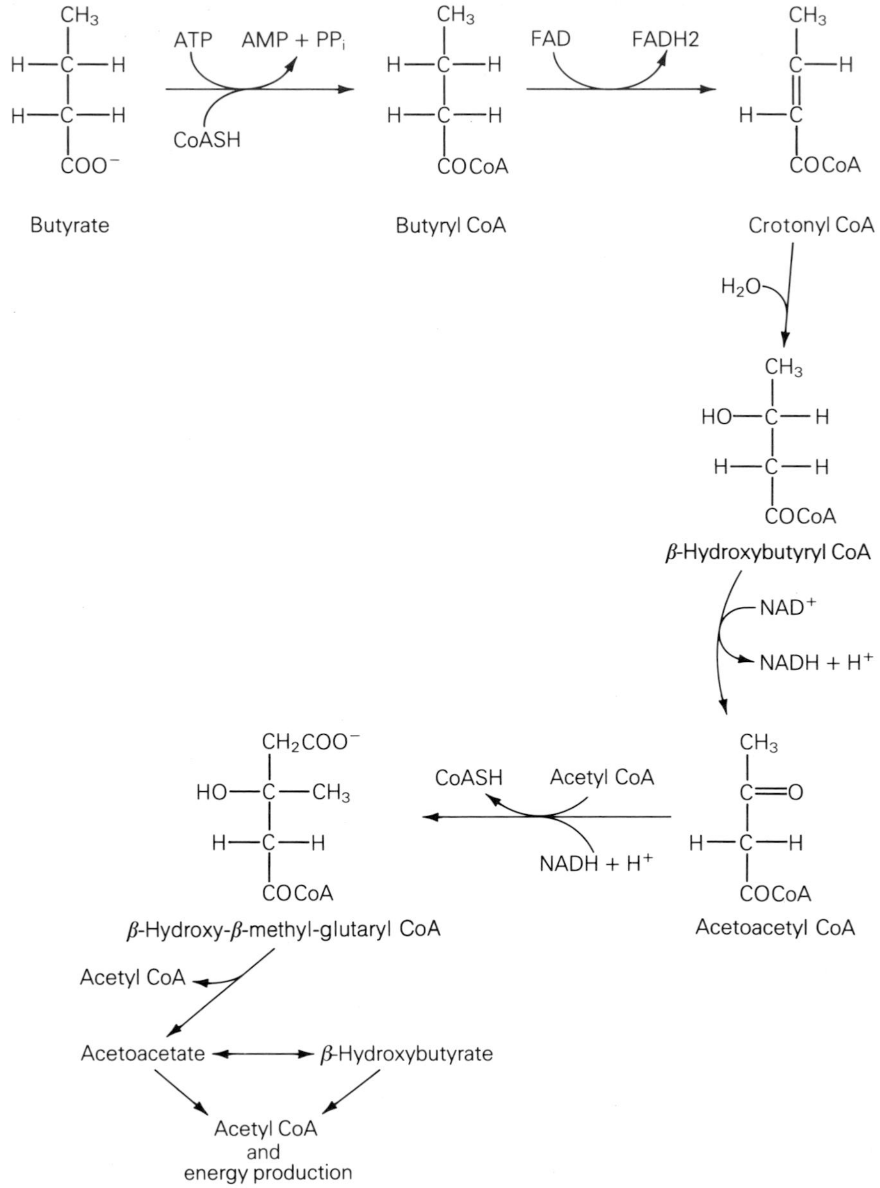

Fig. 6.18 Conversion of butyrate to acetyl CoA.

Reference Garton, G.A. (1977) Fatty acid metabolism in ruminants. In *International Review of Biochemistry*, Vol. 14 (ed. Goodwin, T.W.), University Park Press, London. Although rather old, still a good concise review of ruminant fatty acid metabolism.

6.6 Overview

The highly reduced, water-repellent triacylglycerols are ideal as compact energy-rich stores. This energy is made available to the cell by oxidative processes, of which β-oxidation is the most important. In brown fat this energy is not used primarily to drive biochemical activities, but to produce heat.

Ruminants, like all mammals, lack digestive cellulases but have evolved a unique solution to the problem of digesting cellulose. The development of a four chambered stomach and the use of gut microorganisms allows the anaerobic, that is to say incomplete, digestion of cellulose to short-chain fatty acids. Following their absorption by the ruminant these organic acids are used as the major metabolic fuels.

Answers to exercises

1. Unsaturated fats would bypass the first oxidation step (acyl CoA dehydrogenase reaction) of β-oxidation. Thus, one less $FADH_2$ would be produced for each double bond in the molecule. The oxidation of stearic acid produces 8 $FADH_2$, and oxidation of oleic acid 7 $FADH_2$.
2. If tissue generates 300 W kg^{-1} then 2.8 g lipid would be oxidized.
3. Propionyl CoA is converted to succinyl CoA and enters the TCA cycle. The conversion of succinyl CoA to oxaloacetate by the TCA cycle will generate GTP, $FADH_2$ and NADH, i.e. a total of 6 ATP per molecule of succinly CoA. Two ATP are used in converting propionate to methylmalonyl CoA, therefore *net* yield is 4 ATP.

QUESTIONS

FILL IN THE BLANKS

1. Fatty acids are stored in an anhydrous form as _______ in _______ _______. The hydrolysis of _______ is catalysed by _______ and releases _______ and _______ _______. Hormones, such as _______ and _______ regulate the action of these enzymes. The _______ spiral degrades fatty acids generating _______ _______ and _______ _______. The reactions of the spiral occur in _______ and consist of the cyclic repetition of _______ reactions. These are a _______ requiring the coenzyme _______, a hydrolysis producing an _______ derivative, a second _______ which uses the coenzyme _______ and the release of acetyl CoA which is catalysed by the enzyme _______.
The _______ _______ produced is oxidized by the _______ cycle and the oxidation of the _______ _______ is linked to the formation of _______ in aerobic conditions.

Choose from: acetyl CoA (two occurrences), adipose tissue, adrenalin (or glucagen), ATP, dehydrogenation (2 occurrences), FAD, fatty acids, four, glycerol, L-hydroxyacyl, insulin, lipases, mitochondria, NAD^+, β-oxidation, reduced coenzymes (2 occurrences), TCA, thiolase, triacylglycerols (2 occurrences).

MULTIPLE-CHOICE QUESTIONS

2. Which of the following combinations of coenzymes are necessary for β-oxidation?

A. NAD^+, $FADH_2$, coenzyme A
B. NAD^+, FAD, coenzyme A
C. acetyl CoA, NAD^+, FAD
D. NADH, coenzyme A, FAD
E. acetyl CoA, $FADH_2$, NAD^+

SHORT-ANSWER QUESTIONS

3. What would you expect to be the effects of a deficiency of carnitine?

4. Describe the β-oxidation of palmitoleate (*cis*-9-$C_{16:1}$).

5. Draw a possible sequence of steps to represent the oxidation of a methyl group (CH_3–) to an acid group (–COO^-). Suggest possible hydrogen acceptors involved in these reactions.

ESSAY QUESTION

Write an essay explaining how a meal rich in triacylglycerols ingested in the summer by a small furry animal can eventually be used to generate heat at the end of hibernation the following spring.

7

Amino acid catabolism

Objectives

After reading this chapter you should be able to:

- ☐ illustrate how cells remove the nitrogen from amino acids, and how organisms eliminate it;
- ☐ describe how the amino acid alanine functions as an intermediate in the transport of lactate from skeletal muscles to mammalian liver, where it is reconverted into glucose;
- ☐ describe the diversity of physiologically active products that can result from amino acid catabolism;
- ☐ outline how the transaminases function in the interconversion of amino acids;
- ☐ describe the mechanisms of the urea cycle and its relationship to the general intermediary metabolism of the organism.

7.1 Introduction

Amino acids are commonly thought of in the context of their major role as building units for proteins. When proteins are being synthesized, all 20 of the standard amino acids found in proteins must be available. Half-finished polypeptides cannot be stored and then finished later when the additional amino acids become available. In contrast to plants and many microorganisms that are capable of making their own amino acids, animals need to take in at least half of the standard amino acids preformed in the diet. When dietary amino acids are plentiful and in greater abundance than is necessary for normal protein synthesis and tissue maintenance, their carbon skeletons are degraded and used in part to generate energy during their catabolism. When this happens the nitrogen must be removed and eliminated. These are some of the issues that will be considered in this chapter. Several other interesting and relevant aspects of amino acid catabolism will also be discussed.

7.2 Protein degradation

A series of proteolytic enzymes act upon proteins so as ultimately to release free amino acids. A variety of routes could lead to the complete conversion of a protein to its complement of free amino acids. For example, in the mammalian gastrointestinal tract, or gut, ***endopeptidases*** catalyse the cleavage of peptide bonds at internal points in the polypeptides so as to produce smaller peptides and oligopeptides of varying lengths. Simultaneously ***exopeptidases*** will be working at the amino and/or carboxyl

☐ The length of small peptides which contain 2–10 amino acid residues is generally indicated by Greek prefixes (*di-*, *tri-*, *tetra-*, etc.). The smallest peptide to be called a protein would contain about 50 or more amino acid residues. In between these limits exist a group of intermediate-length (*oligo*) peptides and larger polypeptides that may approach the size of small proteins. It would be acceptable to arbitrarily use the expression oligopeptide for chains containing 6–20 amino acid residues but there is no official designation to be used.

Endopeptidases: *enzymes that hydrolyse amide linkages in the interior of a polypeptide*
Exopeptidases: *enzymes that hydrolyse at either the amino terminal or at the carboxyl terminal ends of peptides.*

Reference Schoenheimer, R. (1942) *The Dynamic State of Body Constituents*, Harvard University Press, Cambridge, MA, USA. Less than 100 pages that puts the last half-century of isotopic metabolic studies in proper historical perspective.

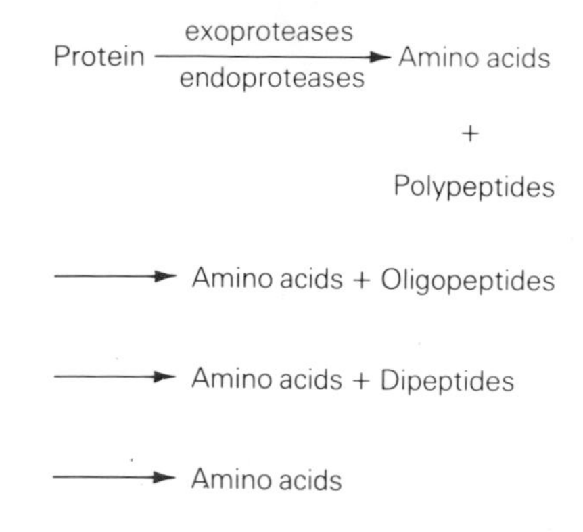

Fig. 7.1 Stages in the digestion of proteins.

terminal ends of the proteins and of their intermediate degradation products. Some enzymes catalyse the hydrolysis of dipeptides from the amino terminal ends of polypeptides. Eventually the entire amino acid complement of the protein is available to be used as needed (Fig. 7.1). The amino acids may then be catabolized, or in animals they may be transported to the liver and other organs for incorporation into protein. Human blood plasma typically contains 350–650 mg dm^{-3} free amino acids. In animals, amino acids are not stored.

Unicellular organisms, such as the bacteria and yeasts and also other fungi (Fig. 7.2) are often capable of extracellular digestion by means of secreted degradative enzymes, including extracellular proteases. These organisms can break down a potential foodstuff upon which they may have been deposited by the whims of fate. The useful degradation products, including the amino acids, are then absorbed and used for the growth of the organism. Multicellular organisms generally have to be more conservative about secreting digestive enzymes, since they must avoid damaging their own cells and tissue. It is common for such organisms to manufacture and transport their digestive enzymes in the form of inactive proenzymes, or **zymogens,** which must be activated in the gut lumen. In many cases the activation of the proenzyme to the enzyme requires selected proteolytic cleavage of an oligopeptide from the amino terminus of the proenzyme.

Box 7.1 Fungi and extracellular digestion

Fungi lack chlorophyll and must live on organic matter. Many fungi, known as saprophytes, decompose various kinds of dead plants and animals (see Fig. 7.2c). Other fungi attack living organisms and are therefore parasites. Still others form a unique association with the roots of higher plants and are called mycorrhizal associates. All three types are capable of extracellular digestion of carbohydrates and proteins, by which they derive at least part of their energy and nutrients. Many cheeses owe their distinctive character to various kinds of moulds (fungi) that are either injected into the cheese or are absorbed naturally from the floors and walls of the caves and buildings where these cheeses are prepared. Stilton is such a cheese. Better known worldwide is Roquefort, whose blue-green veining is due to the mould *Penicillium roqueforti*, common in the humid mountain caves of southern France. During the ripening and ageing of the cheese, the mould digests some of the cheese protein to create the taste and aroma characteristic of this cheese.

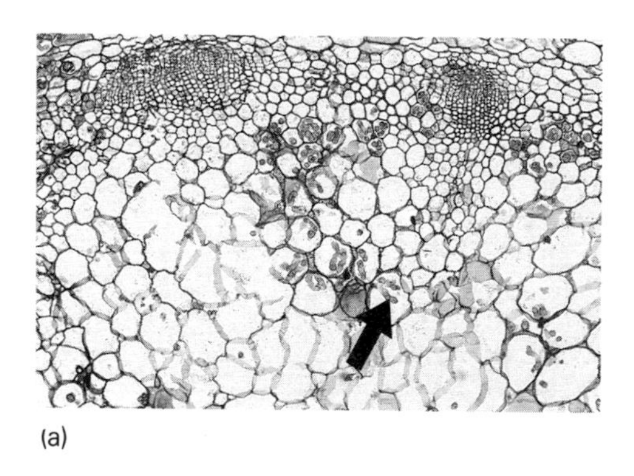

(a)

(b)

Photomicrographs of: (a) the fungus *Peronospora* (arrow) infecting a host plant, and (b) segment of *Dinus* root (dark area) with surrounding ectotrophic mycorrhizal infection (magnification × 180). Courtesy of M.J. Hoult, Department of Biological Sciences, Manchester Metropolitan University, UK.

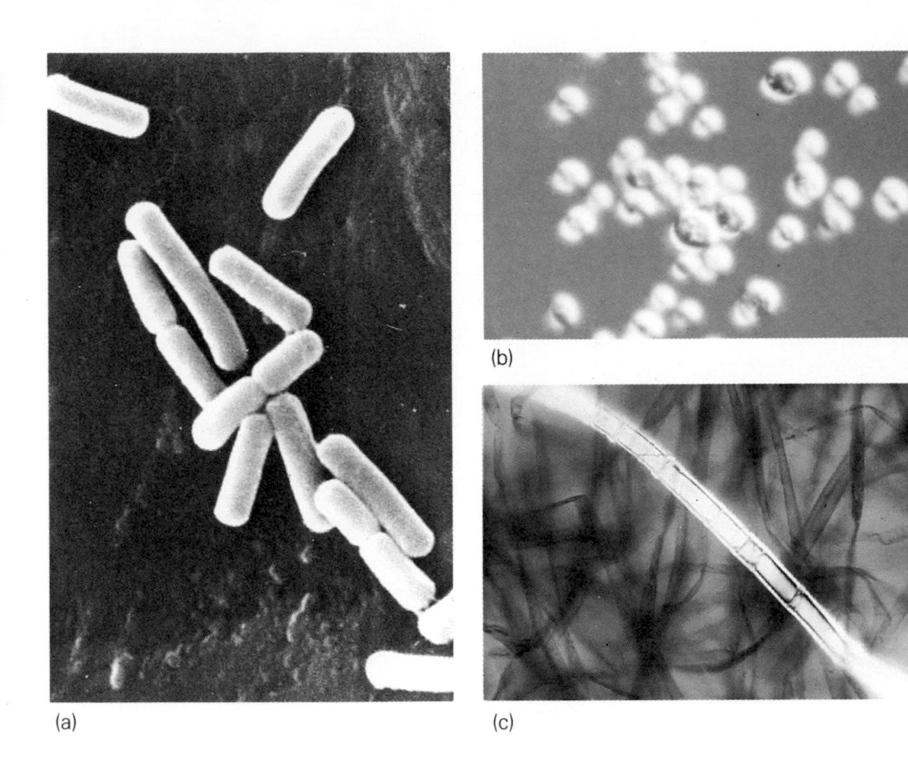

Fig. 7.2 (a) Scanning electron micrograph of *Bacillus subtilis* (magnification × 800). Courtesy of Professor N.H. Mendelson, University of Arizona. Photomicrographs of (b) yeast cells (magnification × 720) and (c) *Mucor* sp. (magnification × 45). Courtesy of M.J. Hoult, Department of Biological Sciences, Manchester Metropolitan University, UK.

7.3 Removal of nitrogen from amino acids

Amino acids typically lose their α-amino groups by one of several possible methods. Some of these are specific to certain amino acids and are not generally applicable. For example, the direct elimination of ammonia from the carbon chain of aspartate produces fumarate (Fig. 7.3a). The enzyme responsible for this conversion is called **aspartase**, and is generally only found in microorganisms. **Serine** and **threonine** can be dehydrated to amino acrylate intermediates, and then these can lose ammonia after hydrolysis (Fig. 7.3b). Elimination of water, followed by a shift of the double bond yields an imino acid; spontaneous hydrolysis releases ammonia and the α-oxoacids, pyruvate and α-oxobutyrate. These two elimination routes are specific for the amino acids mentioned.

More generally, amino acids lose ammonia after first being dehydrogenated in reactions catalysed by enzymes that either use a **flavin** coenzyme or a **nicotinamide adenine dinucleotide** coenzyme. In either case, the intermediate imino acid is subsequently hydrated as in the cases above. The reduced coenzymes that result may be utilized in the generation of ATP for use by the cell. Flavoprotein dehydrogenations are catalysed by enzymes that usually fall into two classes, restricted to those dealing with either the L-amino acids or the D-amino acids. The enzymes are called ***amino acid oxidases***. D-Amino acids are sometimes found in microbial cell walls, so it is perhaps not surprising that such organisms have D-amino acid oxidases (Fig. 7.3c). The explanation for why this type of enzyme can also be found in vertebrates may lie in their need to remove D-amino acids absorbed from gut bacteria. In contrast, the L-amino acid oxidases are more directly involved in normal catabolism of protein amino acids. The reduced flavin may be reoxidized by reacting with molecular oxygen to yield hydrogen peroxide and this is usually detoxified by the action of peroxidase. The α-oxoacids of the carbon chains are further catabolized by several pathways discussed in other chapters.

☐ The pyridine or pyridinium coenzymes include the oxidized and the reduced forms of **nicotinamide adenine dinucleotide** and its close relative **nicotinamide adenine dinucleotide phosphate**. The oxidation–reduction function of both cofactors resides in the structural portion derived from the vitamin **niacinamide**. The first and slightly simpler structure, known now as **NAD^+(NADH)**, was first isolated in 1933 by von Euler in Sweden. The alternative form, known now as **$NADP^+$ (NADPH)**, was isolated in Germany by Warburg and Christian in 1934. Ironically, it was not until 1937 that niacinamide was recognized to be a vitamin for animals. Dogs first, then human beings.

Amino acid oxidases: *flavin-containing enzymes that dehydrogenate amino acids to the intermediate imino acid, which spontaneously gives rise to the corresponding 2-oxoacid.*

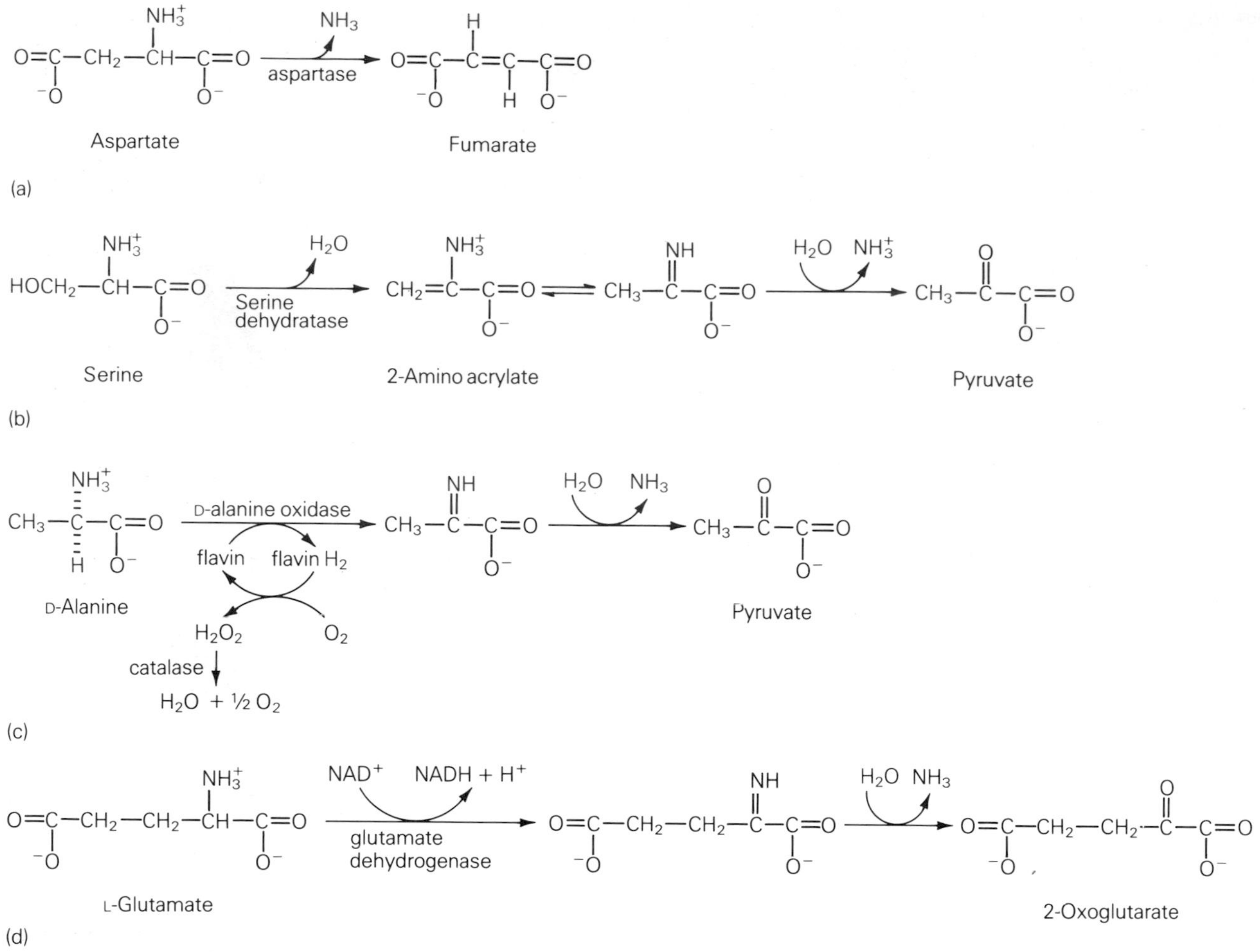

Fig. 7.3 (a) Direct elimination of ammonia from aspartate. (b) Dehydration and deamination of serine. (c) D-Amino acid oxidase action on D-alanine. (d) Glutamate dehydrogenase activity.

□ The liver is the largest organ in the mammalian body. It has many vital functions, such as to produce numerous plasma proteins; bile for secretion into the digestive tract; the storage of fat-soluble vitamins, and to remove toxic substances from body fluids. For the discussion in this section, however, two of the most important functions are to receive, through the portal vein, dietary amino acids and use them to produce serum proteins which are distributed through the central vein; and, to convert excess amino nitrogen into urea for subsequent elimination.

See *Biosynthesis*, Chapter 3

The pyridine coenzyme dehydrogenases, or ***amino acid dehydrogenases***, for **alanine** and for **glutamate** are of special importance (Fig. 7.3d). The reduced coenzymes can be used for energy metabolism (ATP production under aerobic conditions, or indirectly for reductive synthesis). The resulting carbon chain products of alanine and of glutamate are, respectively, pyruvate and α-oxoglutarate, key molecules in intermediary metabolism. They are used for ATP production and also serve as the principal acceptors of the amino groups from other amino acids during transamination reactions (to be discussed below). In many organisms the activities of these two dehydrogenases are the major means by which amino acids lose their α-amino nitrogen, and yet the same enzymes also play an important role in the incorporation of ammonia into amino acids. **Alanine dehydrogenase** is also important in the process of ***gluconeogenesis*** by which muscle lactate transported to the mammalian liver is converted into glucose (Fig. 7.4).

The amide ammonia of the amino acids **asparagine** and **glutamine** also needs to be dealt with. The enzymes **asparaginase** and **glutaminase** hydrolyse them to the corresponding free amino acids, aspartate and glutamate respectively, releasing ammonia (Fig. 7.5).

Both asparagine and glutamine may act as ammonia transfer agents (Fig. 7.6). They provide the major route for the safe transport of toxic

Amino acid dehydrogenases: *enzymes that dehydrogenate amino acids to the intermediate imino acid using niacinamide (nicotinamide) coenzymes.*

Gluconeogenesis: *literally means new synthesis of glucose. It is the process that converts catabolites such as pyruvate and oxaloacetate, which may have been derived from amino acids or other types of molecules, into glucose. In human liver as much as 100 g/day of glucose may be synthesized by this process.*

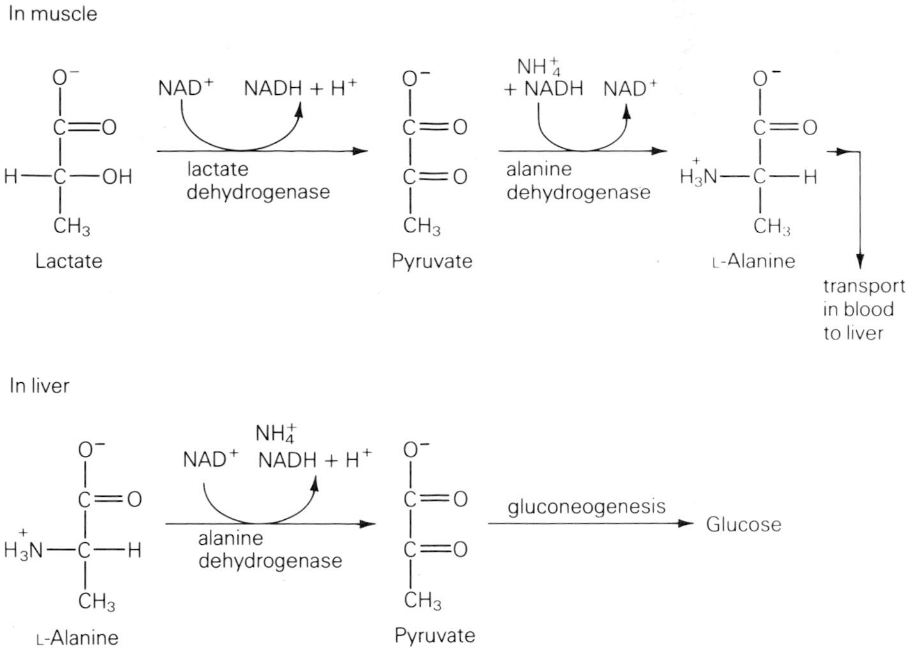

In muscle

Lactate (CH_3–CH(OH)–COO^-) $\xrightarrow[\text{lactate dehydrogenase}]{NAD^+ \rightarrow NADH + H^+}$ Pyruvate (CH_3–CO–COO^-) $\xrightarrow[\text{alanine dehydrogenase}]{NH_4^+ + NADH \rightarrow NAD^+}$ L-Alanine (CH_3–CH($H_3\overset{+}{N}$)–COO^-) → transport in blood to liver

In liver

L-Alanine $\xrightarrow[\text{alanine dehydrogenase}]{NAD^+ \rightarrow NH_4^+ \;\; NADH + H^+}$ Pyruvate $\xrightarrow{\text{gluconeogenesis}}$ Glucose

Fig. 7.4 The role of alanine in the transport of muscle lactate and ammonia to the liver.

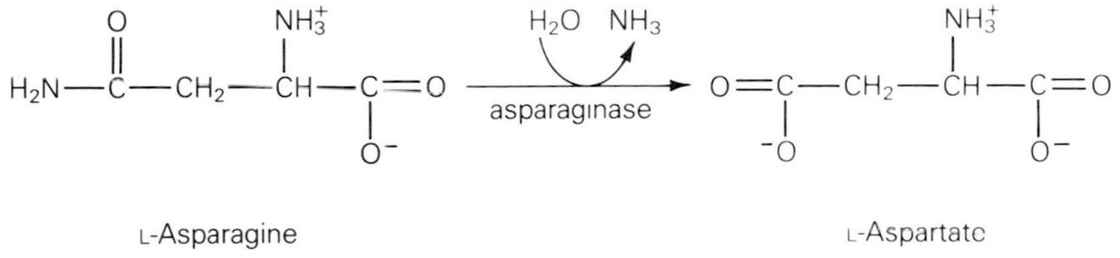

H_2N–C(=O)–CH_2–CH(NH_3^+)–C(=O)–O^- $\xrightarrow[\text{asparaginase}]{H_2O \rightarrow NH_3}$ O=C(^-O)–CH_2–CH(NH_3^+)–C(=O)–O^-

L-Asparagine → L-Aspartate

Fig. 7.5 Deamidation of asparagine.

☐ Other amino acid dehydrogenases exist. For example, the enzyme leucine dehydogenase has been implicated in the germination of bacterial spores of the genus *Bacillus*.

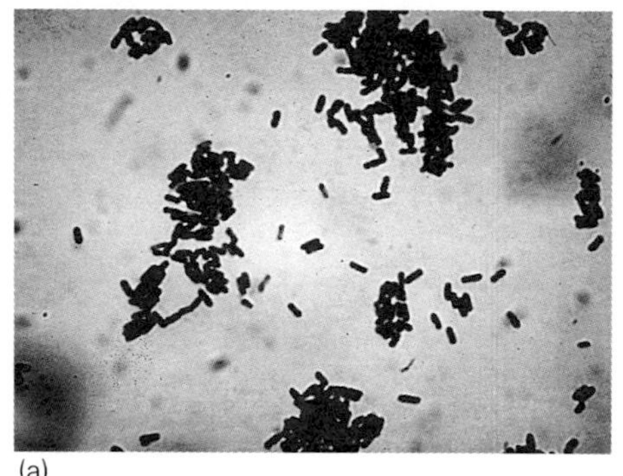

(a)

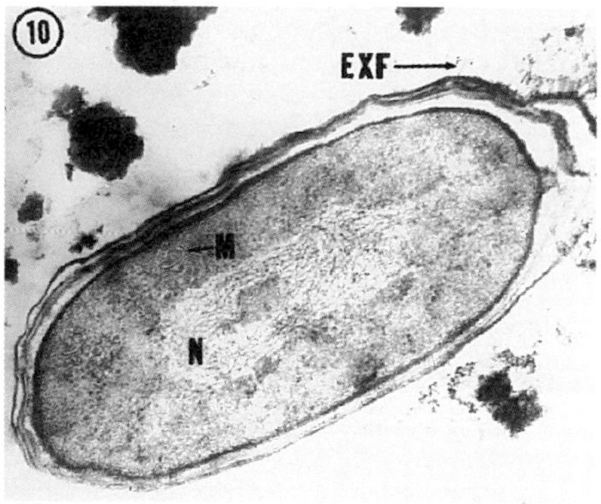

(b)

(a) Photomicrograph of a culture of *Bacillus subtilis* showing spore formation. Courtesy of Dr M.W. Whalley, Department of Biological Sciences, Manchester Polytechnic (magnification ×875).
(b) Electron micrograph of a thin section through a spore of *B. anthracis*, which was germinated in a broth containing ataninc. The nucleoid (N), mesosores (M) and fingerlike extensions of the exosporium (EXF) are visible. Reprinted from Mabberly, B.J., Shafen, F. and Gerhardt, P. (1966) *Journal of Bacteriology*, **92**, 225. Photograph courtesy of Professor P. Gerhardt, Michigan State University.

Box 7.2 Asparaginase

It was noted over a decade ago that some leukaemia cell lines were sensitive to guinea-pig serum. This was eventually shown to be a result of high asparaginase activity in the serum. Apparently leukaemia cells have a greater than normal requirement for asparagine, which explained this sensitivity. Purified bacterial asparaginase was then shown to bring remission in a few test leukaemia patients, but it has not proved to be a practical treatment. The enzyme is expensive as a drug, it must be injected, it is rapidly eliminated from the animal or patient after administration, and it often induces an immune response in the recipient.

Gallagher, M.P., Marshall, R.D. and Wilson, R. (1989) L-Asparaginase as a drug for the treatment of acute lymphoblastic anaemia. In *Essays in Biochemistry*, **24**, 1–40. An up-to-date review of the potential of asparaginase in cancer treatment.

Box 7.3 Why is ammonia toxic?

In animal liver the ammonia concentration is about 5×10^{-4} mol dm^{-3}, and most of this is in the form of the ammonium ion (NH_4^+). Higher concentrations are known to be toxic to the brain, but common postulations do not adequately explain why this is so. It sounds reasonable that brain 2-oxoglutarate is depleted through the action of the enzyme glutamate dehydrogenase, which converts it to glutamate and from that to glutamine via the action of glutamine synthetase. The brain requires 2-oxoglutarate to maintain the operation of the tricarboxylic acid cycle. However, the equilibrium constant for glutamate dehydrogenase would require an extremely high ratio of reduced to oxidized pyridinium coenzymes for extensive conversion to glutamate to occur, and so this theory is probably not correct. High levels of glutamate are known to overstimulate certain nerve populations in the brain, causing their death. Could this possibly be the indirect cause of ammonia toxicity?

□ Different organisms have different synthetic capabilities. Most plants can start from basic inorganic materials and carbon dioxide. Many microorganisms can use simple metabolites, and in some cases can even use diatomic nitrogen. *Escherichia coli* can make all 20 standard amino acids, but *Lactobacillus casei*, which is adapted to grow on milk, requires 16 or so of these preformed.

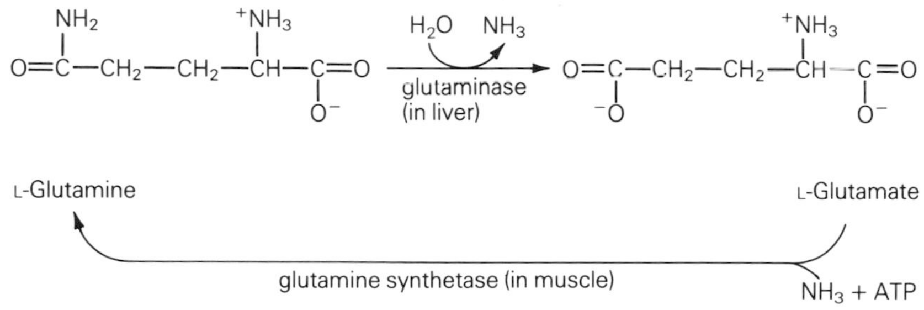

L-Glutamine: $O{=}C(NH_2){-}CH_2{-}CH_2{-}CH(^+NH_3){-}C({=}O)O^-$ — H_2O, NH_3; glutaminase (in liver) → L-Glutamate: $O{=}C(^-O){-}CH_2{-}CH_2{-}CH(^+NH_3){-}C({=}O)O^-$

L-Glutamate + NH_3 + ATP — glutamine synthetase (in muscle) → L-Glutamine

Fig. 7.6 Glutamine in the transport of ammonia to muscle and other tissues.

Table 7.1 Daily adult requirements for the nutritionally essential amino acids

Essential	mg/kg body wt	g/70 kg man
Isoleucine	20	1.4
Leucine	31	2.2
Valine	23	1.6
Tryptophan	7	0.5
Phenylalanine	31	2.2
Methionine	31	2.2
Threonine	14	1.0
Lysine	23	1.6

ammonia from the peripheral tissues of vertebrates to the liver, where asparaginase and/or glutaminase catalyse the release of ammonia for incorporation into molecules that are useful for the cells, or into non-toxic excretory products. Selective deamidation of some asparagine and glutamine residues within certain proteins is known to occur by the action of a set of different enzymes. Rather than being for the purpose of metabolic release of ammonia, this latter process is probably a mechanism for modifying protein structure and function.

7.4 Essential amino acids

Many microorganisms and plants are capable of making all of the amino acids that they require from glucose and other common metabolites and a source of nitrogen. In contrast, most animals are capable of making no more than half of the standard 20 amino acids found in proteins (Table 7.1). The problem lies in their incapacity to make the corresponding α-oxoacids. Providing that these carbon chains are available, they may be transformed into the corresponding α-amino acids by receiving an amino group from a donor such as glutamate or alanine. This process is known as ***transamination***, and the details will be discussed below. However, in the absence of the capability to produce these carbon skeletons, these amino acids are said to be **essential amino acids** and must be supplied in the diet. This designation is dependent upon the species, and even within mammals there is some small variation. The age of the individual animal is also a factor. The requirements for humans include some hydrophobic amino acids (valine, leucine, isoleucine, methionine and

□ Threonine was the last of the 20 protein amino acids to be discovered. According to legend, in 1935 Rose had several University of Illinois research students subsisting on a diet that included only the then known amino acids. When the students began to show signs of poor health (they went into negative nitrogen balance), the experiment was terminated until a further search revealed the existence of a previously unknown essential amino acid. The amino acid isolations had been obtained from acid-hydrolysed casein and the acid-sensitive threonine had been mostly destroyed during the process. Enzymatic hydrolysis allowed the threonine to survive for isolation and identification (and enable the students to survive also!).

Transamination: a reaction type in which an α-amino acid becomes a 2-oxoacid, and its coreactant 2-oxoacid then becomes the new α-amino acid. The coenzyme for this reaction is pyridoxal phosphate, a vitamin B_6 derivative.

Reference Bender, D.A. (1985) *Amino Acid Metabolism*, 2nd edn, John Wiley, New York, USA. Broad coverage of most aspects of amino acid metabolism.

Box 7.4
Kwashiorkor

Human adults are generally able to meet their needs for histidine and for arginine (the latter from the urea cycle, but infants may require dietary sources. A well-fed, hypothetical 70 kg man may have a total metabolic turnover of up to 400 g protein daily. Nevertheless, less than 13 g of essential amino acids are required (see Table 7.1). No specific deficiency syndromes exist for those individual amino acids.

To determine the nitrogen balance of an individual it is necessary to measure the total protein nitrogen ingested and then to subtract from that the total nitrogen excreted in all forms. Positive nitrogen balance indicates adequate dietary intake. A negative nitrogen balance can mean insufficient total protein intake, lack of one or more of the essential amino acids, or possibly an infection or pathological condition.

Humans need 2 g protein per 100 g diet. Cows' milk contains 5 g protein per 100 g, so it is generally adequate with respect to total amino acids. Egg-white protein is somewhat more digestible and has a slightly more favorable distribution of the essential amino acids. Starvation under conditions in which total calories are adequate, but protein intake is not, results in the nutritional deficiency syndrome called kwashiorkor (the term comes from a Ghanian dialect and implies 'displaced from the breast', thus referring to children at age 1–3 who suddenly have to find adult food upon the birth of the next sibling). Symptoms in infants include oedema and a bloated look, lack of skin and hair pigmentation, a general malaise and restricted movement. Neurological development will be impaired or delayed, even if adequate nutrition becomes available in time to prevent death.

Children in rural Mozambique at a CARE food distribution centre. The child on the right shows symptoms of kwashiorkor, including skin and hair depigmentation. Coutesy of G. Wirt with the co-operation of CARE, New York.

phenylalanine), the neutral polar amino acids (threonine and tryptophan), and the basic ionizable amino acids (histidine, lysine and arginine). Cysteine and tyrosine are usually considered as essential, but they can be derived from methionine and phenylalanine, respectively.

7.5 Carbon chain catabolism

See Chapters 4 and 5

The pathways for the biosynthesis of the non-essential amino acids, although diverse, usually originate from either glycolytic or tricarboxylic acid cycle intermediates. In those organisms capable of manufacturing most or all amino acids, the essential groups generally have more complicated pathways for their synthesis. Likewise, the degradative pathways for these amino acids are also more complicated than those for the non-essential amino acids. Nevertheless, all of the amino acid carbon chains can ultimately be broken down to yield either pyruvate or tricarboxylic acid cycle intermediates, or acetyl CoA, or intermediates that can give rise to acetyl CoA (Fig. 7.7). Those amino acids in the first category are called ***glucogenic*** and those in the second category are called ***ketogenic***.

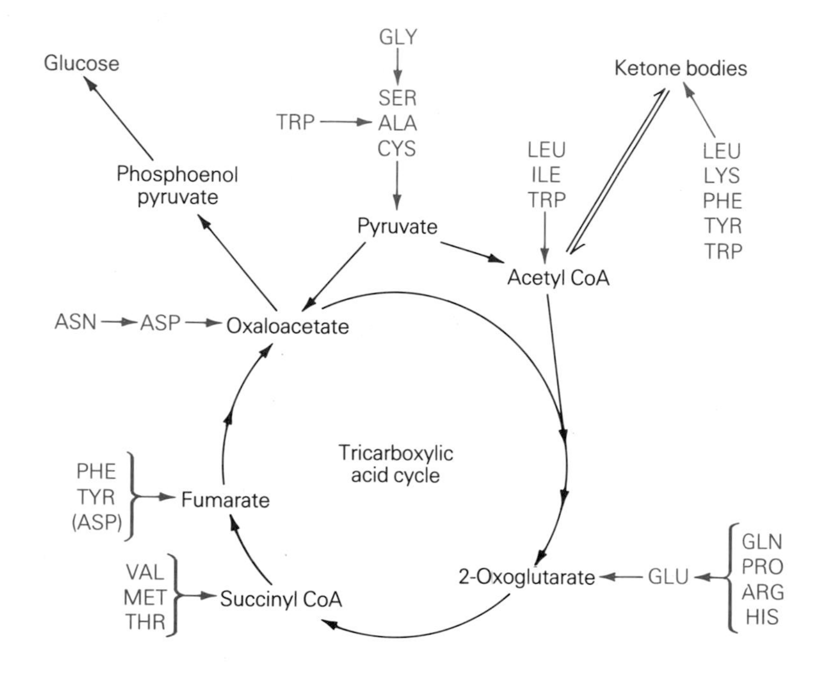

Fig. 7.7 Potential catabolic routes of amino acids.

Exercise 1

Some amino acids, the glucogenic ones, can serve as dietary glucose replacements. Can ethyl alcohol do likewise?

☐ Methylmalonyl CoA is an intermediate in the catabolism of methionine, isoleucine, and valine. Methylmalonyl CoA is enzymatically rearranged to give succinyl CoA, a tricarboxylic acid cycle intermediate that can be readily metabolized further. The rearrangement step uses cobamide as a coenzyme. Inherited disorders are known in which either a functional B_{12} coenzyme cannot be made, or the methylmalonyl CoA mutase apoenzyme (the protein portion) is defective. The resulting condition is known as methylmalonic aciduria and is characterized by excretion of as much as one gram of methylmalonate daily. More importantly, the patient exhibits acidosis with an arterial blood pH of 7.0, rather than the normal value of 7.4. The condition can be lethal, although approximately half of the patients respond to intramuscular administration of cobalamine.

The products of carbon chain catabolism distinguish the amino acids, since animals are unable to effect a net conversion of acetyl CoA into carbohydrate. A fasted animal fed solely on the glucogenic amino acids (alanine, cysteine, glycine, serine, asparagine, aspartic acid, glutamine, glutamic acid, proline, arginine and histidine) would be able to use the amino acid catabolic intermediates to resynthesize glucose. On the other hand, the ketogenic amino acids (leucine, isoleucine, phenylalanine, tyrosine, tryptophan and lysine) give rise to acetyl CoA, and animals cannot use acetyl CoA to make pyruvate or oxaloacetate which can be used to make glucose. As a result, accumulated acetyl CoA is either used for energy production, for fatty acid biosynthesis, or for the production of ketone bodies, ***acetoacetate*** and β-hydroxybutyrate.

Phenylalanine metabolism

See Chapter 6

The essential amino acid phenylalanine can be converted to tyrosine in animals (see below). Plants and microorganisms sometimes synthesize

Glucogenic amino acids: *amino acids that can be catabolized to yield intermediates for gluconeogenesis, the new synthesis of glucose.*
Ketogenic amino acids: *amino acids whose catabolism yields primarily ketone bodies, such as acetoacetic acid or acetyl CoA.*
Acetoacetate: *a ketone body, also known as 3-oxobutyrate.*

Reference Meister, A. (1965) *Biochemistry of the Amino Acids*, 2nd edn, Academic Press, New York, USA). These two volumes discuss many aspects of amino acids, but of especial interest is the large section that surveys amino acid catabolic pathways as they were understood up to that date.

tyrosine in this way, but generally they synthesize tyrosine directly from an intermediate that is on the pathway for phenylalanine biosynthesis. Complete catabolism of tyrosine yields fumarate, a glycogenic precursor, and acetoacetate, a ketogenic precursor (Fig. 7.8). Tyrosine is a potential fuel. It also plays an indirect (regulatory) role in energy production since it gives rise to the hormone ***adrenalin***. Adrenalin, also called epinephrin, is the mammalian hormone that initates the cAMP-mediated phosphorylase cascade that allows muscle glycogen to be metabolized rapidly at times when there is a sudden demand for a lot of energy. In animals tyrosine can be converted into several other physiologically important metabolites (Fig 7.9).

See *Cell Biology*, Chapter 8

The first step in the catabolism of phenylalanine is catalysed by the enzyme **phenylalanine 4-hydroxylase**, producing tyrosine. This enzyme is a ***monooxygenase*** because it uses molecular oxygen which furnishes only one atom of oxygen to the product (Fig. 7.10). The second oxygen atom is used to oxidize a reduced ***tetrahydrobiopterin*** cofactor, producing a molecule of water. The resultant dihydrobiopterin is reduced back to tetrahydrobiopterin using two electrons furnished by NADH or NADPH. If an individual inherits defective genes for the phenylalanine hydroxylase, a serious genetic disorder results known as phenylketonuria (PKU). There is then the distinct possibility that tyrosine could become a dietary limiting amino acid.

□ Tyramine, the tyrosine decarboxylation product, occurs in the normal diet, a typical daily consumption amounting to 1 mg or more. Aged cheeses contain tyramine and red wine contains 2 mg dm^{-3}, so tyramine has been implicated as a potential cause of migraine headaches in those persons for whom the response is triggered by wine or brandy consumption. The actual culprit may more likely be a flavonoid inhibitor of one or more of the enzymes involved in catecholamine inactivation.

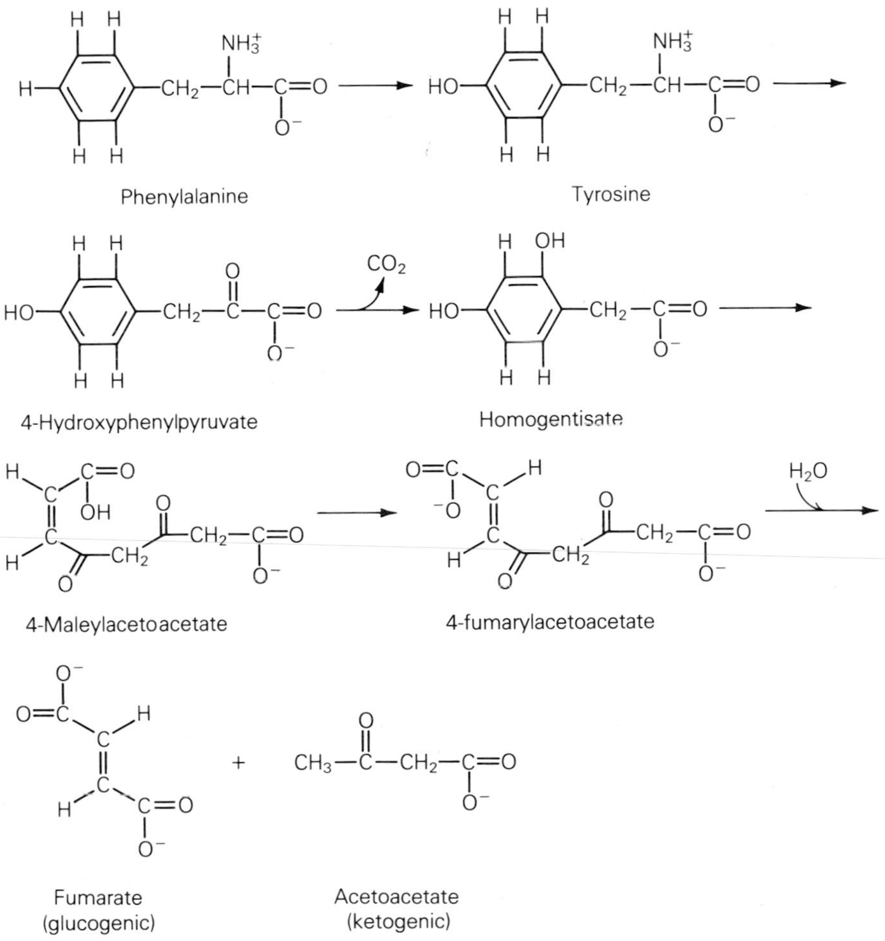

Phenylalanine → Tyrosine → 4-Hydroxyphenylpyruvate → (CO_2) → Homogentisate → 4-Maleylacetoacetate → 4-fumarylacetoacetate → (H_2O) → Fumarate (glucogenic) + Acetoacetate (ketogenic)

Fig. 7.8 Pathway of phenylalanine and tyrosine catabolism.

Exercise 2

Phenylalanine and tyrosine are both catabolized to homogentisate, which is further degraded to fumarate plus acetoacetate. What would be expected when starved animals are fed either of these two amino acids in excess?

Adrenalin: *a hormone whose synthesis is the result of chemical modification of tyrosine.*
Monooxygenase: *an enzyme for which only one of the oxygen atoms from diatomic oxygen is incorporated into the product of the reaction. The second oxygen atom is converted to HOH as it oxidizes a coreactant, usually a reduced coenzyme.*

Tetrahydrobiopterin: *a cofactor for the phenylalanine hydroxylase reaction. It is structurally related to folic acid and to a group of natural product molecules that are sometimes found in such locations as the wing pigments of butterflies.*

Box 7.5
Cocaine

The habituating drug cocaine is thought to exert its supposed 'pleasure' sensation by blocking the re-uptake at nerve terminals of the neurotransmitter dopamine (step 4 in the illustration). Consequently, the dopamine receptors on the adjacent neuron remain saturated with the neurotransmitter for a long time period and the signal remains steady, rather than responding to repeated releases of the neurotransmitter.

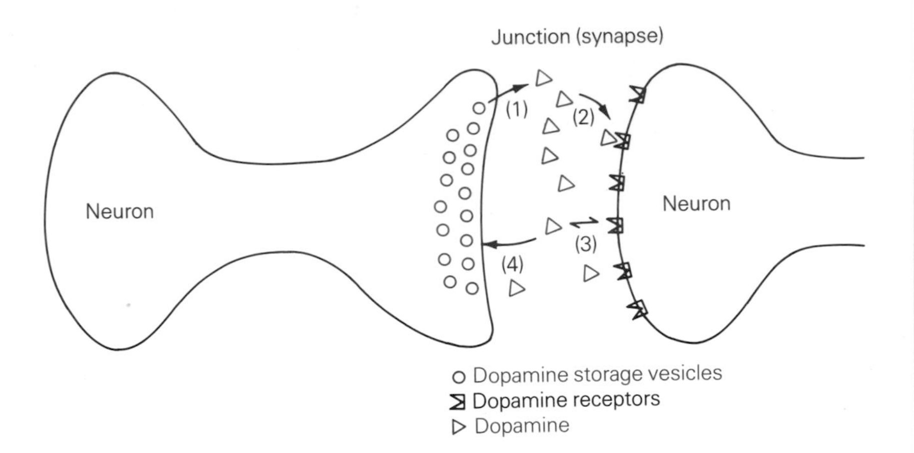

(1) Release of dopamine from receptors; (2) binding by receptors; (3) dissociation from receptors; (4) re-uptake into vesicles.

Box 7.6
Phenylketonuria

A serious problem arises for those who lack phenylalanine hydroxylase, usually referred to as phenylketonurics. The condition also arises in about 3% of infants who are phenylketonuric who have normal phenylalanine hydroxylase but are defective in either the synthesis or reduction of biopterin. They accumulate excess phenylalanine in their blood, and this amino acid apparently interferes with the absorption of several hydrophobic amino acids across the blood–brain barrier. The infant then shows pathological brain development. Although physical development of the afflicted can be essentially normal, mental development is severely impaired in an irreversible manner. The incidence of phenylketonuria in Western European populations is estimated to be 1 in 15 000–20 000. Undiagnosed cases accounted for a large fraction of patients in mental wards, yet the diagnosis of PKU is now fairly simple. A drop of the newborn baby's blood is cultivated with a bacterial strain that requires phenylalanine for its growth, a so-called *phe*$^-$ mutant: growth of the culture is proportional to serum phenylalanine levels. Treatment of a positively diagnosed individual consists of maintenance on a low phenylalanine diet for the first seven or so years of the child's life. By that time brain development is almost complete and no additional damage results when a normal diet is then assumed. However, such females must be cautious, since pregnancy might expose their developing fetus to intolerable levels of placental phenylalanine.

Reference Garrod, A.E. (1923) *Inborn Errors of Metabolism*, 2nd edn, Oxford University Press, Oxford, UK. Introduces the term, which serves also as the title, that is still used today to describe genetic biochemical disorders. The author, showing great insight, discusses the biochemical and genetic concepts involved, way ahead of his time.

Reference Benson, P.F. and Fensom, A.H. (1985) *Genetic Biochemical Disorders*, Oxford University Press, UK. Excellent survey of salient aspects of classical inborn errors of metabolism, arranged into sections covering the major molecular groups.

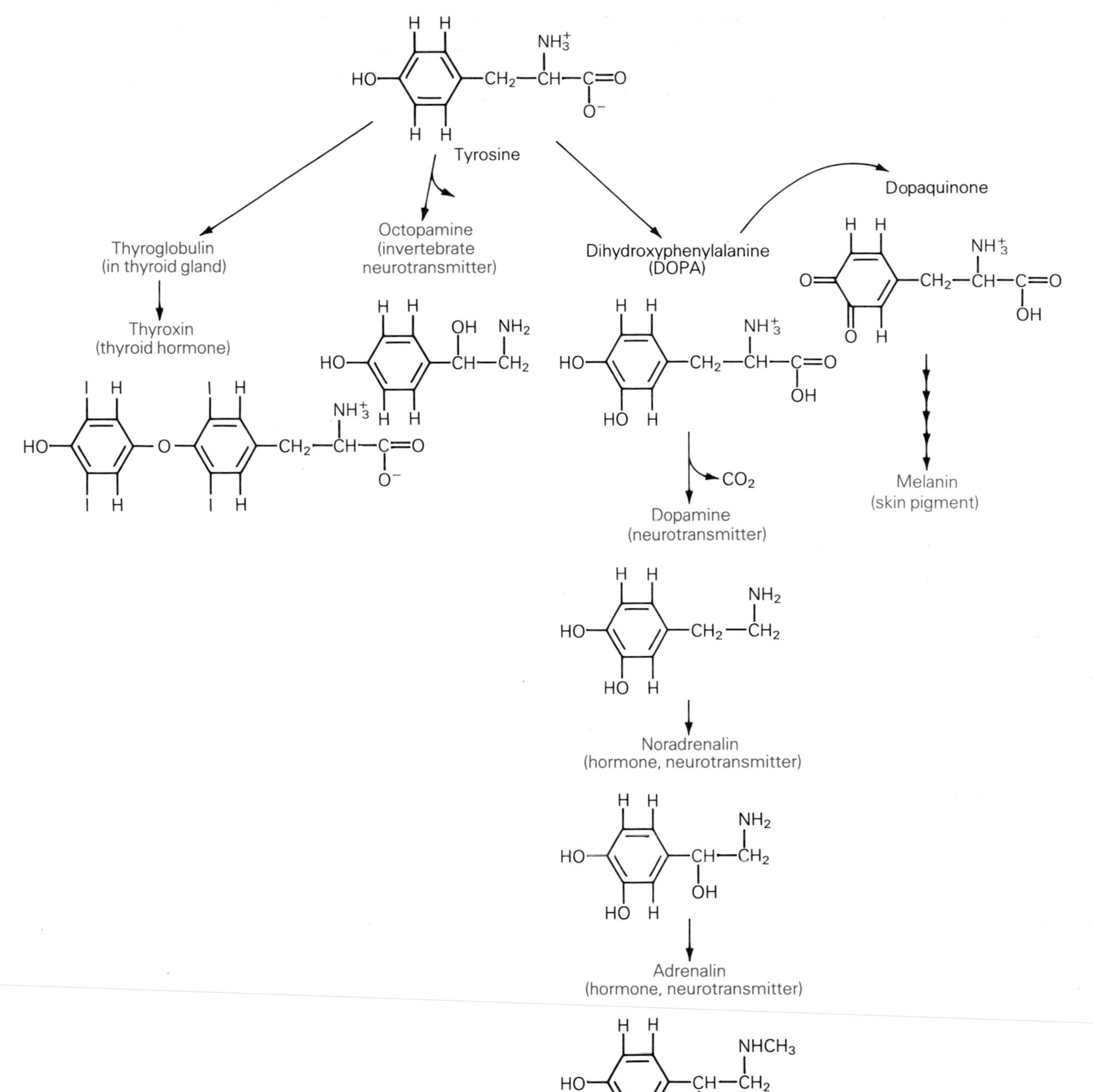

Fig. 7.9 Derivation of physiologically important molecules from tyrosine.

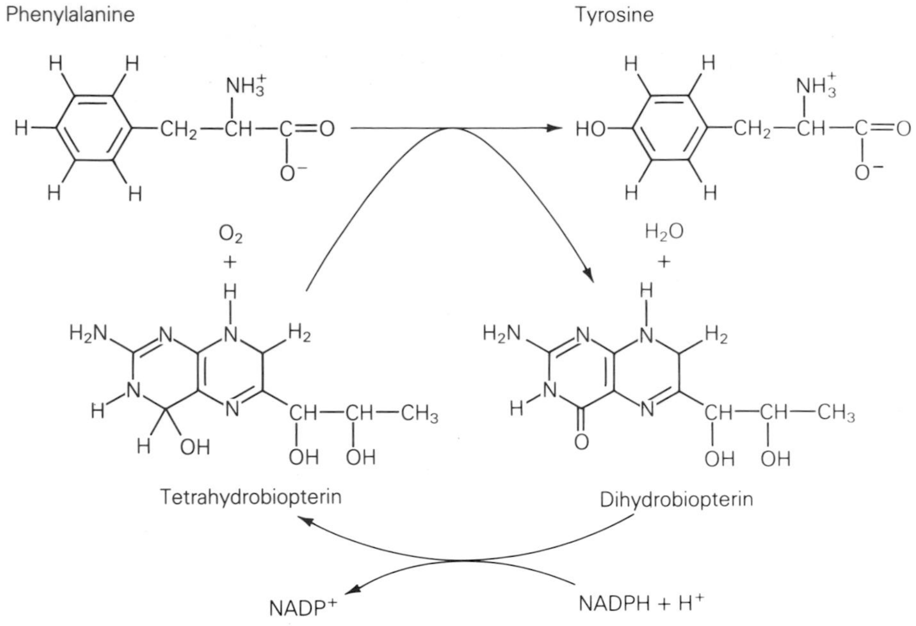

Phenylalanine

Tyrosine

O_2 +

H_2O +

Tetrahydrobiopterin

Dihydrobiopterin

$NADP^+$

$NADPH + H^+$

Fig. 7.10 Conversion of phenylalanine to tyrosine.

7.6 *Interconversion of amino acids and transamination*

The elimination of nitrogen from amino acids by oxidases, dehydrogenases and deaminases has been discussed above. Another vital process in amino acid metabolism involves the effective transfer of ammonia from an amino acid to an acceptor oxoacid, frequently 2-oxoglutarate. This process is referred to as **transamination** (Fig. 7.11). The enzymes that catalyse these transfers are known as **transaminases** or **aminotransferases**. These enzymes are apparently ***constitutive*** and do not respond to allosteric regulation: such behaviour is a general indication that a steady demand exists for their services. Transaminases require **pyridoxal phosphate**, a form of vitamin B_6, as their coenzyme. The aldehyde form of pyridoxal phosphate receives the amino group from the donor amino acid, which then becomes an oxoacid. The resultant amine form of the coenzyme pyridoxamine phosphate transfers the amine to the *acceptor* oxoacid and converts it into the new amino acid. The coenzyme has now been regenerated to its original form and is ready to begin a new round of activity. The mechanism of the reaction is illustrated in Fig. 7.12.

☐ Pyridoxal phosphate is synthesized in animals from one of three precursors, **pyridoxal**, **pyridoxol** or **pyridoxamine**, all are forms of vitamin B_6.

☐ The common amino acid proline is really an *imino* acid, and a reaction such as transamination cannot be applied to it until the pyrrolidine ring is first opened. This is accomplished via a distinctive oxidase reaction, followed by conversion of the product to glutamate-γ-semialdehyde, and thence to glutamate.

Box 7.7 Blood enzyme levels change in disease

Damage to certain tissues such as heart muscle or liver results in the release into the blood of enzymes characteristic of those tissues of origin. Clinical practitioners take advantage of this observation for diagnostic purposes. Blood chemistry analyses frequently include tests for enzymes such as **serum glutamate oxaloacetate transaminase** (SGOT) and **serum glutamate pyruvate transaminase** (SGPT). Levels of enzyme activity elevated above the norm can be indicative of a mild cardiac infarction or of a diseased liver, amongst other conditions. Alternative names for these two enzymes are glutamate–aspartate transaminase and glutamate–alanine transaminase, respectively.

Constitutive enzymes: *enzymes whose presence does not vary much with diet.*
Inducible enzymes: *enzymes that may not be made by the organism until it is exposed to some inducer substance in its medium or diet.*

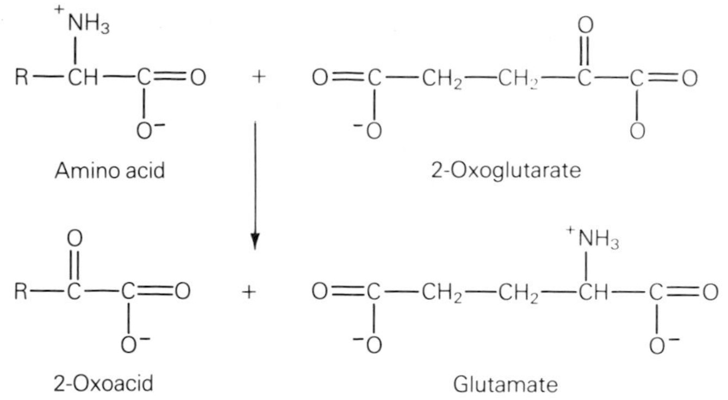

$$\underset{\text{Amino acid}}{R-CH(\overset{+}{N}H_3)-C(=O)O^-} + \underset{\text{2-Oxoglutarate}}{{}^-O-C(=O)-CH_2-CH_2-C(=O)-C(=O)O^-} \longrightarrow \underset{\text{2-Oxoacid}}{R-C(=O)-C(=O)O^-} + \underset{\text{Glutamate}}{{}^-O-C(=O)-CH_2-CH_2-CH(\overset{+}{N}H_3)-C(=O)O^-}$$

Fig. 7.11 Transamination scheme.

The oxoacids corresponding to the amino acids can all be degraded further to yield common metabolic intermediates and may account for about 15% of the total energy production from the diet for a human. The major oxoacid recipient is 2-oxoglutarate, and consequently glutamate is the most common amino acid product. Glutamate in animals can be converted back to oxoglutarate by the action of liver glutamate dehydrogenase (Fig. 7.3d). The released ammonia can be used to synthesize urea (section 7.8) which is subsequently transported to the kidneys for excretion. This overall scheme allows for the nitrogen of any amino acid to be removed and its carbon chain to be utilized for energy.

Exercise 3

The calcium salt of 2-hydroxy-4-(methylthio)butyrate is sometimes used in poultry feeds as an inexpensive way of raising the equivalency of methionine, a limiting essential amino acid. Explain the likely route of transformation that occurs.

Other coenzyme functions of vitamin B_6

Pyridoxal readily reacts with an amino acid to form a Schiff base, as illustrated in Fig. 7.12. The pyridine ring possesses a weakly basic nitrogen atom that is almost fully protonated at physiological pH (Fig. 7.13). The positive charge is positioned with respect to the conjugated double bond system in the Schiff base so as to be able to act as an electron sink. Electrons can be withdrawn from designated bonds adjacent to the α-carbon of the amino acids (Fig. 7.12). Several bonds are eligible for this electron withdrawal, so the actual choice will be influenced by the functional groups of the enzyme that are in the vicinity of the substrate–coenzyme complex. Therefore, amino acid reactions of several types have pyridoxal phosphate as their cofactor. These reactions include dehydration and subsequent deamination of serine, as illustrated in Fig. 7.3b), amino acid decarboxylations, amino acid racemization/optical inversion, and the hydroxymethyl transfer from serine (Figs 7.14–7.16). It is not clear what the significance of the racemization reaction is for animals, but in bacteria it is the most likely route for making the D-amino acids that are used

Box 7.8 Folic acid

Tetrahydrofolic acid or pteroylglutamic acid consists of three components: glutamate, *p*-aminobenzoate and a base, 2-amino-4-hydroxy-6-methylpterin. In animals, polyglutamate derivatives with two to seven glutamate residues are found and folic acid is stored in the liver as a polyglutamate derivative. Deficiency is characterized by low purine and dTMP levels. The symptoms of lack of folic acid are growth failure, anaemia, and deficiency of many types of white blood cell. **Megaloblastic anaemia** due to failure of DNA synthesis is the principal disease associated with folic acid deficiency. Megaloblastic anaemia is characterized by the appearance of mycrocytes, the percursors of erythrocytes, in the blood. Deficiency of folic acid is uncommon because of its production by intestinal microoganisms.

Reference Christen, P. and Metzler, D.E. (eds) (1985) *Transaminases*, John Wiley, New York, USA. A detailed reference work in which mechanisms of transamination reactions are thoroughly described.

Amino acid

Pyridoxal phosphate

2-Oxoacid

Pyridoxamine phosphate

2-Oxoglutarate

Glutamate

Pyridoxal phosphate

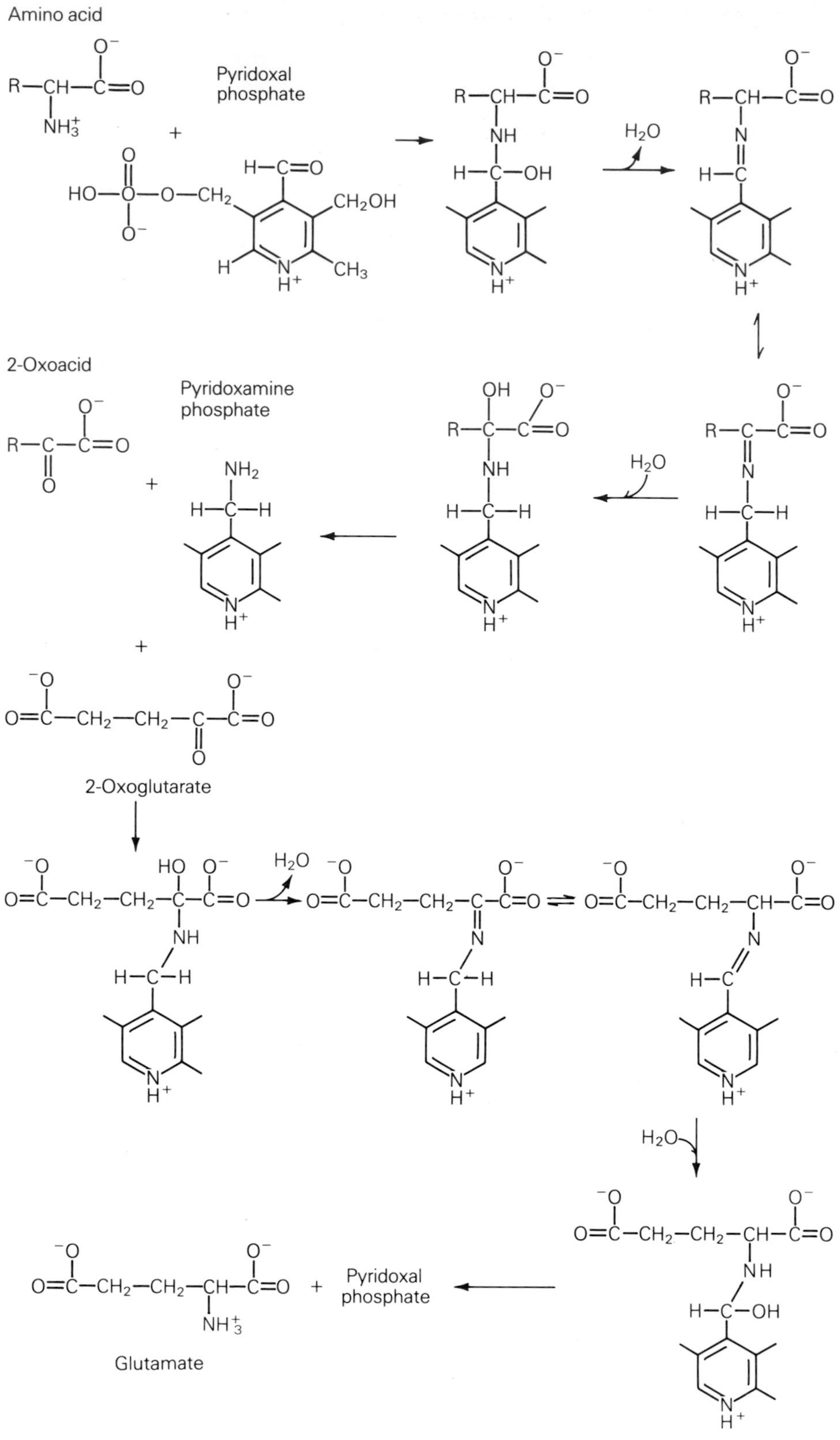

Fig. 7.12 Schiff base formation between pyridoxal phosphate and an amino acid: transamination mechanism.

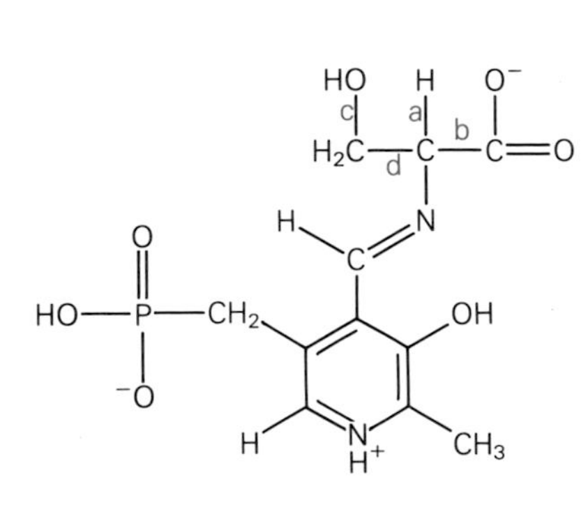

Fig. 7.13 Pyridoxal catalysis in several amino acid catabolic reactions. The positive nitrogen centre serves as an electron sink. Electrons are withdrawn through a conjugated double-bond system, thus facilitating the breaking of bonds some distance removed. (a) Racemase: removal of H^+, followed by readdition in the opposite stereochemical sense. (b) Decarboxylase: loss of CO_2. (c) Dehydratase: serine dehydratase yields the Schiff base of 2-amino acrylate; refer to Fig. 8.3b. (d) Serine hydroxymethyl transferase: the $HOCH_2$-group is transferred to tetrahydrofolic acid as an equivalant of formaldehyde.

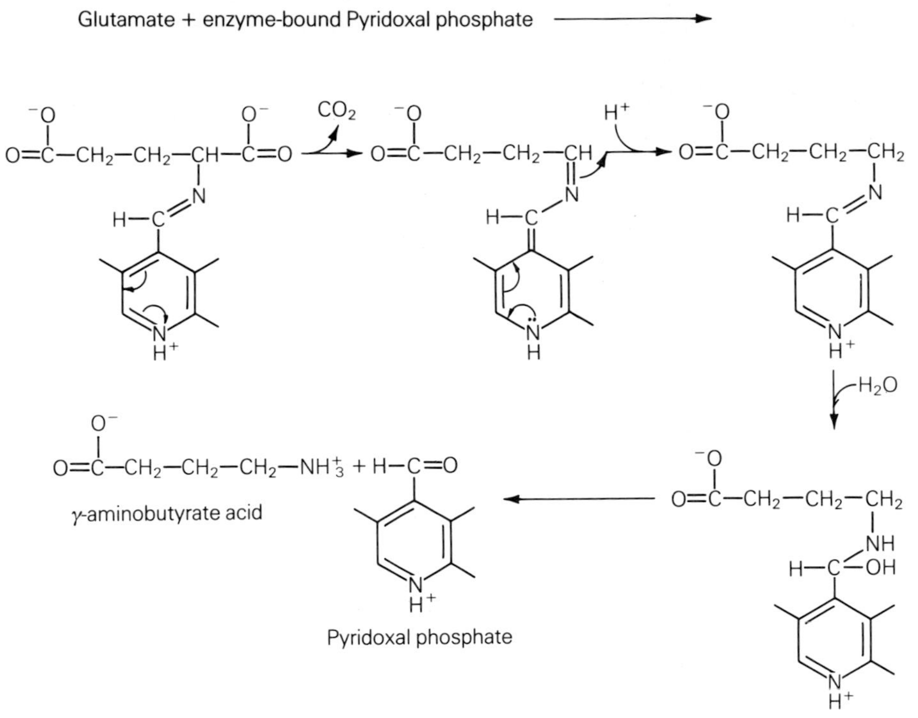

Glutamate + enzyme-bound Pyridoxal phosphate
CO_2
H^+
H_2O
γ-aminobutyrate acid
Pyridoxal phosphate

Fig. 7.14 Amino acid decarboxylation reaction: glutamate.

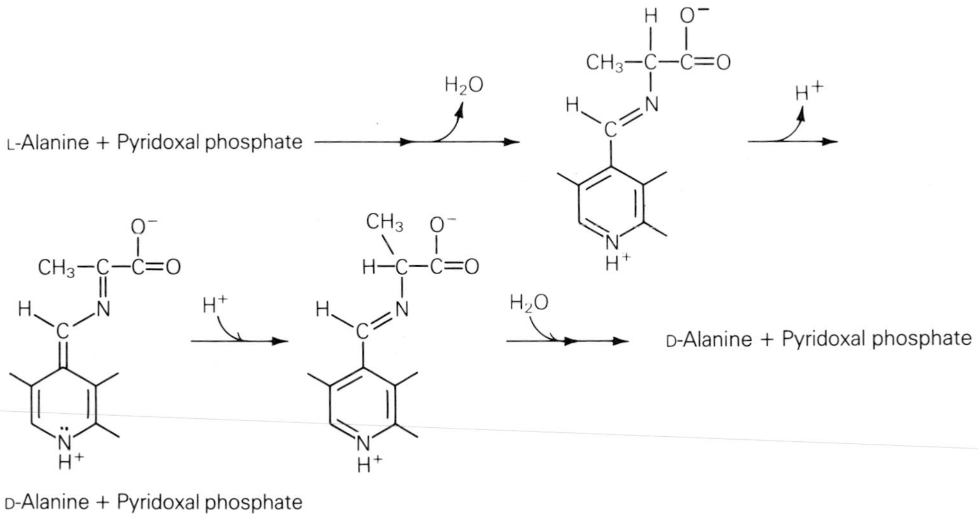

L-Alanine + Pyridoxal phosphate
H_2O
H^+
H^+
H_2O
D-Alanine + Pyridoxal phosphate
D-Alanine + Pyridoxal phosphate

Fig. 7.15 Conversion of L-alanine to D-alanine for bacterial cell wall synthesis: an abbreviated scheme.

Serine
Glycine
Pyridoxal phosphate
Tetrahydrofolic acid
5,10-Methylenetetrahydrofolic acid

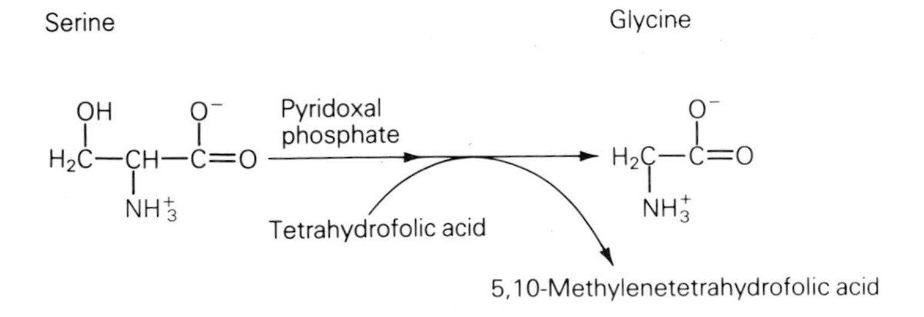

Fig. 7.16 Serine hydroxymethyl transferase and glycine production.

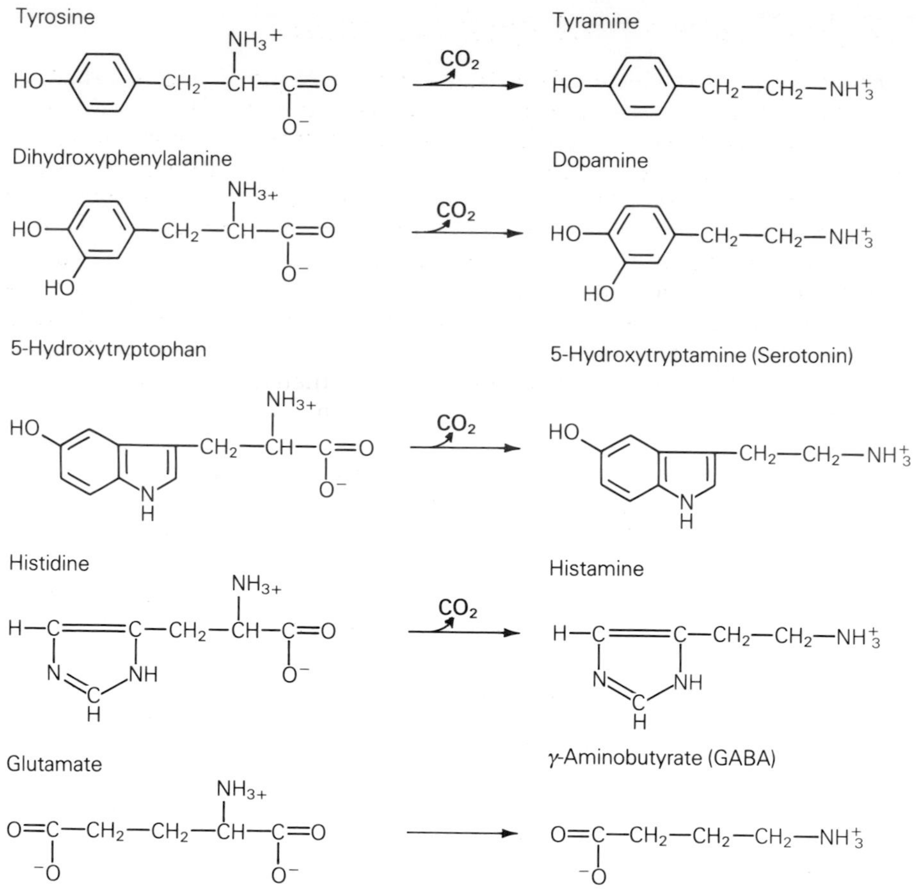

Tyrosine → Tyramine (CO_2)

Dihydroxyphenylalanine → Dopamine (CO_2)

5-Hydroxytryptophan → 5-Hydroxytryptamine (Serotonin) (CO_2)

Histidine → Histamine (CO_2)

Glutamate → γ-Aminobutyrate (GABA)

Fig. 7.17 Structures of precursor amino acids and their neurochemical messenger amines derived via decarboxylation reactions.

Exercise 4

The mechanism for the serine hydroxymethyl transferase reaction was not shown in Fig. 7.16. Study Figs 7.13–7.16 and then see if you can complete a simple mechanistic pathway for this reaction.

in the synthesis of cell wall peptidoglycans. The decarboxylation reaction produces several amines that are neurochemical messengers (Fig. 7.17): tyramine from tyrosine, dopamine from dihydroxyphenylalanine, serotonin from 5-hydroxytryptophan, histamine from histidine and γ-aminobutyrate (GABA) from glutamate. The serine hydroxymethyl transferase reaction (Fig. 7.16) is the biosynthetic route to glycine in animals. Equally important, this reaction converts the B vitamin **folic acid**, in its coenzyme structure of tetrahydrofolic acid, into a form that is necessary for synthesis of deoxythymidine nucleotides. Thus, DNA synthesis is indirectly dependent upon a pyridoxal phosphate-requiring amino acid reaction. Clearly, vitamin B_6 is especially well adapted for reactions involving amino acids, and life appears to have evolved in such a way as to take advantage of the chemical properties of this coenzyme.

Box 7.9
The B-group vitamins

See Box 6.3

It is interesting to note the number of different B vitamins that have been mentioned in the discussions on amino acid catabolism: riboflavin and niacinamide in the oxidase–dehydrogenase reactions, vitamin B_{12} for valine catabolism, folic acid in serine catabolism, and of course pyridoxal. In addition, not specifically mentioned, but implied, were pantothenic acid as part of coenzyme A, biotin as a participant in gluconeogenesis, and thiamin in the conversion of alanine to pyruvate via CoA. That completes the list of B vitamins!

7.7 *Pathways for nitrogen excretion*

Gardeners know that the fertilizers they apply are rich in nitrogen, usually in the form of ammonium and/or nitrate salts. The ammonia is converted into nitrate by a process known as **nitrification**. The ammonia is first oxidized to nitrite by one group of bacteria, and then the nitrite is further oxidized to nitrate by a second group. Nitrogen fixation, in which atmospheric molecular nitrogen (diatomic nitrogen) is converted into usable compounds by some prokaryotic organisms, plays a vital role in replenishing the total biotic pool of available nitrogen. These nitrogen-fixers include free-living blue-green algae (cyanobacteria), bacteria such as *Azotobacter* and *Clostridium* spp. and other bacteria such as *Rhizobium* spp. that live in symbiosis in the root nodules of legumes and other plants. Only prokaryotes, and then not all of them, appear to have evolved the mechanism for fixing nitrogen.

See *Biosynthesis*

Plants that are not nitrogen-fixers must have an external source of usable nitrogen. Most other organisms, and especially multicellular animals, generally have the reverse problem: how safely to eliminate the excess nitrogen derived primarily from protein and amino acid catabolism. It has already been mentioned that ammonia, itself, is toxic to animals. Animals that live in an aqueous environment have adapted by eliminating the ammonia directly into their environment, since they have sufficient water to dilute the ammonia below toxic levels as it is being excreted. These animals are called ***ammonotelic***, and include fish, amphibians and some protozoa. In contrast, there are animals for whom water is precious. For example, animals that fly and so cannot afford to carry extra weight in the form of water and animals whose environment is normally arid. These organisms must adopt some other metabolic strategy; that is, they must convert the nitrogen to some non-toxic product that can be eliminated without at the same time requiring the loss of water for the elimination process. The molecule that is frequently used is the purine derivative, uric acid (Fig. 7.18). Animals that excrete most of their excess nitrogen in this form are called ***uricotelic***. They include birds, most reptiles, some amphibians that have a terrestrial stage in their life cycle, and insects.

☐ Uric acid is almost insoluble in water, and generally crystallizes as the dihydrate, the predominant constituent of bird faeces. Coastal cliffs and islands off Peru, which host thousands of nesting shore birds, have been a rich source of **guano**, a bird manure that is an excellent nitrogen fertilizer.

Uric acid has a low solubility in water, for example, less than 66 mg dm^{-3} at 20°C, and usually appears in excrement as highly pure, crystalline uric acid dihydrate. Uric acid is also the major end-product of purine catabolism. Humans, higher apes, and a few other animals such as the dalmatian coach hound excrete uric acid derived from this source. Other animals can further degrade the uric acid from purines into more soluble products, such as allantoin, allantoic acid, and even urea. However, the excess amino acid nitrogen that is eliminated as uric acid must be synthesized as such in the liver of uricotelic organisms. The initial steps in the synthetic route are the same as those for the synthesis of purines of the nucleic acid.

The third way of excreting nitrogen is the conversion of ammonia into the relatively non-toxic material, urea (Fig. 7.18). These animals are said to be

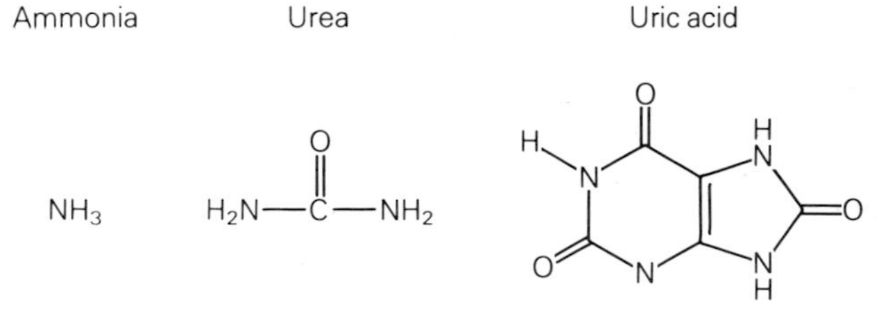

Fig. 7.18 Nitrogen excretory molecules.

Ammonotelic: *organisms (such as fish) that can excrete their excess nitrogen directly in the form of ammonia.*
Uricotelic: *organisms that must excrete their excess nitrogen in the form of the insoluble uric acid.*

Box 7.10
Urea

Urea has extensive industrial uses, such as in the manufacture of plastics. It also has medicinal and agricultural uses. Urea is essentially non-toxic for animals in moderate quantities and is recommended as a diuretic for cats. It has been added (as much as 10%) to skin emolients and lotions. It can be used as a source of ammonia in fertilizers, and it is added to some animal feeds as a partial source of dietary nitrogen for ruminants.

Exercise 5

The enzyme **urease** is found in many plants and microorganisms. The most common commercial source is the jack bean, *Canavalia ensiformis*. When well-fed rabbits are injected with purified urease, they rapidly go into convulsions and die. Why?

ureotelic, and 90% of their total nitrogen excretion takes the form of urea. They include the mammals and some other terrestrial vertebrates in the reptiles. The solubility of urea approaches 1 kg dm^{-3}, and so it is eliminated along with some water, but much less than would be needed for the dilution and elimination of the highly toxic ammonia. This accounts for one of the major needs for continual water intake by mammals. The relative nuisance of having constantly to seek water is offset by the fact that urea is non-toxic. Uricotelic animals, such as chickens, have little urea in their blood in comparison. Blood urea levels in humans are normally about 250 mg dm^{-3}. An adult human on an ordinary diet produces about 30 g, or 0.5 moles, of urea a day. On a high-protein diet a human may produce three times as much urea.

7.8 Biosynthesis of urea

The biosynthesis of one equivalent of urea requires two equivalents of NH_3, one of CO_2, and the energy input equivalent to the hydrolysis of four ATP molecules. Urea biosynthesis is a cyclic process requiring four enzymes and in addition an enzyme required for the synthesis of **carbamoyl phosphate** (Fig. 7.19). One of the two nitrogens in the urea is obtained from NH_3. The indirect source of the NH_3 is glutamate, acted upon by liver glutamate dehydrogenase, or from the glutamine amide nitrogen, released by the enzyme glutaminase. The second nitrogen atom arises from an asparate

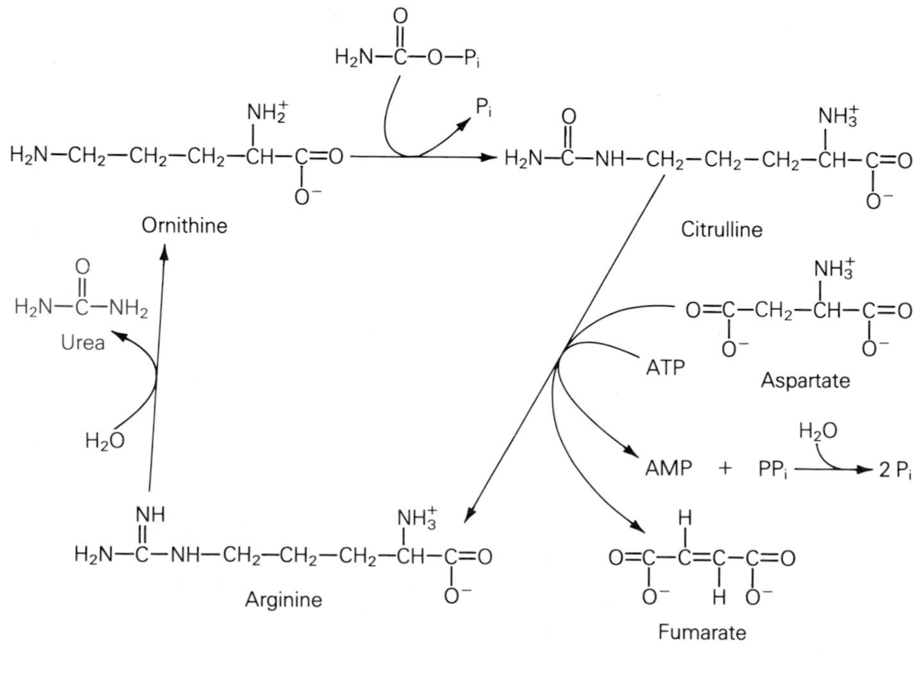

Fig. 7.19 Overview of the urea cycle.

***Ureotelic:** organisms (such as mammals) that convert their nitrogen to urea. The ending 'telic' in these three terms comes from the Greek* telos, *or end.*

molecule. The molecule that cycles, acting essentially like a catalyst, is **ornithine**; this compound gradually accumulates the two nitrogen atoms and the carbon atom that ultimately end up in urea. When the urea is finally released by hydrolysis at the end of the cycle by the action of arginase, ornithine is reformed. The whole process is generally referred to as the **urea cycle**. The enzymes of the cycle are identical to those that synthesize arginine in many microorganisms and higher organisms. The enzymes of arginine biosynthesis are widely distributed in nature because arginine is needed for the synthesis of protein and it seems likely that in the evolution of ureotelic organisms, the arginine biosynthesis pathway was simply exploited for nitrogen excretory purposes.

□ Arginine is a critical metabolite for invertebrates, because arginine phosphate is their energy storage molecule for the rapid regeneration of muscle ATP from ADP. This type of molecule is called a muscle phosphagen. The vertebrate equivalent is creatine phosphate.

Details of the urea cycle

In the pathway one NH_3 and one CO_2 are combined with a phosphate to yield carbamoyl phosphate. This reaction, which takes place in the mitochondrial matrix, requires the cleavage of two ATP molecules and is catalysed by carbamoyl phosphate synthetase. The 'first' enzyme active in the cycle itself, is ornithine transcarbamoylase, and is also found in the mitochondrial matrix. The reaction product is an amino acid, citrulline (Fig. 7.20a), which diffuses into the cytosol for further transformation by the second enzyme of the cycle,

□ A carbamoyl phosphate synthetase isoenzyme is also found in the cytosol. This enzyme provides carbamoyl phosphate that is needed for the biosynthesis of pyrimidines.

Box 7.11 Krebs' first cycle

It is worthwhile to put the study of the urea cycle in historical perspective. Krebs studied medicine in Germany and then worked in the laboratory of Warburg. Warburg introduced the technique of using tissue slices 0.2–0.4 mm thick in his studies on respiration. In Krebs' first independent post in the Department of Internal Medicine at Freiburg University in 1931 he decided to use the liver slice technique for biosynthesis studies. He chose urea as his target because of the simplicity of its structure and the large amount in which it was produced by mammals. Krebs developed a new saline medium that corresponded better to the ionic composition of blood than did media then in use. He adapted an enzymatic analysis for urea using urease and a manometric procedure that measured the amount of released CO_2 in apparatus which came to be called the Warburg respirometer. This method enabled Krebs accurately to detect as little as 0.05 mg urea. These studies were carried out with the assistance of a medical student, Henseleit. Together they found that urea synthesis was high in the presence of both NH_4^+ and amino acids, and it was highest with the presence of the non-protein amino acid, ornithine. They then observed that the ornithine need only be in small amounts relative to the urea produced, so ornithine appeared to act like a catalyst. The high arginase activity in liver that had been reported almost two decades earlier was again apparent. They therefore assumed that the synthesis of urea involved the conversion of ornithine with CO_2 and two ammonium ions into arginine, which was then 'hydrolysed' by arginase to yield urea and to regenerate the ornithine. The non-protein amino acid, citrulline, which had previously been identified in watermelon (*Citrullus* sp.), was demonstrated by Krebs and Henseleit to be an intermediate in urea formation. It was not until much later, 1946–57, that the entire cyclic scheme was completed. Nevertheless, by 1932 Krebs had earned some reputation amongst international biochemists for his discovery of the first metabolic cycle, which he referred to as the 'ornithine cycle'.

Krebs, of Jewish heritage, was dismissed from his Freiburg position after the Nazi rise to power in 1933. He moved to Cambridge first and from there to Sheffield in 1935, where he was given his own laboratory. His studies in Sheffield led to the concept of the tricarboxylic acid cycle, for which he is perhaps best known. Krebs made many additionally important discoveries in metabolism up until his death in 1981 aged 81. In 1972 Krebs and Henseleit were reunited in Germany at a scientific meeting.

Krebs, H. with Martin, A. (1981) *Reminiscences and Reflections*, Clarendon, Oxford, UK. Autobiography of one of the most influential scientists in the field of medical studies. The book gives historical insight into a half-century of biochemistry, as well as world and scientific politics.

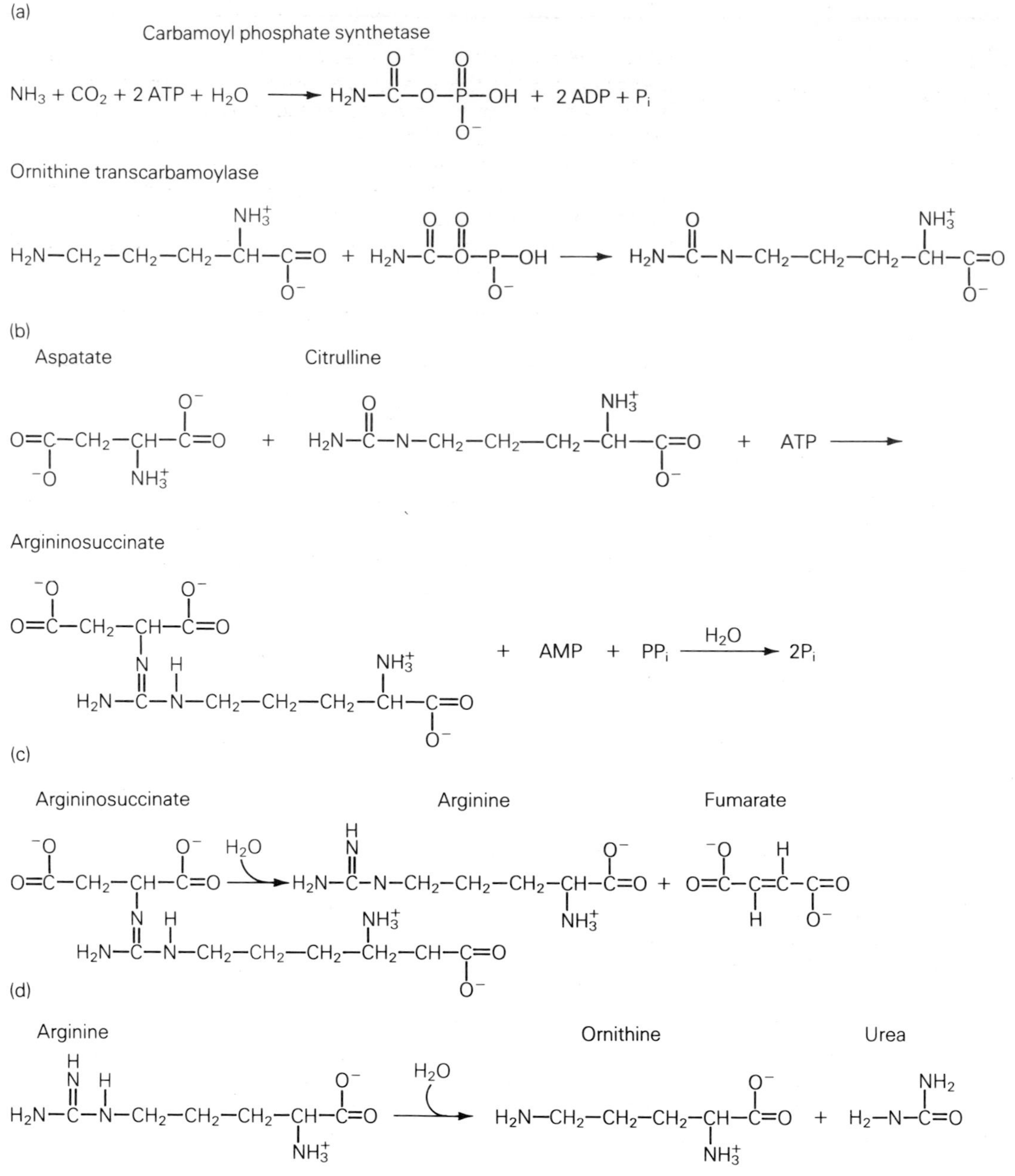

(a)

Carbamoyl phosphate synthetase

$$NH_3 + CO_2 + 2\,ATP + H_2O \longrightarrow H_2N-\overset{O}{\overset{\|}{C}}-O-\overset{O}{\overset{\|}{P}}(O^-)-OH + 2\,ADP + P_i$$

Ornithine transcarbamoylase

$$H_2N-CH_2-CH_2-CH_2-CH(NH_3^+)-C(=O)O^- + H_2N-\overset{O}{\overset{\|}{C}}-O-\overset{O}{\overset{\|}{P}}(O^-)-OH \longrightarrow H_2N-\overset{O}{\overset{\|}{C}}-N-CH_2-CH_2-CH_2-CH(NH_3^+)-C(=O)O^-$$

(b)

Aspatate | Citrulline

$$O{=}C(^-O)-CH_2-CH(NH_3^+)-C(O^-){=}O + H_2N-\overset{O}{\overset{\|}{C}}-N-CH_2-CH_2-CH_2-CH(NH_3^+)-C(=O)O^- + ATP \longrightarrow$$

Argininosuccinate

$$O{=}C(^-O)-CH_2-CH(N{=})-C(O^-){=}O \quad H_2N-\overset{N}{\overset{\|}{C}}-N(H)-CH_2-CH_2-CH_2-CH(NH_3^+)-C(=O)O^- + AMP + PP_i \xrightarrow{H_2O} 2P_i$$

(c)

Argininosuccinate | Arginine | Fumarate

$$O{=}C(^-O)-CH_2-CH(N{=})-C(O^-){=}O \quad H_2N-\overset{N}{\overset{\|}{C}}-N(H)-CH_2-CH_2-CH_2-CH(NH_3^+)-CH-C(=O)O^- \xrightarrow{H_2O} H_2N-\overset{NH}{\overset{\|}{C}}-N-CH_2-CH_2-CH_2-CH(NH_3^+)-C(O^-){=}O + O{=}C(^-O)-C(H){=}C(H)-C(O^-){=}O$$

(d)

Arginine | Ornithine | Urea

$$H_2N-\overset{NH}{\overset{\|}{C}}-N(H)-CH_2-CH_2-CH_2-CH(NH_3^+)-C(O^-){=}O \xrightarrow{H_2O} H_2N-CH_2-CH_2-CH_2-CH(NH_3^+)-C(O^-){=}O + H_2-N-C(NH_2){=}O$$

Fig. 7.20 (a) The mitochondrial phase of urea biosynthesis. (b) Argininosuccinate synthetase reaction. (c) Generation of arginine from argininosuccinate by argininosuccinate lyase. (d) The arginase reaction: generation of urea.

argininosuccinate synthetase. This step is energetically driven by the cleavage of ATP to AMP and inorganic pyrophosphate (PP_i). Since a cellular pyrophosphatase hydrolyses PP_i to two inorganic phosphates (P_i), the energy input is that of two pyrophosphate bonds, equivalent to two ATP being hydrolysed to two ADP (Fig. 7.20b). In this step, the second urea nitrogen atom is obtained from asparate, which is incorporated into the product, argininosuccinate. In the following step, catalysed by argininosuccinate lyase, the carbon chain of aspartate is released as fumarate. The carbon skeleton originally contributed by ornithine is now part of the arginine that is formed (Fig. 7.20c).

□ Inorganic pyrophosphate, abbreviated PP_i, is rapidly hydrolysed to two molecules of inorganic phosphate, or $2P_i$ by the inorganic pyrophosphatases that seem to be ubiquitous in animal tissues. Conversion of PP_i to $2P_i$ enhances the thermodynamic drive to completion for the reaction; consequently, a sequence of nucleoside triphosphate to nucleoside monophosphate plus PP_i is a common motif found in biosynthetic activation reactions of lipids, amino acids, and nucleic acids.

Box 7.12 Metabolic defects of the urea cycle

Inherited abnormalities of the urea cycle have been discovered. Defects in all four of the enzymes within the urea cycle have been characterized and are listed in the table. Ammonia and/or cycle intermediates accumulate in the body fluids and in the urine of these patients. If any of these enzymes are totally deficient, the infant usually dies shortly after birth. Partial deficiencies can result in coma, lethargy and mental retardation. In some cases where the defective enzyme is still minimally active, the patient can be maintained on a restricted protein intake, evenly distributed throughout the day. This prevents the liver from being overloaded with ammonia and the patient can survive.

Defect	Symptom
Ornithine transcarbamoylase	Hyperammonaemia (high blood ammonia)*
Argininosuccinate synthetase	Accumulation of citrulline
Argininosuccinate lyase	Accumulation of argininosuccinate
Arginase	Hyperargininaemia**

* Normal human blood plasma ammonium concentration is less than 50 μmol dm^{-3}, and neurological disturbance is common with levels greater than 200 μmol dm^{-3}.
** Typical human blood plasma arginine is 12–30 mg dm^{-3}.

Lowenthal, A., Mori, A. and Marescau, B. (eds) (1982) *Urea Cycle Diseases*. Advances in Experimental Medicine and Biology, Vol. 153, Plenum Press, New York, USA. Discusses a variety of clinical screening methods and abnormalities related to the urea cycle.

The final step of the cycle is the hydrolytic cleavage of arginine by arginase regenerating ornithine and releasing urea (Fig. 7.20d). The net stoichiometry of the urea cycle reactions is as follows:

$$NH_3 + CO_2 + \text{aspartate} + 3\,ATP + 4\,H_2O \rightarrow \text{urea} + \text{fumarate} + 2\,ADP + AMP + 4\,P_i$$

The fumarate can diffuse into the mitochondrion where it can be converted to malate and then to oxaloacetate by the enzymes of the tricarboxylic acid cycle. The oxaloacetate can be converted back to aspartate by transamination with one of several other amino acids. Thus, the second urea nitrogen is indirectly derived from almost any of the amino acids. This general scheme, shown in Fig. 7.21, links together the urea cycle, the tricarboxylic acid cycle, and transamination. The urea cycle was elucidated in 1931 by Krebs and Henseleit, some years before the Krebs tricarboxylic acid cycle was proposed. The linkage of two metabolic cycles has stimulated some to refer to this scheme as the 'Krebs bi-cycle'. It clearly illustrates how metabolism involves interaction between different cellular compartments and seemingly unrelated metabolic pathways.

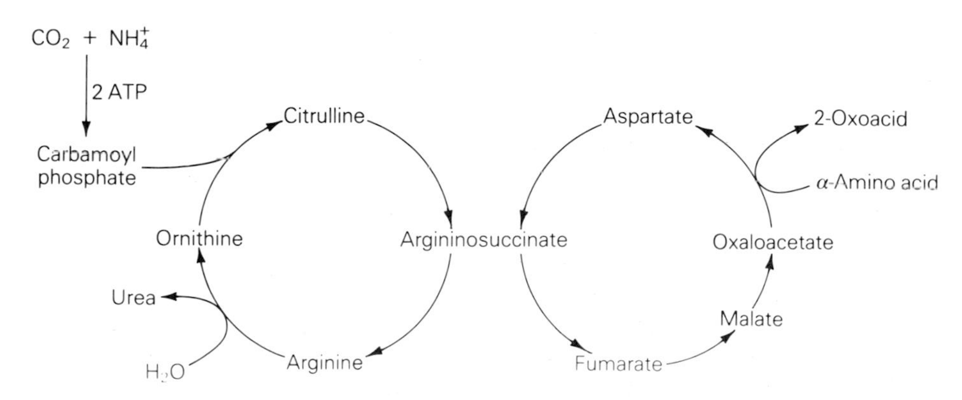

Fig. 7.21 Interaction of the urea cycle with the tricarboxylic acid cycle.

7.9 Overview

Amino acids are utilized in building proteins but their carbon skeletons can also serve as an energy source. Some amino acids also can be converted to specific hormones and/or neurotransmitters. Amino acid catabolism poses a problem, however, since the ammonia generated must be eliminated in a manner that does not allow toxic levels to accumulate. Animals can be categorized according to how they accomplish nitrogen excretion, into ammonotelic, ureotelic and uricotelic types. Mammals are ureotelic. Inherited deficiencies for urea synthesis and for amino acid metabolism occur in humans. The syndromes are well characterized, and in some cases can be alleviated by dietary regulation.

Answers to exercises

1. Animals depleted of carbohydrate cannot use alcohol for replenishment. The alcohol is oxidized to acetate and thence to acetyl CoA, which can be used to produce calories or fat. Athletes cannot carbohydrate-load by drinking beer alone; they could do so to a limited extent, however, by consuming the appropriate amino acids (or, of course, glucose).
2. The animals are able to produce glucose, because the fumarate can be converted to oxaloacetate and thence to phosphoenolpyruvate and on to glucose. Acetoacetate can be used by the brain and the muscles for energy. The acetoacetate could also give rise to acetyl CoA for conversion to fat. Thus, phenylalanine and tyrosine can be categorized as both glucogenic *and* ketogenic.
3. The 2-hydroxy group must first be converted into a 2-oxo group, and the likely enzyme would be a non-specific alcohol dehydrogenase. The product of this reaction would be 2-oxo-4-(methylthio)butyrate, and after catalysis by a transaminase it would be converted into methionine.
4. Form the serine–pyridoxal Schiff base, have electron withdrawal from the appropriate bond(s), show loss of a formaldehyde equivalent without worrying about the transfer to tetrahydrofolic acid, reverse the complex formation and pick up a proton from the milieu, and finally obtain the release of glycine.

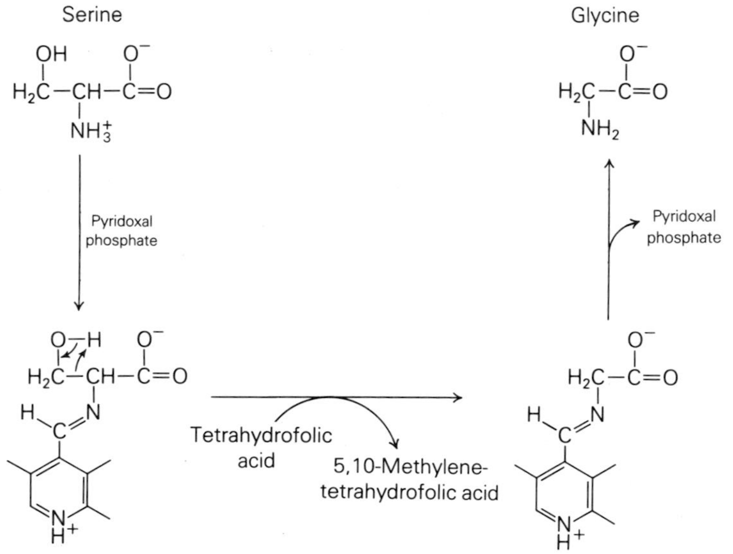

Serine

Glycine

Pyridoxal phosphate

Pyridoxal phosphate

Tetrahydrofolic acid

5,10-Methylene-tetrahydrofolic acid

5. Urease hydrolyses urea to two molecules of toxic ammonia.

QUESTIONS

FILL IN THE BLANKS

1. Amino acids are not just used in the building of protein structure, but they can also be ______ in order to produce energy. In addition, some amino acids can give rise to physiologically active messenger molecules: ______ and ______. Protein breakdown is initiated by ______ and ______, plus a variety of other hydrolytic enzymes. The first step in their catabolism is the removal of the α-______ group from the amino acids. This can be accomplished by a variety of enzymatic reactions using ______,

dehydrogenases or _______ . Enzymes specific for certain amino acids may also remove the nitrogen directly as NH_3, or in the case of serine and threonine, _______ may remove HOH as the first step. Although some organisms are capable of making all of the amino acids that they need, other organisms must be provided with some, referred to as _______ _______ _______ . Amino acids that can be degraded to metabolites that can be resynthesized into glucose are said to be _______ ; the alternative category are those that are _______ , and give rise to acetoacetate and related molecules.

The classification of animals according to their excretory pathway for excess nitrogen is _______ , _______ or _______ . Humans are classified as _______ organisms.

Choose from: amino, ammonotelic, catabolized, dehydratases, endoproteases, essential amino acids, exoproteases, glucogenic, hormones, ketogenic, neurotransmitters, oxidases, transaminases, ureotelic (2 occurrences), uricotelic.

MULTIPLE-CHOICE QUESTIONS

2. Which of the following statements is correct?

A. All mammals require the 20 amino acids in their diet.
B. Most mammals require about 10 amino acids in their diet.
C. All bacteria can make all of the amino acids.
D. Humans must have the acidic amino acids in their diet.
E. Albino rats have amino acid requirements vastly different from humans.

3. The hormone or neurotransmitter listed below that is not derived from the catabolism of phenylalanine is:

A. adrenalin
B. thyroxin
C. dopamine
D. insulin
E. noradrenalin

4. Vitamin B_6 functions as a coenzyme for many enzyme reactions in amino acid metabolism because it can:

A. help in the dehydrogenation between the α-amino and the α-carbon atoms.
B. form a Schiff base with the enzyme.
C. form a Schiff base with the amino acid and that intermediate facilitates the migration of electrons away from bonds to be cleaved.
D. act as an electrophile.
E. act as a nucleophile.

5. Which of the following statements relating to the urea cycle is correct?

A. It takes place entirely within the mitochondria.
B. It takes place entirely within the cytosol.
C. It results in the conversion of CO_2 and free ammonia (exclusively) into urea.
D. It involves no amino acids.
E. It requires the participation of aspartate and, indirectly, glutamate.

SHORT-ANSWER QUESTIONS

6. List several vitamins that play a role as coenzymes in reactions of amino acids.

7. Divide the amino acids into two groups based upon their dietary needs in humans.

8. Draw the structural formulae for all of the metabolites listed in Fig. 7.21.

Answers to questions

Chapter 1

1. In a redox reaction the molecule that loses electrons becomes oxidized , and the molecule that gains electrons becomes reduced .

At pH 7.0 ATP occurs as a multiple charged anion because its phosphate groups are almost completely ionized at this pH.

One of the principal reasons why the hydrolysis of ATP has a high ΔG^0 is that at pH 7.0 ATP molecules have four closely spaced negative charges. ATP functions as an energy-carrying common intermediate in living cells.

ATP provides free energy for the major forms of biochemical work, namely, biosynthesis of cell components, muscular contraction and active transport of numerous molecules and ions.

2. B.
3. A.
4. E.
5. A and C.
6. A. True.
B. False.
C. True.
D. False.
E. False.
F. True.
G. False.
H. False.
I. True.
J. True.

7. (a) Non-spontaneous because of result in (b).
(b) Since $\Delta G^{0'} = RT \ln K'_{eq}$

$$\begin{aligned} &= 8.314\ \text{J mol}^{-1} \times 2.3 \\ &= 298\ \text{K} = \log 10^{-3} \\ &= 5698\ \text{J mol}^{-1} \times -3 \\ &= 17095\ \text{J mol}^{-1} \\ &= 17.1\ \text{kJ mol}^{-1} \end{aligned}$$

8. Mg^{2+} ions will neutralize two negative charges on the triphosphate arm of ATP thus decreasing the charge repulsion between the O^- atoms. This should reduce the intramolecular strain and cause a decrease in ΔG, i.e. ΔG will become less negative.

9. $\Delta G^{0'}$ is defined at standard conditions of 1 mol dm^{-3} reactant and products, pH 7.0, 298 K and a pressure of 101 kPa while ΔG refers to the free energy change under any other conditions of temperature and reactant and product concentrations.

10. If heat is to do work, it must pass from a region of high to a region of low temperature. Since most animals operate at constant body temperature (homotherms) or at least under isothermal conditions during activity, they cannot use heat energy to do work.

11. Assume from Box 1.3 that the ΔG for ATP is -55 kJ mol^{-1}. Then the minimum amount of free energy required to form ATP under physiological conditions must be $+55$ kJ mol^{-1}, i.e. the free energy of formation of ATP. Thus, a minimum of 129 mol $\times$ 55 kJ mol^{-1} = 7095 kJ of free energy is stored as ATP when one mole of palmitate is oxidized.

12. The $\Delta G^{0'}$ of 24 kJ mol^{-1} is the free energy change at 1 mol dm^{-3} concentrations of reactant and products at pH 7.0. While intracellular pH is close to 7.0, the concentrations of D-fructose 1,6-bisphosphate, dihydroxyacetone phosphate and D-glyceraldehyde 3-phosphate will be in the millimolar concentration range and far removed from the *standard* concentrations. Thus, it is necessary to know G for the reaction to answer the question.

13.

$$\begin{aligned} G^{0'} &= -mF\Delta\psi + RT\ln \frac{[1]_o}{[I]_i} \\ &= -1 \times 96\,000 \\ &\quad \times -0.110 + 8.314 \\ &\quad \times 310 \times \ln[140/5] \\ &= 10\,615 + 8588 \\ &= 19.2\ \text{kJ mol}^{-1}. \end{aligned}$$

Chapter 2

1. Light that drives photosynthesis is absorbed primarily by the green pigment chlorophyll . All green plants have chlorophyll but photosynthetic bacteria have a distinctive type of pigment called bacteriochlorophyll . Halobacteria have no chlorophyll or bacteriochlorophyll but capture light using the protein bacteriorhodopsin . Oxygen is evolved in aerobic photosynthesis. The oxygen comes from water . Purple sulphur bacteria carry out photosynthesis by oxidizing reduced sulphur compounds, such as hydrogen sulphide , to sulphur or sulphate . In eukaryotes photosynthesis occurs in the chloroplasts . During photosynthesis, reducing power in the form of NADPH is generated. ATP is produced during photosynthetic electron transport , which generates a proton gradient . This proton gradient activates an ATPase , which catalyses the phosphorylation of ADP to produce ATP.
2. A. Sulphur is an environmental source of electrons for sulphur respiration .
B. Sulphate can accept electrons during sulphate respiration .
C. Methane is the ultimate reduced state of carbon in fermentation .
D. Water donates electrons in aerobic photosynthesis .
E. Oxygen is the electron acceptor for aerobic respiration .
F. A net reducing potential is not generated in fermentation .
3. A. Fatty acids, hexose sugars and amino acids are the principal fuel molecules for heterotrophs.
B. In plants and animals, NADPH is catabolically generated in the pentose phosphate pathway.
C. The three principal products generated during the complete catabolism of glucose are ATP , NADPH and carbon dioxide .
4.
A. (i).
B. (i).
C. (ii), (iii).
D. (i), (ii).
E. (i), (ii).
F. (i), (ii), (iii).
5. A. False
B. False
C. False
D. True
E. True.
6. The relative values of [ATP], [ADP] and [P_i] in figures (a) and (b) are found by measuring the heights of the relevant peaks. In this answer, the peak height of ATP was taken as a measure of [ATP]. The [P_i] in the cytosol is used in the calculation of the phosphorylation potential. The [P_i] in vacuole represents free phosphate in the aqueous vacuolar contents. The phosphorylation potential is calculated by the method shown on page 31.

The phosphorylation potential of maize root cells in the presence of oxygen is 0.89 compared to 0.17 for cells in nitrogen. The [ATP]/[ADP] ratio of oxygenated cells is 3.38 compared to 1.00 for cells in nitrogen.

In cells perfused with nitrogen less ATP is present. Such cells metabolise anaerobically (page 40 to 43) and can generate ATP only at substrate level (page 42). In oxygenated cells, oxidative phosphorylation takes place generating large amounts of ATP.
7. Direct involvement of a proton gradient: flagellar rotation of the bacterium *Escherichia coli*; ion transport in many processes in plants; heat production in plants and animals (pages 32 and 133). Involvement of a membrane bound ATPase; establishment of proton gradients in cells; Na^+/K^+ exchange pumps in nerve cells (Fig. 2.11); muscular movement in vertebrates; bioluminescence in microorganisms.
8. This is an investigative question which encourages you to read beyond this text (eg see the texts recommended on page 18). The following are some examples extracted from Chapter 2.

Table 2.5

Prokaryotic autotrophs, *Chlorobium* sp. and *Thiocystis* sp. page 36.
Prokaryotic heterotrophs, *Thiovulum* sp. page 38
Eukaryotic autotrophs, *Passiflora caerulea*, figure 2.1f.
Eukaryotic heterotrophs, *Amoeba proteus*, figure 2.1c.

Table 2.6

Photoautotrophs, *Molinia caerulea*, figure 2.1g
Chemolithotrophs, *Thiovulum* sp. page 38
Photoheterotrophs, *Halobacterium*, page 38
Heterotrophs, *Penicillium* sp. page 40

9. A solution similar to that in Section 2.5. Fermentation is required here.
10. Catabolism of glucose realises energy primarily in the form of proton gradients. Oxidation-reduction reactions and electron transfer play a key role in the establishment of such gradients. Section 2.2 and 2.3 provide the essential discussion.
11. Heat can only be converted into useful work in a heat engine where heat is transferred from a region of higher temperature to lower temperature. However, organisms are not heat engines. They operate more or less at constant temperature and pressure while the heat engine depends on the exchange of heat and the compression of gases to do work. Heat generated in catabolism can only be used to generate body heat. GLUCOSE CATABOLISM, Pages 24 and 25 and Page 32: Catabolism generates heat.
12. (a) Digestion is the hydrolysis of food molecules prior to absorption in the gut while catabolism is the degradation of food molecules after absorption. Only catabolism leads to the formation of ATP.
(b) Anaerobic respiration is the partial catabolism of organic food molecules in the absence of oxygen when electron acceptors other than oxygen enable ATP to be generated by oxidative as well as substrate level phosphorylation. Fermentation is the partial oxidation of food in the absence of any external electron acceptors. ATP is produced only by substrate level phosphorylation.
(c) Photoautotrophs are autotrophs which utilise inorganic sources of electrons to photosynthetically reduce carbon dioxide. Photoheterotrophs use organic molecules as electron donors in photosynthesis.
(d) Aerobic photosynthesis takes place in the presence of oxygen. Molecular oxygen is generated when electrons are removed from water. Anaerobic

photosynthesis takes place in the absence of oxygen. Electron donors other than water are used and molecular oxygen is not produced.

13. Photoautotrophs capture light energy in photosynthetic reactions. They pass this energy on as food to heterotrophs which turn the energy into useful work during catabolism. The important biochemical processes involve oxidation and reduction. The essay should include: (1) a description of the generation of NADPH by photosynthesisers and its use in reduction of carbon dioxide (Section 2.4); (2) the use of the products of photosynthesis to generate reducing potential in heterotrophs (Section 2.5); (3) a brief comment on chemolithotrophy as an exception to the idea that the sun is the ultimate source of ALL energy in the biosphere (Section 2.4, CHEMOTROPHS and LITHOTROPHS).

Chapter 3

1. The synthesis of ATP requires the establishment of a proton gradient across a biomembrane. In mitochondria, the protons are pumped across the inner membrane making the intermembrane space more acidic. Protons flow back across the membrane through the F_1F_0-ATPase. This enzyme catalyses the synthesis of ATP. It has a quaternary structure and consists of five subunits α, β, γ, δ and ε. The α and β subunits bind nucleotides (ADP and ATP). The γ subunit allows the movement of protons across the membrane, while the δ subunit attaches the F_1 particle to the F_0 portion embedded in the membrane. The ε subunit has a regulatory role.

According to Mitchell's chemiosmotic hypothesis, three protons move across the membrane for each ADP molecule phosphorylated. An alternative hypothesis, proposed by Boyer, suggests that protons are involved in initiating a conformational change in the F_1 particle. This alters the affinity of the protein for ATP and P_i.

2.
A. False.
B. True.
C. False.
D. False.
E. False.
F. True.
G. False.
H. False.

3. Using the equation:

$$\Delta G^{0'} = -nF\Delta E^{0'}$$

$$151\,200 = -2 \times 96487 \times \Delta E^{0'}$$

and so

$$\Delta E^{0'} = \frac{151\,200}{-2 \times 96487} = 0.78\,V$$

Therefore the O_2/H_2O couple has a redox potential 0.78 V greater than the succinate/fumarate couple.

4. Again using the equation:

$$\Delta G^{0'} = -nF\Delta E^{0'}$$
$$= -2 \times 96487 \times (-0.185 - (-0.32))$$
$$= -2 \times 96487 \times 0.135$$
$$= -26.05\,kJ\,mol^{-1}$$

The $\Delta E^{0'}$ value is determined by the relationship $\Delta E^{0'} = E^{0'}$ (oxidizing couple) $- E^{0'}$ (reducing couple).

5. The relationship used is:

$$\Delta G^{0'} = -RT\ln K_{eq}$$

The first step is to calculate the $\Delta G^{0'}$ using:

$$\Delta G^{0'} = -nF\Delta E^{0'}$$
$$= -2 \times 96487 \times (-0.185-(-0.32))$$
$$= -2 \times 96487 \times 0.135$$
$$= -26.051\,kJ\,mol^{-1}$$

Substituting this value in the first equation:

$$26051 = -8.314 \times 298 \times K_{eq}$$

$$\frac{26051}{8.314 \times 298} = -\ln K_{eq}$$

$$K_{eq} = 3.16 \times 10^{-5}$$

6. H_2, NADH, cytochrome *c* reduced, lactate

7. Using the equation:

$$\Delta G^{0'} = -nF\Delta E^{0'}$$

substitute values of $E^{0'}$ in each case:

for NADH/uQ $\Delta G^{0'} = -52.10\,kJ\,mol^{-1}$
for cyt *b*/cyt *c* $\Delta G^{0'} = -42.45\,kJ\,mol^{-1}$
for cyt *a*/O_2 $\Delta G^{0'} = -51.14\,kJ\,mol^{-1}$

8. (a) two electrons, one proton
(b) two electrons, two protons
(c) neither
(d) two electrons, no protons
(e) one electron, no protons

9. The uncouplers can dissipate transmembrane proton gradient by shuttling backwards and forwards across the membrane carrying protons.

Chapter 4

1. The TCA cycle is the major pathway for the aerobic catabolism of many types of compounds. Some of its intermediates also form starting points for the biosynthesis of carbohydrates, lipids, many amino acids and other compounds, such as porphyrins.

The TCA cycle consists of eight enzyme-catalysed reactions in which the acetyl residue of acetyl CoA is oxidised to carbon dioxide. Acetyl CoA combines with oxaloacetate to form the C_6 tricarboxylic acid, citrate. The reactions of the cycle operate to reform oxaloacetate, and produce one molecule each of GTP and $FADH_2$ and three molecules of NADH. Thus the cycle produces the equivalent of twelve molecules of ATP for each acetyl residue oxidized.

The key regulating enzyme of the cycle is isocitrate dehydrogenase. Other enzymes which exert control are pyruvate carboxylase and pyruvate dehydrogenase. All three enzymes are subject to allosteric regulations.

2. A. False.
B. True.
C. True.
D. False.
E. True.
F. True.
G. False.

3. The point to be made here is that there intermediates are intermediates in an oxidative pathway. The most reduced state of carbon in a multicarbon compound is $-CH_3$, and the most oxidized is $-COOH$; at the next oxidation step the $-COOH$ can be released as CO_2. In the TCA cycle we see the gradual change of reduced to oxidized states of carbon atoms.

4. Complete oxidation of pyruvate generates 4 NADH, 1 $FADH_2$ and 1 GTP. In electron transport/oxidative phosphorylation, each NADH generates 3 ATPs in principle and the reduced FAD produces 2 ATPs. The GTP, of course, produces 1 ATP. So the answer is

$(4 \times 3) + (1 \times 2) + 1 = 15$.
5. 2-chlorosuccinate. This imaginary experiment would in principle work just like the radio-isotope experiment, though in practice the chlorinated derivative would probably inhibit everything in sight, beginning with pyruvate dehydrogenase.
6. (a) Cessation, because the reduced cofactors cannot be reoxidized. (b) Acceleration (decreased allosteric inhibition of isocitrate dehydrogenase by ATP). (c) Cessation, as (a). (d) Inhibition. (e) Inhibition. (f) Inhibition: although ATP production is prejudiced, the oxidation of reduced cofactors is also greatly retarded.
7. (a) C; (b) D; (c) F; (d) A; (e) B; (f) E.
8. Combine data from Figures 4.11 and 4.22–4.25 (or an alternative presentation of the same information) to answer this question.
9. This question could be used for tutorial discussion. The crucial point is that in this imaginary situation, we would expect to find hydrogenation reactions in place of dehydrogenation reactions. We would also expect molecular reshufflings that would facilitate hydrogenations, and the appearance of terminal methyl (not carboxylic acid) groups that would be released as methane when the next hydrogenation occurred. Any student who grasps these general points has understood the main principles of the TCA cycle.
10. Acetate is converted (i) to acetyl CoA which is then converted to (ii) glycolyl, (iii) glyoxalyl, (iv) oxalyl CoA, this in turn is decarboxylated (v) to formyl CoA which is finally oxidized (vi) to CO_2 and free CoA. Two ATP equivalents are used in reaction (i) and, at best, 3 NAD^+s are reduced at each of reactions (iii), (iv) and (vi), giving a net gain of 7 ATPs. This may be optimistic, given that reaction (ii) is likely to be energy-requiring. Discussion of this problem could include some interesting guesses about enzyme reaction mechanisms (e.g. and electrophilic hydroxyl donor in reaction (ii), a disulphide bridge at the active site in reaction (vi).

Chapter 5

1. Carbohydrates and lipids are the main fuel molecules in organisms. Glucose is degraded by a series of reactions called glycolysis . The first enzyme active in the pathway is hexokinase , which requires the cofactors ATP and Mg^{2+} . Feedback control of the regulatory enzyme phosphofructokinase decreases glycolysis if the concentration of ATP is high. Hexokinase is inhibited by glucose 6-phosphate. However, the enzyme glucokinase is not inhibited so strongly by this metabolite. The latter enzyme also differs from hexokinase in having a high Michaelis constant for glucose .
Two molecules of ATP per glucose molecule are required to initiate glycolysis. In the absence of oxygen a net gain of only two molecules of ATP for each glucose molecule is possible. In an absence of oxygen the end product of glycolysis is lactate , in yeast ethanol and carbon dioxide .
2. C.
3. B.
4. (a) Creatine kinase is raised when skeletal or cardiac muscle is damaged, e.g. after myocardial infarction, in dermatomyositis and certain other skeletal muscle disorders, and after trauma that entails widespread damage to muscle tissue. (b) Lactate dehydrogenase is raised slightly in a wide variety of conditions (myocardial infarction, various liver disorders including acute and chronic hepatitis, disseminated carcinomatosis, leukaemias, etc.) and very markedly raised in cases of megaloblastic anaemia, severe haemolytic anaemia and acute solvent poisoning (the last because of the liver damage). Discussion of this question can usefully include a mention of the various isoforms of lactate dehydrogenase.
5. There will be no label in the CO_2 in either case. The ethanol will be labelled on (a) C-2, (b) C-1.
6. (a) The control step (phosphofructokinase) of glycolysis is bypassed because most of the fructose is phosphorylated at the 1-position and the product is cleaved by aldolase to trioses that include dihydroxyacetone phosphate. Therefore, glycolysis is accelerated. (b) Fructose infusion causes metabolic acidosis (and consequent electrolyte and water depletion of the extracellular compartment) because of the enhanced lactate production.
7. (a) Von Gierke's disease (Type I glycogen storage disease). (b) Glucagon and cortisol are already raised in this condition (homoeostatic responses to fasting hypoglycaemia). Both exacerbate the symptoms, glucagon because it accelerates glycolysis and therefore lactate production, and cortisol because it promotes gluconeogenesis and therefore glycogen synthesis. (Further increases of these hormones might have little or no effect in practice because the receptors will already be more or less fully occupied.)
8. If the individual's metabolism is normal, a low carbohydrate diet will tend to decrease the fasting blood glucose level, thus promoting glucagon production and therefore fat mobilization. Also, acetyl CoA production will not greatly exceed the rate of oxidation through the TCA cycle; therefore relatively little citrate will enter the cytosol, and fatty acid biosynthesis will be decreased.
9. Refer to a range of modern methods, such as: subcellular fractionation; enzyme purification by chromatographic and electrophoretic techniques; use of radioactively labelled substrates; purification of intermediates by techniques such as HPLC; use of NMR and X-ray crystallograpy to characterize the chemical structures of intermediates.
10. Of the many possible alternatives, the simplest by far is to add the reduction of lactate to propionate as an additional step to the mammalian glycolytic pathway. In this case , the anaerobic situation could lead to, example, reoxidation to lactate, perhaps by activating the carbon-2 by forming a thioester of the propionate, which would make the net ATP yield zero. More complicated possibilities include an altered reaction sequence after 3-phosphogglycerate.
11. A good answer would include the production and effect of 2,3-bisphosphoglycerate, the role of ATP in ion pumping, and the various roles of NADH, NADPH and glutathione in the reduction of methaemoglobin, membrane-associated peroxides and intracellular peroxides. If a clinical orientation is encouraged, appropriate references to malaria, glucose-6-phosphate dehydrogenase deficiency

and the induction of haemolytic anaemias by various toxins or dietary deficiencies (e.g. selenium deficiency) would be expected.

Chapter 6

1. Fatty acids are stored in an anhydrous form as triacylglycerols in adipose tissue. The hydrolysis of triacylglycerols is catalysed by lipases and releases glycerol and fatty acids. Hormones, such as insulin and adrenalin (or glucagon) regulate the action of these enzymes. The β-oxidation spiral degrades fatty acids generating acetyl CoA and reduced coenzymes. The reactions of the spiral occur in mitochondria and consist of the cyclic repetition of four reactions. These are a dehydrogenation requiring the coenzyme FAD, a hydrolysis producing an L-hydroxyacyl derivative, a second dehydrogenation which uses the coenzyme NAD^+ and the release of acetyl CoA which is catalysed by the enzyme thiolase.
The acetyl CoA produced is oxidised by the TCA cycle and the oxidation of the reduced coenzymes is linked to the formation of ATP in aerobic conditions.
2. Carnitine is required to transport fatty acids into the mitochondrion before β-oxidation. Deficiency of carnitine would lead to an inability to oxidize fats.
3. B.
4. See text (p. 129) for details if reactions.
5. $R{-}CH_3 \rightarrow R{-}CH_2OH \rightarrow R{-}CHO \rightarrow R{-}COO^-$
In the cell hydrogen acceptors would be NAD^+ or FAD.
6. Plan

Ingested food $\xrightarrow{1}$ Lipids emulsified digested absorbed $\xrightarrow{2}$ liver $\xrightarrow{3}$ storage tissues including BAT

$\xrightarrow{4}$ stores $\xrightarrow{5}$ β-oxidation electron transport heat production

1, 2, 3 : summer
3, 4 : winter
4, 5 : spring

Chapter 7

1. Amino acids are not just used in the building of protein structure, but they can also be catabolized in order to produce energy. In addition, some amino acids can give rise to physiologically active messenger molecules: hormones and neurotransmitters. Protein breakdown is initiated by endoproteases and exoproteases, plus a variety of other hydrolytic enzymes. The first step in their catabolism is the removal of the α-amino group from the amino acids. This can be accomplished by a variety of enzymatic reactions using oxidases, dehydrogenases or transaminases. Enzymes specific for certain amino acids may also remove the nitrogen directly as NH_3, or in the case of serine and threonine, dehydratases may remove HOH as the first step. Although some organisms are capable of making all of the amino acids that they need, other organisms must be provided with some, referred to as essential amino acids. Amino acids that can be degraded to metabolites that can be resynthesized into glucose are said to be glucogenic; the alternative category are those that are ketogenic, and give rise to acetoacetate and related molecules.
The classification of animals according to their excretory pathway for excess nitrogen is ammonotelic, ureotelic, or uricotelic. Humans are classified as ureotelic organisms.
2. B.
3. D.
4. C.
5. E.
6. All of the B vitamins function in the structure of coenzymes utilized for amino acid catabolism. These would include niacinamide, riboflavin, pyridoxal, B_{12}, folic acid, and (less directly), thiamin, pantothenic acid and biotin.
7. The essential amino acids for adult humans include: isoleucine, leucine, valine, tryptophan, phenylalanine, tyrosine, methionine, cysteine, threonine and lysine. Infants require, in addition, arginine and histidine. The requirement for tyrosine may be met by phenylalanine, and that for cysteine may be fulfilled by methionine. The remained eight amino acids can generally be biosynthesized and are called non-essential: glycine, serine, glutamate, glutamine, aspartate, asparagine, alanine and proline.
8. The structural formulas for the metabolites in Fig. 7.21:

$H_2N{-}C(=O){-}O{-}P_i$
Carbamoyl phosphate

$H_2N{-}CH_2{-}CH_2{-}CH_2{-}CH(NH_2){-}C(=O){-}O^-$
Ornithine

$H_2N{-}C(=O){-}NH{-}CH_2{-}CH_2{-}CH_2{-}CH(NH_2){-}C(=O){-}O^-$
Citrulline

$O{=}C(O^-){-}CH_2{-}CH(C(=O)O^-){-}N{=}C(NH_2){-}NH{-}CH_2{-}CH_2{-}CH_2{-}CH(NH_2){-}C(=O){-}O^-$
Argininosuccinate

$H_2N{-}C(=NH){-}NH{-}CH_2{-}CH_2{-}CH_2{-}CH(NH_2){-}C(=O){-}O^-$
Arginine

$H_2N{-}C(=O){-}NH_2$
Urea

$O{=}C(O^-){-}CH_2{-}CH(NH_2){-}C(=O){-}O^-$
Asparate

$O{=}C(O^-){-}C(H){=}C(H){-}C(=O){-}O^-$
Fumarate

$O{=}C(O^-){-}CH_2{-}CH(OH){-}C(=O){-}O^-$
Malate

$O{=}C(O^-){-}CH_2{-}C(=O){-}C(=O){-}O^-$
Oxaloacetate

Glossary

A

Acetoacetate: *a ketone, also known as 3-oxobutyrate.*
Acidophile: *an organism that can live at low pH.*
Adrenalin: *a hormone whose synthesis is the result of chemical modification of tyrosine.*
Aerobic respiration: *involves reactions which bring about the complete oxidation of food molecules, such as glucose.*
Aerobic: *metabolic processes that take place in the presence of oxygen.*
Amino acid dehydrogenases: *enzymes that dehydrogenate amino acids using niacinamide coenzymes as intermediates.*
Amino acid oxidases: *flavin-containing enzymes that dehydrogenate amino acids to the intermediate imino acid, which spontaneously gives rise to the corresponding 2-oxoacid.*
Ammonotelic: *organisms (such as fish) that can excrete their excess nitrogen directly in the form of ammonia.*
Anabolism: *that part of metabolism concerned with synthesis, from the Greek* anabole, *to heap up.*
Anaerobic (anoxic): *metabolic processes that take place in the absence of oxygen.*
Anaerobic respiration: *the partial catabolism of organic food molecules in the absence of oxygen. Electron transport to acceptors, other than oxygen, as well as substrate level phosphorylation, enables ATP to be generated.*
Anaplerotic: *from the Greek ana,* up, and pleroma, *full. It refers to reactions that 'fill up' the supply of intermediates that tend to become depleted during metabolism.*
Anoxia: *a deficiency of O_2.*
Autotroph: *from the Greek* troph, *to feed, and* auto, *self. 'Self-feeders' need an environmental source of energy. Heterotroph means feeding from sources other than onself.*

B

Biosphere: *that part of the Earth which is living.*

C

Canonical forms: *two or more contributing structures which approximate to the real hybrid molecule and are written according to certain rules.*
Catabolism: *that part of metabolism concerned with the chemical breakdown of organic molecules with the release of energy. From the Greek* katabole, *to throw down.*
Chloroplast: *sites of photosynthesis in green plants, from the Greek* chloros, *green, and* plasma, *to mould.*
Constitutive enzymes: *enzymes whose presence does not vary much with diet.*
Cytochromes: *electron-carrying proteins that contain haem. From the Greek* kutos, *vessel, and* khroma, *colour.*

D

Decomposer: *an organism that uses dead organic matter as a source of food. It breaks the food (proteins, polysaccharides, etc.) down extracellularly, absorbing smaller carbon compounds and releasing ions and molecules to the environment. Decomposers are heterotrophs.*

E

Electromotive force: *the free energy, available to do work, arising from an unequal distribution of charged particles.*
Endopeptidases: *enzymes that hydrolyse amide linkages in the interior of a polypeptide.*
Endotherm: *an organism that uses an internal (metabolic) source of heat to maintain body temperature.*
Epistemology: *the field of philosophy that investigates the nature and limits of knowledge. From the Greek* episteme, *knowledge.*
Exopeptidases: *enzymes that hydrolyse at either the amino terminal or at the carboxy terminal end of peptides.*

F

Fermentation: *the partial breakdown of food molecules in the absence of external electron acceptors such as oxygen. Reducing potential is conserved in the process and ATP is produced only by substrate level phosphorylation.*

G

Glucogenic amino acids: *amino acids that can be catabolized to yield intermediates for gluconeogenesis, the new synthesis of glucose.*

Gluconeogenesis: *The process that converts catabolites such as pyruvate and oxaloacetate, which may have been derived from amino acids or other types of molecules, back into glucose. In human liver as much as 100 g/day of glucose may be resynthesized. From glucose, and the Greek* neos, *new, and* genis, *produce.*
Glycogen: *from the Greek* glykr, *sweet, plus* gen *from the verb* gennaein, *to produce. This alludes to the production of sweet-tasting glucose upon hydrolysis of glycogen.*
Glycogenolysis: *glycogen, plus the Greek* lysis, *dissolution.*
Glycolysis: *literally the breakdown of sugar. It refers specifically to the process where one molecule of glucose is cleaved to produce two molecules of pyruvate. From the Greek* glykys, *sweet, and* lysis, *dissolution.*

H

Haem: *the iron–porphyrin prosthetic group of a number of proteins (haemoglobin, myglobin and the cytochromes). From the Greek* aima, *blood.*
Half-cell: *one electrode of an electrolytic cell and the solution or electrolyte with which it is in contact.*
Hepatomegaly: *enlarged liver. From the Greek* hepar, *liver, and* megas, *large.*
Hypoglycaemia: *low serum concentration of glucose. From the Greek* hypo, *under.*
Hypophosphataemia: *low serum concentration of phosphate. From the Greek* hypo, *under.*
Hypoxia: *low* O_2 *intake.*

I

Inducible enzymes: *enzymes that may not be made by the organism until it is exposed to some inducer substance in its medium or diet.*
Isothermal: *literally the same temperature (throughout the body).*

K

Ketogenic amino acids: *amino acids whose catabolism yields primarily ketone bodies, such as acetoacetic acid or acetyl CoA.*
Ketosis: *from ketone, and the suffix -osis, which can mean 'process' or 'condition'.*

L

Lithotrophs: *autotroph that uses inorganic sources of energy from the environment. From the Greek* litho, *rock.*

M

Matrix: *the part of the mitochondrion enclosed by the inner membrane. From the Latin* matrix, *womb.*
Metabolism: *the total chemical reactions in an organism from the Greek* metabole, *change. Cells are in a state of continual chemical change.*
Monooxygenase: *an enzyme for which only one oxygen atoms from diatomic oxygen is incorporated into the product of the reaction. The second oxygen atom is converted to HOH as it oxidizes a coreactant, usually a reduced coenzyme.*

N

Nutrients: *environmental chemicals from which organisms are built. From the Latin* nutrire, *to nourish.*

O

Ontology: *the science of reality. From the Greek* ontos, *being; i.e. the theoretical and the practical.*
Organelle: *a membrane-bound structure in cells which performs specific biochemical functions, e.g. the nucleus, endoplasmic reticulum, mitochondrion.*
Oxidative phosphorylation: *the production of ATP coupled to an electron transport chain transferring electrons from a reductant to an oxidant* (O_2).
Oxygen debt: *the physiological condition produced during temporary anoxia. Intermediary metabolism is switched to an anaerobic mode, producing compounds that can be stored until sufficient* O_2 *becomes available to complete oxidative processes.*

P

Phosphoanhydride bond: *the P-O-P linkage formed when two phosphate molecules or groups condense with the elimination of a water molecule.*
Photoautotrophs: *autotroph that uses light as an environmental source of energy. From the Greek* photo, *light.*
Primary productivity: *the production of biomass by autotrophs.*

R

Redox pair/couple: *an electron-donating molecule and its oxidized form.*
Resonance stabilization: *the stabilization of organic molecules mediated by the delocalization or 'sinking' of certain electrons to a lower energy level, thus forming a hybrid structure between two or more canonical forms.*

S

'Squiggle' or ~: *the symbol introduced by Lipmann in 1941 to denote a 'high-energy' bond in biochemical compounds.*
Substrate-level phosphorylation: *the generation of ATP from ADP in a flow whereby a substrate that is an intermediate in a metabolic pathway, becomes phosphorylated and passes its phosphate group on to ADP.*

T

Transduction: *the conversion and transmittance of energy in one form to another.*
Thioesters: *esters containing an S rather than an O atom* ($-\overset{\overset{O}{\|}}{C}-S-$). *From the Greek,* theion, *sulphur.*
Tetrahydrobiopterin: *a cofactor for the phenylalanine hydroxylase reaction. It is structurally related to folic acid and to a group of natural product molecules that are sometimes found in such locations as the wing pigments of butterflies.*
Thermogenesis: *i.e origin of heat from the Greek* thermos *and* genesis *for heat and origin.*
Transamination: *a reaction type in which an* α*-amino acid becomes a 2-oxoacid, and its coreactant 2-oxoacid then becomes the new amino acid. The coenzyme for this reaction is pyridoxal phosphate, a vitamin* B_6 *derivative.*
Transducer: *a device for transmitting the energy from one system to another. This often involves changing the form of the energy.*

U

Ureotelic: *organisms (such as mammals) that can convert their nitrogen to urea. The 'telic' in these three terms comes from the Greek* telos, *end.*
Uricotelic: *organisms that excrete their excess nitrogen in the form of the insoluble uric acid.*

Z

Zymase: *collection of enzymes extractable from yeast. From the Greek* zyme, *leaven.*

Index

Page references to Tables are in *italic* and those to Figures are in **bold**. References to Boxes and Side-notes are indicated by (B) and (S) after the page numbers respectively.

Acetate 136
Acetoacetate 96, 128–9, 146
Acetone 96, 128–9
Acetyl CoA (B), 82, 127
Active transport 9–11
Activity (of reactants) 5 (S)
Acyl CoA derivatives 85–6, 126–7
Adenosine triphosphate, *see* ATP
Adenylate cyclase 126
Adenylate energy charge 13 (B), 29
Adenylate kinase 28
Adenylate pool (ATP, ADP, AMP) 28–9
Adrenalin 110, 125, 147, **149**
Alanine dehydrogenase 142
Albumin, serum 126
Alcohol dehydrogenase 118 (B)
Allosteric regulation 24, 29, 88–9, 108–9
Amino acid decarboxylations 151, **153**, 154
Amino acid dehydrogenases 142
Amino acid oxidases 141
Amino acids
  α-amino group removal 141–3
  essential 144–5
  glucogenic 146
  ketogenic 146
  racemization 151, 153, 154
Aminotransferases (transaminases) 96 (B), 150
Ammonia 142, 144 (B), 155
Ammonotelic organisms 155
Anabolism 20, 22, **26**, 76
Anaplerotic reactions 87
Anoxia 108
Antibiotics (ionophores) 59 (B)
Arginine 157
Arum, thermogenesis 32
Asparaginase 142, 143 (B)
ATP
  and active transport 9–11
  cellular concentration 28–31
  chemical structure 12
  in coupled reactions 12–15
  high-energy (phospho-anhydride) bonds 11 (B), 12
  synthesis
    models (Mitchell/Boyer) 70–1
  see also Electron transport; Phosphorylation
ATPases
  F_1F_0 22, 28, 49, 61, 65, 70
  Na^+/K^+-transporting 9
Autotrophs 33, 34–9
Bacteria
  electron transport systems 61–5
  extracellular digestion 39, 140
  fermentation 42
  lithotrophic 38–9
  nitrogen fixation 155
  photosynthetic 36–8, 48, 69–70
  in ruminant metabolism 133–4
Bacteriochlorophylls 36, 69
Bacteriopheophytin 69
Bacteriorhodopsin 38, 70
Bioluminescence 28
  and ATP assay 30 (B)
Blue-gree algae, *see* Cyanobacteria
Bomb colorimeter 23 (B)
Boyer's model (ATP synthesis) 71
Brown adipose tissue 32, 129–30, 132–3
Butyrate 136
Ca^{2+} transport 61 (B)
Calorific value 23 (B)
Carbamoyl phosphate 156–7
Carbohydrate metabolism
  early studies 102–3
  and fat metabolism 94–5
  and protein catabolism 97
  see also Gluconeogenesis; Glycolysis
Carnitine 92, **93**, 126
Catabolism 2, 20, 22, 26–32, 40, **43**, 76
Catecholamines 132
Cellulose 133
Chemiosmotic theory (Mitchell's model) 59–60, 70–1
Chemoautotrophs (lithotrophs) 33, 38–9, 62
Chlorophylls 34, 37, 65–8
Chloroplasts 65
Citrate 83–5, 86, 93, 109
Citrate synthetase 86
Citric acid cycle, *see* Tricarboxylic acid cycle
Cobalamin (vitamin B_{12}) 119, 135 (B), 146 (S)
Cobamide 146 (S)
Cocaine 148 (B)
Coenzyme A (CoA) 81 (B), 82
Coenzyme Q (ubiquinone) 50
Coenzymes (group-transfer molecules) 21
  see also named types
Colourless sulphur bacteria 38–9
Combustion 23 (B), 24 (S)
Common 'activated' reaction intermediates 12–15
Cori cycle 112–14
Corticosteroids 96 (B)
Coupled reactions 12–14
Creatine 10 (B), 115
Creatine kinase 115
Creatine phosphate 48 (S), 114–15
Cyanobacteria 34–5, 68, 155
Cyclic AMP cascade 110, 114, 125, 147
Cytochrome *bf* complex 66–7
Cytochrome *c* oxidase 53
Cytochrome P450 27 (B), 62 (B)
Cytochromes
  bacterial 62, 63–4
  mammalian 52–3
Decomposers 39, 140
Dehydrogenases (electron transport) 49–50, 62–3
Detoxification 27 (B), 62 (B)
Diabetes mellitus 95 (B)
Difference spectra, electron transport chain 55
Dihydroxyacetone phosphate 126
Dinitrophenol 58, 59 (B), 133
Dopamine 148 (B), **149**
Ectotherms 32 (S)
Electromotive force, *see* Redox potentials
Electron transport
  bacterial 61–5
  in catabolism 41–2
  and cation transport 61 (B)
  microsomal 62 (B)
  mitochondrial 49–54
  sequence determination 54–6

Embden–Meyerhof pathway, *see* Glycolysis
Endergonic reactions 5 (S)
Endopeptidases 139–40
Endotherms 32
Enthalpy (H) 3, 22, 23, 24
Entropy (S) 2, 3, 4, 23
Epinephrin, *see* Adrenalin
Equilibrium (biochemical) 4
Equilibrium constants
  K_{eq} 24
  K'_{eq} 5–6
Erythrocytes, reductive machinery 119 (B), 122 (B)
Essential amino acids 144–5
Esters 86, 87 (S)
Exergonic reactions 5 (S)
Exopeptidases 139
Extracellular digestion 39, 140

FAD 50 (B)
Fat metabolism
  links with other pathways 92–3, 97, 109
  β-oxidation 42, 92, 126–8, 129
  ω-oxidation 129
  in ruminants 133–6
  and thermogenesis 32, 129–30, 132–3
  triacylglycerol breakdown 125–6
Fermentation 42–3, 102–3, 134–6
Ferredoxins 37, 52, 67
Flavin coenzymes 50 (B)
Flavin-linked dehydrogenases 49–50, 62–3
FMN 50 (B)
Folic acid 151 (B), 154
Free energy (G) 3, 4, 5, 6, 14, 23, 24
  of active transport 9–11
  and redox potentials 8, 56 (B)
Fructosaemia 118 (B)
Fructose 117, 118 (B)
Fructosuria 118 (B)
Fungi, extracellular digestion 140 (B)

Galactosaemia 118 (B)
Galactose 116
Galactosuria 118 (B)
Gibbs free energy, *see* Free energy (G)
Glucagon 109, 110, 125
Glucocorticoids 114 (S)
Glucogenic amino acids 146
Glucokinase 112
Gluconeogenesis 93, 112–14
Glucose 6-phosphate dehydrogenase 119
Glucuronide **110**
Glutamate dehydrogenase 142, 151
Glutaminase 142
Glycerol metabolism 109, **110**, 126
Glycogen metabolism 110–15
Glycogen phosphorylase 110, 114
Glycogen storage diseases 112 (B)
Glycolipids **116**
Glycolysis
  in aerobic catabolism 41, 42, **75**, 76
  in anaerobic catabolism 107–9
  free energy of 24–5
  links with other pathways 93, 109, 121
  reactions of 104–7
  regulation 108–9
Glycoproteins 116
Glyoxylate cycle 90–1, 96
Green sulphur bacteria 36, 37, 69
Greenhouse effect 43 (B)
Group-transfer molecules (coenzymes) 21

H^+ (proton) pump 9, 21–2, 48, 59–60, 133
Haem, in cytochromes 52–3
Halobacteria 38, 70
Hangovers 118 (B)
Heat energy 3, 5
  see also Thermogenesis
Heterotrophs 33, 38, 39–40
Hexokinase 112
Hexose monophosphate shunt, *see* Pentose phosphate pathway
Hexose sugars 115–16
Hibernation 129–30
Hill reaction 66
Honey 118 (B)
β-Hydroxybutyrate 96, 128–9, 146
Hypophosphataemia 118 (B)
Hypoglycaemia 118 (B)
Hypoxia 108

Immunosuppression 96 (B)
Inhibitors
  of electron transport chain 55–6
  of glycolysis 105–7
Insulin 126
Ion gradients 27–8, 59 (B)
Ionophores 59 (B)
Iron–haem proteins 52–3
Iron-sulphur proteins 51–2
Isocitrate dehydrogenase 88
Isocitrate lyase 90
Isothermal systems 2

Joule 1 (S)

Ketogenic amino acids 146
Ketone bodies 96, 128–9, 146
Ketosis 95 (B)
Krebs cycle, *see* Tricarboxylic acid cycle
Kwashiorkor 145 (B)

Lactate 103, 107–8, 114, 142
Lipase 125, 132
Lipids, *see* Fat metabolism
Lipoic acid 86
Lithotrophs (chemoautotrophs) 33, 38–9, 62
Liver, and metabolism 142 (S)
Luciferase 30 (B)

Malate shuttle 92
Mass Action, Law of 24 (S)
'Mechanism' 102 (B)
Megaloblastic anaemia 151 (B)
Membrane potential 10
Membrane transport 9–11, 91–3, 97 (S), 126
Membranes, and electron transport 21–2, 47–9, 59–60, 65, 67
Metabolic control 24–5, 87–90
  see also named m. pathways
Metabolic disorders 25 (B)
Metabolic pathways 20
  see also named types
Methane 43
Methylmalonyl CoA 146 (S)
Microsomes, electron transport 52 (B)
Mitchell's model (ATP synthesis) 59, 70–1
Mitochondria
  electron transport 21–2, 47–9
  fat metabolism 92–3, 126–7, 132–3
  TCA cycle 91–2
Monooxygenase 147
Movement, energy for 27
Muscle contraction 48 (S), 75, 102–3, 114

Na^+/K^+-transporting ATPase 9
NAD^+ 7 (B), 141 (S)
NAD^+/NADH redox couple 8, **20–1**
NAD^+-linked dehydrogenases **7 (B)**, 49–50
$NADP^+$ 7 (B), 141 (S)
Nernst equation (redox potentials) 9, 56 (B)
Neurochemical messengers 138 (B), 154
Niacinamide 141 (S)
Nitrification 155
Nitrogen excretion 155–6
Nitrogen fixation 155
Noradrenalin 130, 132, **149**
Nuclear magnetic resonance (NMR), phosphate assay 31 (B)
Nucleoside triphosphates, as common intermediates 12–15
Nutrition, modes of 33–40, **43**

Ornithine 157
Oxaloacetate 86–7
β-Oxidation 42, 92, 126–8, 129, 131 (B)
ω-Oxidation 129
Oxidation–reduction reactions, *see* Redox reactions
Oxidative phosphorylation 22, 34, 48, 56–60, 64–5
α-Oxoacids 85, 86, 141, 150–1
β-Oxoacids 85 (S)
Oxygen debt 114

P/O ratios 56
Palmitate 127–8
Pasteur effect 108–9
Pentose phosphate pathway 42, 118–21
Peptidoglycans 154
Peroxisomes 131 (B)
Phenylalanine metabolism 146–9
Phenylketonuria 147, 148 (B)
Pheophytin 66
Phosphoanhydride bonds (ATP) 11 (B), 12
Phosphoenolpyruvate 104 (S)
Phosphofructokinase 108
Phosphoglucomutase 114 (S)
Phosphogluconate pathway, *see* Pentose phosphate pathway
Phosphorylation
  assay 31
  oxidative 22, 34, 48, 56–60, 64–5

photosynthetic 22, 34, 48, 68–70
substrate level 48
Phosphorylation potential 31
Photoautotrophs 33–7
Photoheterotrophs 38
Photophosphorylation 22, 34, 48, 68–70
Photosynthesis 1–2, 34–7, 65–70
Phytanic acid storage disease 131 (B)
Plastoquinone (PQ) 66
Pognophoran (tube) worms 38–9
Polysaccharides 101, **116**
Porphyrin (cytochrome) 52–3
Primary productivity 38
Proenzymes (zymogens) 140
Proline 150 (S)
Propionate 135
Protein catabolism
and carbohydrate catabolism 97
and glycolysis 109, 146
nitrogen removal 141–4
protein degradation 139–40
and stress 96 (B)
and TCA cycle 94, 114 (S), 146
transamination 142, 144, 150–4
Proteoglycans **116**
Proteolytic enzymes 139–40
Proton (H^+) pump 9, 21–2, 48, 59–60, 133
Protophorphyrin IX 52
Purple non-sulphur bacteria 37–8
Purple sulphur bacteria 36, 37, 69
Pyridine coenzyme dehydrogenases 142
Pyridoxal phosphate (vitamin B_6) 150, 151, 154
Pyrophosphate 126, 127 (S), 158
Pyruvate carboxylase 89
Pyruvate dehydrogenase **85**, 86, 89

Q cycle (electron transport) 60

Redox potentials 7–8, 20–1, 47, 54–8
Redox reactions 20–1
Refsum's disease 131 (B)
Resonance stabilization 11 (B)
Respiration
aerobic 41–2
anaerobic 42, 107–9
Respiratory fractions (I–IV) 49, **54**
Riboflavin 50 (B)
Ribose 117
Rubredoxins 51
Ruminant metabolism 133–6

Schiff base 151, **152**
Serine 151
Serine hydroxymethyl transferase 151, **153**, 154
Serum albumin 126
Serum glutamate oxaloacetate transaminase (SGOT) 150 (B)
Serum glutamate pyruvate transaminase (SGPT) 150 (B)
Spontaneous reactions 2, 4, 5
Squiggle (high-energy bond) 11 (B)
Starch 133
Stress, and protein catabolism 96 (B)
Substrate-level phosphorylation 48
Superoxide dismutases 54 (B)
Surgical staples 41 (B)
Szent-Györgyi model (TCA cycle) 77–8

Tetrahydrobiopterin 147
Thermodynamics, general definitions 2–3
Thermogenesis 32, 129–30, 132–3
Thermogenin 133
Thiamin pyrophosphate 86
Thiazole ring 86
Thioesters 85–6, 126
Threonine 144 (S)
Thylakoid membrane 65, 67
Transaldolase 119
Transaminases (aminotransferases) 96 (B), 150
Transamination 142, 144, 150–4
Transketolase 119
Triacylglycerols 125–6
Tricarboxylic acid cycle
acyl CoA derivatives 85–6
Krebs model 78–80
links with other pathways 41, 42, 76, 90–4, 97 (S), 159
Ogston concept of citrate 83–4
oxaloacetate in 86–7
regulation 87–90
Szent-Györgyi model 77–8
Trophosome tissue (tube worm) 38
Tube worms 38–9
Tyramine 147 (S), 154
Tyrosine 146–9

Ubiquinone (coenzyme Q, CoQ, 'Q') 50
Uncoupling agents 58, 59 (B), 133
Urea metabolism 97 (S), 155–6, 156–9
Ureotelic organisms 155–6
Uric acid 155
Uricotelic organisms 155

Vitalism 102 (B)
Vitamin B group 154 (B)
Vitamin B_6 (pyridoxal phosphate) 150, 151, 154
Vitamin B_{12} (cobalamin) 119, 135 (B), 146 (S)

Yeast fermentation 42–3, 75, 102–3

Ylid 86 (S)

Z-scheme (photosynthesis) 67
Zymase 102
Zymogens (proenzymes) 140